Lecture Notes on Genetics

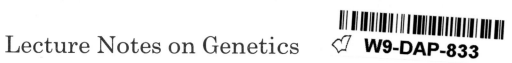

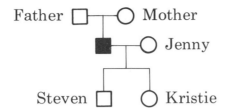

Lecture Notes on Genetics

PETER J. RUSSELL

BSc, PhD
Associate Professor of Biology
Reed College, Portland
Oregon, USA

BLACKWELL SCIENTIFIC PUBLICATIONS
OXFORD LONDON EDINBURGH
BOSTON MELBOURNE

First published 1980

Set in Monophoto Century Schoolbook
by Santype International Ltd,
Salisbury, Wiltshire
Printed and bound in Great Britain
by Morrison and Gibb Ltd, Edinburgh

DISTRIBUTORS

USA
  Blackwell Mosby Book Distributors
  11830 Westline Industrial Drive
  St Louis, Missouri 63141

Canada
  Blackwell Mosby Book Distributors
  120 Melford Drive, Scarborough,
  Ontario MIB 2X4

Australia
  Blackwell Scientific Book
  Distributors
  214 Berkeley Street, Carlton
  Victoria 3053

British Library
Cataloguing in Publication Data

Russell, Peter J
  Lecture notes on genetics.
  1. Genetics
  I. Title
  575.1   QH430

  ISBN 0-632-00609-9

Contents

Preface, vi

1 The Genetic Material, 1

2 The Genetic Material and Chromosome Structure, 8

3 DNA Replication: Prokaryotes, 18

4 DNA Replication and the Cell Cycle in Eukaryotes, 31

5 Mitosis and Meiosis, 39

6 Mutation, Mutagenesis and Selection, 47

7 Transcription, 61

8 Protein Biosynthesis (Translation), 82

9 The Genetic Code, 99

10 Phage Genetics, 106

11 Bacterial Genetics, 118

12 Recombinant DNA, 132

13 Eukaryotic Genetics: Mendel and his Laws, 142

14 Eukaryotic Genetics: Meiotic Genetic Analysis in Diploids, 150

15 Eukaryotic Genetics: Fungal Genetics, 164

16 Eukaryotic Genetics: an Overview of Human Genetics, 181

17 Extrachromosomal Genetics, 195

18 Biochemical Genetics (Gene Function), 204

19 Gene Regulation in Bacteria, 213

20 Regulation of Gene Expression in Eukaryotes, 232

21 Population Genetics, 243

Index, 254

Preface

This book is intended to accompany a general genetics course being given to undergraduate or medical students. It is assumed that the readers of this book have taken elementary courses in biology and chemistry but beyond that no special background is necessary.

Genetics has come to have a central role in all biological disciplines. It is a highly conceptual subject which can either be approached historically, starting with Mendel, or molecularly, with a description of DNA. My approach here is the latter, so that students can feel at ease with molecular genetics before relating those concepts to classical genetics. I have been using this approach in teaching a genetics course for the past seven years and it has been very well received by my students. The topics themselves, though, have been written so as to be self-contained and hence they can be selected in any order desired to complement an instructor's preference. Further, I have tried to make the text readable and up to date, with a reasonable balance between facts and the experimental strategies used to obtain the facts. The liberal number of figures in the book have been carefully prepared to complement and augment the text discussion. The figures have been drawn to give a simple visual expression of the information. To aid the student in further exploration of the subject, an extensive list of references is given at the end of each topic.

The completion of this book has fulfilled an ambition I have had for a number of years. I acknowledge the interest in this project of the students in my Genetics course at Reed College over the last few years. In addition, I wish to thank my wife, Jenny, for her support and encouragement throughout this endeavour, also Robert Campbell and those members of Blackwell Scientific Publications who have been involved with the production of this book for their helpful comments and co-operation.

Peter J. Russell
Reed College 1980

Topic 1
The Genetic Material

OUTLINE
Requirements for the genetic material
Nucleic acid structure
  DNA and RNA
  nucleotides
  polynucleotide organization
  the double helix
Evidence that DNA is the genetic material (historical).

The central theme of this book is the genetic material; its nature, structure, organization, replication, expression, etc. The approach used will be to discuss the salient facts in the context of the current literature and the analytical methods used in the areas under discussion. The emphasis of the book will be on the interrelationships between genetics, molecular biology and biochemistry.

REQUIREMENTS FOR THE GENETIC MATERIAL

The genetic material is of central importance to cell function and therefore must fulfil a number of basic requirements.

1. It must contain the information for cell structure, function and 'reproduction' in a stable form. This information is encoded in the sequence of basic building blocks of the genetic material.
2. It must be possible to replicate the genetic material accurately such that the same genetic information is present in descendant cells and in successive generations.
3. The information coded in the genetic material must be able to be decoded to produce the molecules essential for the structure and function of cells.

4. The genetic material must be capable of (infrequent) variation. Specifically, mutation and recombination of the genetic material are the foundations for the evolutionary process.

The nucleic acids, deoxyribonucleic acid (DNA) and ribonucleic acid (RNA) meet all these requirements.

NUCLEIC ACID STRUCTURE

Both DNA and RNA are linear polymeric macromolecules. The monomeric unit is called a nucleotide; deoxyribonucleotide in the case of DNA and ribonucleotide in RNA. A nucleotide consists of three components: a nitrogenous base (which is a derivative of either purine or pyrimidine), a pentose sugar, and one to three phosphate groups (Fig. 1.1).

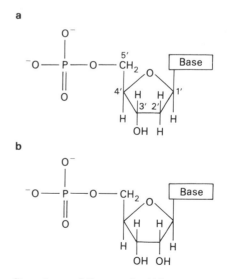

Fig. 1.1 Structure of the nucleotide components of DNA and RNA. (a) Deoxyribonucleoside 5'-monophosphate (monomeric unit of DNA). (b) Ribonucleoside 5'-monophosphate (monomeric unit of RNA).

The carbon positions in the pentose sugar ring are indicated by 1' to 5' to distinguish them from positions in the ring structure of the bases. The phosphoryl groups may be attached to any hydroxyl group of the sugar and the nucleotides with phosphates on the 5' carbon are of particular importance for the structure and function of DNA and RNA.

The four different deoxyribonucleotides are the major components of DNA. These are distinguished by the type of nitrogenous base they contain. The four bases characteristic of the deoxyribonucleotides are the purine derivatives, adenine (A) and guanine (G), and the pyrimidine derivatives thymine (T) and cytosine (C) (Fig. 1.2).

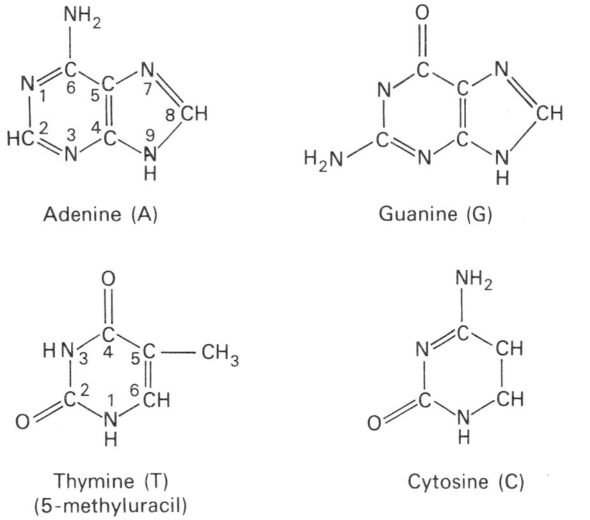

Uracil (U)

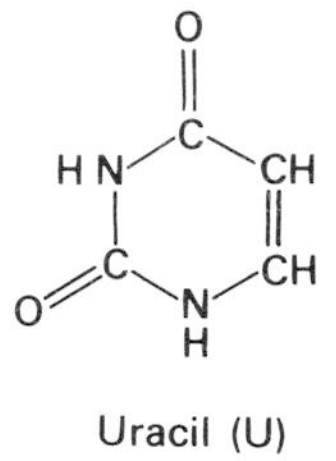

Fig. 1.3 The structure of uracil, the nitrogenous base found in RNA instead of thymine.

Adenine (A)   Guanine (G)

Thymine (T)   Cytosine (C)
(5-methyluracil)

Fig. 1.2 Purine and pyrimidine nitrogenous bases found in DNA.

Similarly, RNA is characterized by four different ribonucleotides which, like the monomeric units of DNA, contain the bases adenine, guanine, and cytosine. However, instead of thymine, RNA contains the pyrimidine derivative uracil (U), which has chemical and physical properties similar to those of thymine (Fig. 1.3).

The bases are attached to the pentose moiety by a covalent bond between the 1' carbon of the sugar and the 9-position nitrogen of the purines or the 1-position nitrogen of the pyrimidines.

Another distinction between DNA and RNA is the nature of the pentose sugar each contains. Specifically, deoxyribonucleotides contain 2-deoxy-D-ribose, whereas ribonucleotides contain ribose. This results in the two nucleic acids having different chemical properties which are biologically important (e.g. enzymes can be specific for DNA or RNA) and which may be exploited to separate the two molecules in the laboratory.

In DNA and RNA the mononucleotides are linked together by 3'-5'-phosphodiester bonds. Thus the backbone of both molecules consists of alternating phosphate and pentose groups. The bases are not part of the backbone structure. An example of an oligodeoxyribonucleotide is shown in Fig. 1.4.

Polynucleotides have polarity. The pentose sugar at one end of the chain has a 5'-hydroxyl or phosphoryl group (5' end), and the sugar at the other end has a 3'-hydroxyl group (3' end). A shorthand way to represent a polynucleotide strand is depicted in Fig. 1.5.

The DNA double helix

In 1953 James D. Watson and Francis H.C. Crick proposed that most DNA is in the form of a double-stranded, right-handed helix. The evidence for their hypothesis was as follows:

1. The DNA molecule consists of bases, sugars and phosphoryl groups linked together as a polynucleotide chain as discussed above.

2. E. Chargaff analysed the nucleotides released by chemical hydrolysis and found that the total amount of purines present is always equal to the total amount of pyrimidines present. More specifically, adenine always equals thymine (A = T), and

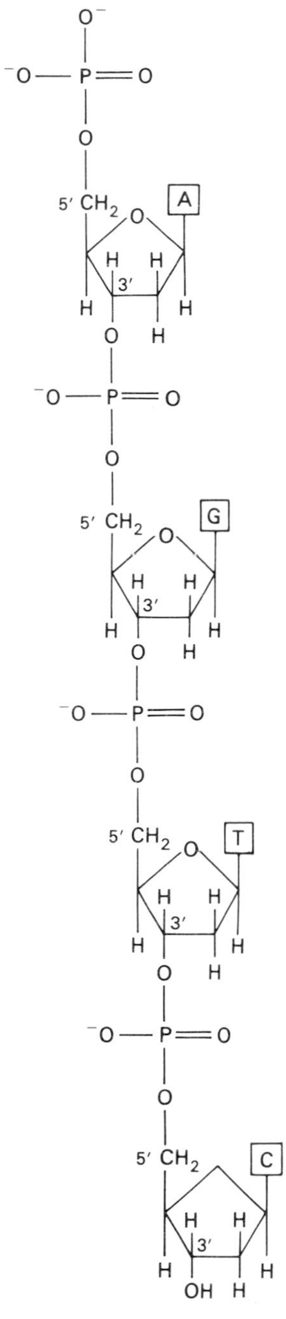

Fig. 1.4 An oligodeoxyribonucleotide chain showing the linkages between the monomeric units in a single DNA chain.

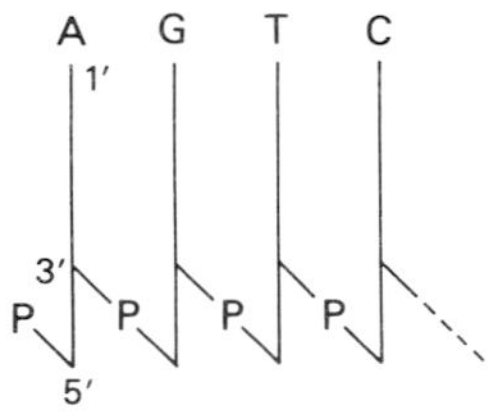

Fig. 1.5 A shorthand way to represent a polynucleotide chain.

guanine always equals cytosine (G = C). Thus the following equations hold for double-stranded DNAs:

$$A + G = C + T$$
$$A + G/C + T = 1$$
$$A + T/G + C \text{ does not} = 1 \text{ (in most cases).}$$

The latter is called the base ratio of the DNA and this is often expressed as % GC. The base ratio varies widely among organisms but it remains constant for any one species.

3. R. Franklin, M.H.F. Wilkins and co-workers analysed fibres of DNA by X-ray diffraction. The patterns they obtained indicated the DNA was a helical structure consisting of two or more chains wound round each other.

Watson and Crick fitted the chemical and physical data into a symmetrical structure that was compatible with all the facts and that also possessed the properties one would expect of genetic material. The Watson-Crick model of DNA involves two polynucleotide chains which are wound around each other to form a double helix (Fig. 1.6). The two chains are joined together by hydrogen bonding between the bases which are flat structures stacked like coins and arranged at right angles to the long axis of the polynucleotide chain. The sugar-phosphate backbones are on the outside of the helix.

From the model it is possible to show that there are ten base pairs per each complete turn of the polynucleotide chain. Since the distance between adjacent base pairs is 0.34 nm, then it follows that the DNA helix has one turn each 3.4 nm of length.

The most important feature of the model is the specific pairing of the bases. Only two complementary base pairs, A-T and G-C, can form stable bonds

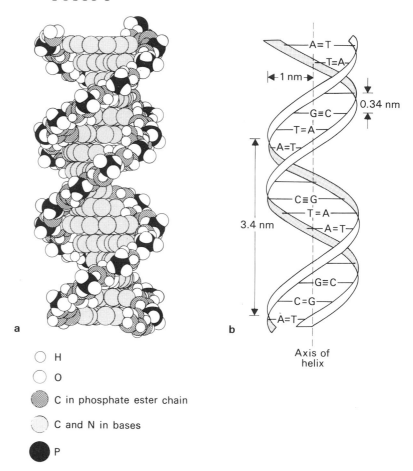

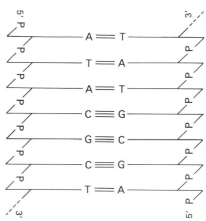

Fig. 1.6 The Watson-Crick model of DNA. (a) Molecular model of DNA double helix. (After M. Feughelman *et al*, 1955. *Nature,* **175** 834, courtesy of M. H. F. Wilkins.) (b) Diagrammatic representation of the DNA double helix.

○ H

○ O

◉ C in phosphate ester chain

○ C and N in bases

● P

in the double-helical structure. As a result of this, the nucleotide sequence in one strand dictates the nucleotide sequence of the other. In other words, the two strands are complementary. The A-T base pair has two hydrogen bonds and the G-C pair has three hydrogen bonds. This specific complementary base pairing is of central importance for many functions of the nucleic acids, for example, DNA replication, transcription, and translation.

Another property of the model is that the two chains of the double helix are oriented with opposite (antiparallel) polarity in terms of the 3′-5′-phosphate-deoxyribose linkages (Fig. 1.7).

Fig. 1.7 A diagrammatic representation of double-stranded DNA showing the opposite polarity of the two strands.

EVIDENCE THAT DNA IS THE GENETIC MATERIAL (historical)

Many lines of evidence strongly indicate that DNA is the genetic material in many organisms. Five examples are given here:

1. A. Mirsky and H. Ris in 1949 demonstrated that all cells of an organism contained the same amount of DNA whereas cells of different types within an organism contained different amounts and kinds of protein. They cited these data to suggest that DNA was the genetic material. We know now that the DNA content of a cell of an organism can vary depending on the tissue of origin. In general, though, with the exception of spontaneous chromosome loss or breakage, the DNA content is usually a multiple of the DNA content of the haploid gamete of the organism. Thus, for example, the DNA content of the root nodules of leguminous plants such as the pea is characteristically double that of the rest of the plant.

2. The amount of DNA per cell is related to the complexity of the cell (e.g. cells of higher organisms contain about 60 times as much DNA as bacterial cells).

3. The nucleic acids show maximal absorbance of ultraviolet light at a wavelength of 260 nm (Fig. 1.8) and this correlates exactly with the wavelength at which maximal mutagenesis of cells can be achieved by ultraviolet irradiation. This observation provided further evidence that nucleic acids and not proteins (which show maximal absorbance of light at 280 nm) are the genetic material.

4. In 1928 F. Griffith discovered that one strain of the bacterium *Pneumococcus,* the S strain, when injected into mice causes death by septicaemia (blood poisoning). Another strain, the R strain, had no effect on the same mice. The distinction between the two strains lies in the fact that the S strain bacteria have a polysaccharide capsule around them resulting in a smooth colony appearance when they grow on solid medium in a culture dish (hence the S designation.) The R strain produces rough-appearing colonies owing to the lack of the capsule. F. Griffith showed that the S bacteria could mutate spontaneously to give rise to the R type. Further, he showed that when mice were

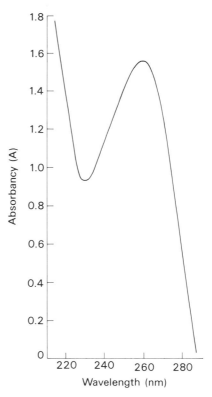

Fig. 1.8 Ultraviolet light absorbancy spectrum of DNA showing the maximal absorbancy at 260 nm.

injected with a combination of live R bacteria and heat-killed S bacteria, the mice died from septicaemia and live S bacteria could be isolated from their blood (Fig. 1.9). Thus something from the dead bacteria converted the R bacteria into S-type cells; this process is called *transformation.*

This phenomenon received further scrutiny from O.T. Avery, C.M. Macleod and M. McCarty in 1944. In some classic experiments they set out to determine the chemical nature of the substance (the so-called *transforming principle*) that induced the specific transformation of the pneumococcal types. They showed that a DNA fraction isolated from S strain was capable of transforming unencapsulated R-type bacteria into fully encapsulated S-type cells. None of the other cell fractions, such as RNA, protein, lipid or carbohydrates, was able to effect the transformation. Further, the transforming activity of the DNA fraction could be

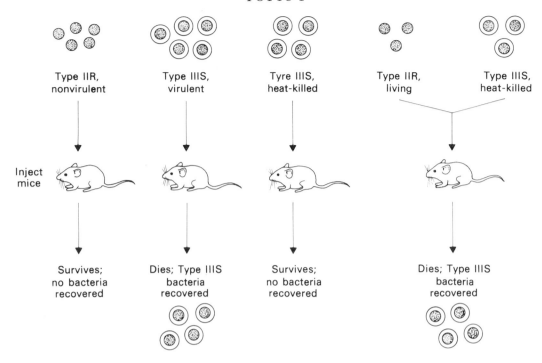

Fig. 1.9 Transformation experiment of F. Griffith. (After M. W. Strickberger, 1976. *Genetics*. Macmillan, New York.)

abolished by treatment with deoxyribonuclease (DNase), a DNA-degrading enzyme. These results, then, strongly indicated that DNA was the genetic material.

5. A.D. Hershey and M. Chase in 1952 studied the replication of bacteriophages (bacterial viruses) in their bacterial hosts. Bacteriophages consist of two components: DNA and protein. Hershey and Chase prepared bacteriophages that were radioactively labelled either in the DNA or in the protein. They then used these phages to infect the bacterial host and found that phage DNA enters the bacteria whereas phage protein does not. Moreover, when progeny phage were released from the bacteria, they contained some of the radioisotope used to label DNA and none of the isotope used to label protein (Fig. 1.10).

We now know that the genetic material in most organisms is DNA. In some viruses, however, the genetic material is RNA, as will be discussed later.

REFERENCES

Avery O.T., C.M. Macleod & M. McCarty. 1944. Studies on the chemical nature of the substance inducing transformation of pneumococcal types. Induction of transformation by a desoxyribonucleic acid fraction isolated from pneumococcus type III. *J. Exp. Med.* **79**: 137–158.

Chargaff E. 1950. Chemical specificity of nucleic acids and mechanism of their enzymatic degradation. *Experientia*, **6**: 201–209.

Chargaff E. 1951. Structure and function of nucleic acids as cell constituents. *Fed. Proc.* **10**: 654–659.

Davidson J.N. 1972. *The Biochemistry of the Nucleic Acids*, 7th edn. Chapman and Hall, London.

Hershey A.D. & M. Chase. 1952. Independent functions of viral protein and nucleic acid in growth of bacteriophage. *J. Gen. Physiol.* **36**: 39–56.

Watson J.D. & F.H.C. Crick. 1953. A structure for desoxyribose nucleic acids. *Nature*, **171**: 737–738.

Wilkins M.H.F., A.R. Stokes & H.R. Wilson. 1953. Molecular structure of deoxypentose nucleic acids. *Nature*, **171**: 738–740.

a. Preparation of radioactively-labelled T2 bacteriophage

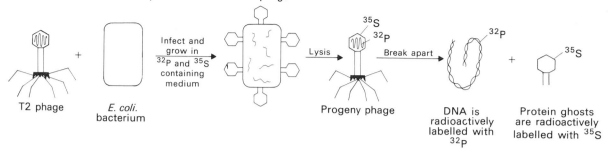

b. Demonstration that DNA is the genetic material in T2

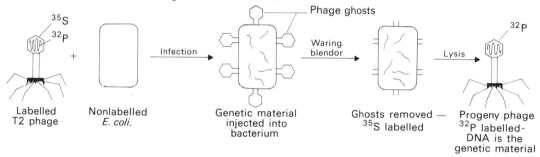

Fig. 1.10 The Hershey and Chase experiment. (After M. W. Strickberger, 1976. *Genetics*. Macmillan, New York.)

Topic 2
The Genetic Material and Chromosome Structure

OUTLINE

Prokaryote, eukaryote definitions
Phage chromosomes
    T phages, phage lambda
Bacterial chromosomes
Eukaryotic chromosomes
    organellar DNA
    karyotype
    chromatin composition
    nucleosomes and chromatin
    euchromatin and heterochromatin
    repetitive DNA sequences

Before beginning discussion of this topic, some terms must be defined in a general (not necessarily complete) way:

Eukaryote—organism with cells that possess a 'true' nucleus, that is, one with a nuclear membrane.

Prokaryote—bacteria and blue-green algae which do not have a 'true' nucleus.

Virus—an 'organism' which can reproduce only after infecting a living cell. Viruses that infect bacteria are called bacteriophages (phages). Most viruses have a 'coat' of protein surrounding the genetic material (DNA or RNA).

PHAGE CHROMOSOMES

A number of phages can infect the intestinal bacterium *Escherichia coli* (*E. coli*). The T series of phages all have double-stranded DNA as their genetic material (Fig. 2.1). When they are added to a culture of *E. coli*, they attach to the outer surface of the bacterium and then inject a single molecule of DNA into the host. Once inside, the DNA is replicated, progeny phages are assembled and eventually the bacterium lyses releasing progeny into the medium (Fig. 2.2).

The chromosomes of the T phages are naked DNA, that is, no proteins are attached. Also, there are interesting differences in the arrangement of genes along the linear chromosome. In both the T2 and T4 phages, the chromosomes are longer than the complete genome. This results from the chromosomes being terminally redundant and circularly permuted as if each phage particle has a standard

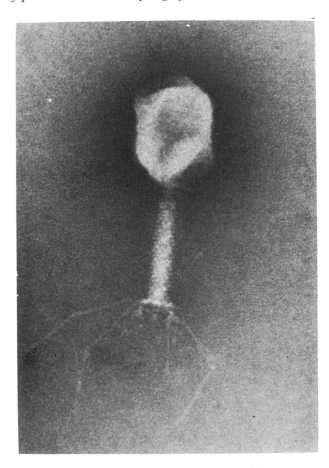

Fig. 2.1 (a) EM photograph of phage T4 (×264 000, negatively stained. Courtesy of M. Wurtz).

8

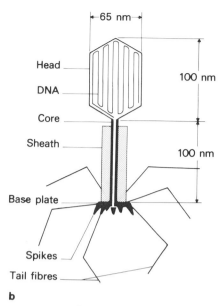

(b) Diagrammatic representation of a T4 phage. (From W. H. Hayes, *Genetics of Bacteria and Their Viruses*. Copyright © 1968, Blackwell Scientific Publications Ltd., with permission from W. H. Hayes and Blackwell Scientific Publications, Oxford.)

length of DNA randomly excised from a long chain of chromosomes joined end to end. On the other hand, the T3, T5, and T7 phages have terminally redundant, but not circularly permuted chromosomes. These types of chromosome arrangements are shown in the following:

1 2 3 4 5 6	Complete genome
1 2 3 4 5 6 1 2	Terminally redundant chromosome
1 2 3 4 5 6 4 5 6 1 2 3 3 4 5 6 1 2	Circularly permuted chromosomes
1 2 3 4 5 6 1 2 3 4 5 6 1 2 3 4 6 1 2 3 4 5 6 1	Terminally redundant and circularly permuted chromosomes

The experimental evidence for these types of chromosome arrangements is as follows:

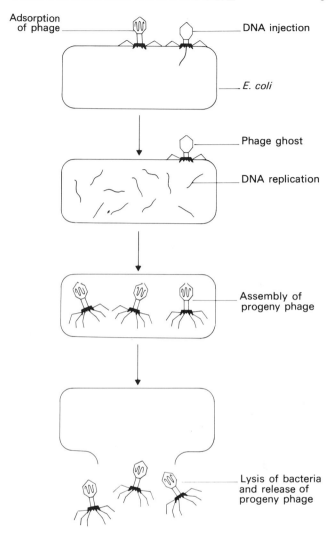

Fig. 2.2 Life cycle of phage T2.

Terminal redundancy

If a chromosome is treated with the enzyme exonuclease III, nucleotides are removed from the 3′ end of each strand of the DNA. If the chromosome is terminally redundant, this treatment will expose complementary 5′-ended strands at the two ends of the linear chromosome. These ends may then pair by hydrogen bonding to form circles which may be visualised by electron microscopy (Fig. 2.3).

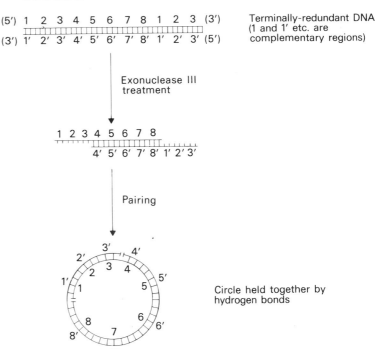

Fig. 2.3 Demonstration of terminally redundant DNA by exonuclease treatment. (After W. H. Hayes, 1968, *Genetics of Bacteria and Their Viruses*. Blackwell Scientific Publications, Oxford.)

Circular permutation

If a double-stranded DNA is heated, then eventually the two strands will separate (denature) by breakage of the hydrogen bonds between the base pairs. If allowed to cool, complementary strands of DNA can come together again (re-anneal) to form double-stranded molecules. If a population of circularly permuted DNA molecules is denatured and then allowed to re-anneal, it is possible that the terminal region of one strand will be complementary to the central part of a second strand. If the complementary regions anneal, then a molecule is produced which is double-stranded in the middle and which has complementary, single-stranded ends. This condition can lead to the formation of circles of double-stranded DNA as discussed for terminally redundant chromosomes (Fig. 2.4).

Another phage that infects *E. coli* is lambda (λ) (Fig. 2.5). Its chromosome is double-stranded DNA and in the phage particles it is a linear molecule. If the chromosome is heated and then cooled, circular structures are seen under the electron microscope. The explanation for this is that the two ends of the chromosome have complementary single-stranded regions which can therefore form hydrogen bonds between them to produce circles. Indeed when the lambda DNA is injected into the host bacterium, the DNA rapidly becomes a covalently bonded circle as a result of the cohesive ends and with the aid of specific enzymes. This event is necessary to enable the phage chromosome either to replicate or to integrate into the bacterial chromosome.

Other viruses present us with examples of a variety of chromosome structures: for example, phage φ × 174 has a single-stranded, circular DNA chromosome, phage Q beta and the plant virus tobacco mosaic virus (TMV) have single-stranded RNA chromosomes, and polyoma viruses have double-stranded RNA for their genetic material.

BACTERIAL CHROMOSOMES

The chromosomes of bacteria are circular, naked, double-stranded DNA molecules. The DNA is

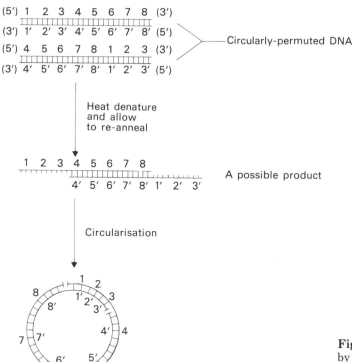

Fig. 2.4 Demonstration of circularly permuted DNA by heat denaturation and reannealing. (After W. H. Hayes, 1968, *Genetics of Bacteria and Their Viruses.* Blackwell Scientific Publications, Oxford.)

usually found attached to the cell membrane at some point or points. Although bacteria do not possess a nucleus, the DNA is localised in a distinct area within the cell called the nucleoid region. Within this region, the DNA is highly convoluted and folded, and very few cytoplasmic particles are found there. There is no membrane around the nucleoid region.

The chromosomes of bacteria can be isolated in a highly folded conformation by gentle lysis at room temperature with nonionic detergents in 1.0 M NaCl. Electron microscopic analysis of the chromosomes reveal extensive packing of the DNA, which are folded into loops (10–80 per chromosome) and supercoils (Fig. 2.6). The folded chromosomes have been shown to contain all of the nascent RNA chains of the cell, and the enzyme for RNA synthesis, RNA polymerase, but no ribosomes. Hence protein synthesis apparently does not occur in the close proximity of the DNA.

EUKARYOTIC CHROMOSOMES

In eukaryotes, most of the DNA is located in the chromosomes found in the nucleus, and the following discussion concentrates on these. However, some DNA is found in mitochondria and chloroplasts and in these instances it is naked, double-stranded, and circular, that is, very similar to the organization of the genetic material in bacteria.

In eukaryotic cells, the chromosomal complement is called its karyotype and is characterised both by the number and morphology of the chromosomes and the positions of the centromeres (where spindle fibres attach during the cell division processes). In animals the karyotype is generally different in males and females owing to the presence of different complements of X and Y (sex) chromosomes. Barring chromosomal aberrations, the karyotype for the autosomes (that is, the set of chromosomes other than the sex chromosomes) is

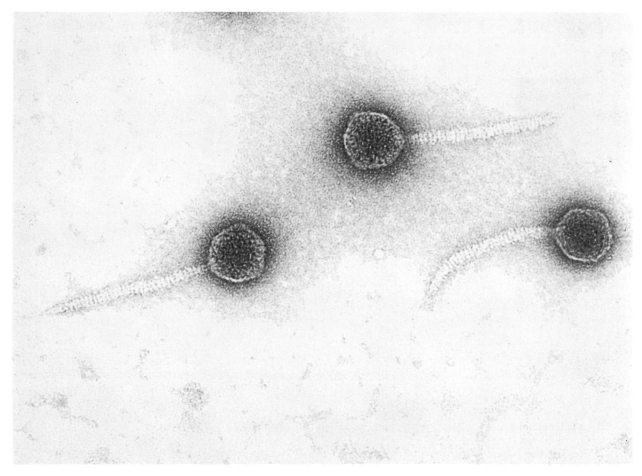

Fig. 2.5 Electron micrograph of phage lambda (λ).
(× 258 000; negatively stained with 2% uranyl acetate.
Courtesy of R. B. Luftig.)

invariant within species, but differs from species to species. A karyotype for a normal human male is shown in Fig. 2.7.

As can be seen, there are 46 chromosomes which, because man is diploid, involve 22 pairs of homologous autosomes and two sex chromosomes, one X and one Y. In this organism the two sex chromosomes are quite distinct morphologically. Until recently it was possible to differentiate the chromosomes only into seven groups, designated A to G, and ordered according to their size. Within each group the chromosomes were almost indistinguishable morphologically. New staining techniques (called banding) have been developed that give rise to specific patterns of bands along the chromosomes that serve to differentiate each chromosome in the karyotype (Fig. 2.8). As an example, one of the methods of staining, Q banding, involves treating the chromosomes with quinacrine mustard and visualising bands under ultraviolet illumination.

Most biochemical studies of chromosome structure and function have been done with interphase chromosomes. When nuclei are isolated and then lysed, the chromosomes are released. Each chromosome contains a single, unbroken, double-stranded DNA running through its length. This DNA is

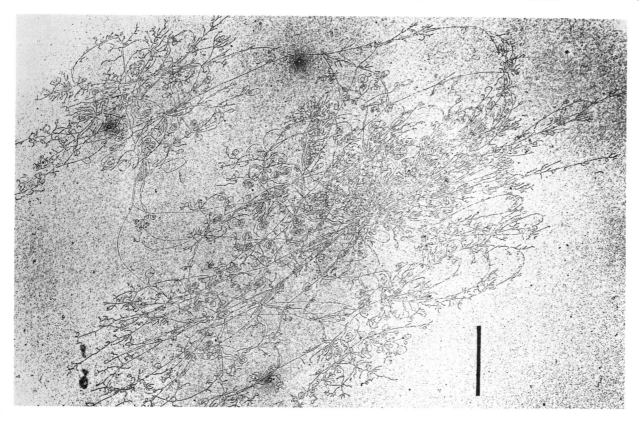

Fig. 2.6 Electron micrograph of the chromosome extracted from *E. coli*. Note the extensive supercoiling of the DNA. The bar is 2 μm. (Courtesy of A. Worcel. With permission from H. Delius and A. Worcel, 1974, *Electron microscope visualization of the folded chromosome of Escherichia coli*. Journal of Molecular Biology **82**: 107–109. Copyright © 1974, Academic Press (London) Ltd.)

associated with proteins in a complex called chromatin. When purified chromatin is analysed, it is found that both basic proteins (histones) and acidic proteins (nonhistones) are associated with the DNA and this is characteristic of eukaryotic nuclear chromosomes. In all eukaryotes there are five distinct histone proteins attached to the DNA in a DNA–histone complex: H1, which is very rich in the basic amino acid lysine; H2A and H2B, which are lysine rich; and H3 and H4 which are arginine rich (Table 2.1). These molecules have been conserved through evolution, as would be expected for these proteins since they have such a basic and universal role in determining chromosome structure. The nonhistone acidic proteins are found more or less firmly attached to the DNA–histone

complexes. In contrast to the histones, these molecules are numerous and heterogeneous. In most organisms that have been examined there are more

Table 2.1 Characteristics of calf thymus histones. (After S.C.R. Elgin and H. Weintraub, 1976, Annu. Rev. Biochem. **44**: 725).

Type	Characteristics	No. of amino acids	Molecular weight
H1	Very lysine rich	~215	~21 500
H2A	Lysine rich	129	14 000
H2B	Lysine rich	125	13 775
H3	Arginine rich	135	15 320
H4	Arginine rich	102	11 280

of nonhistones varies throughout the cell cycle and from one differentiated cell to another. This is in contrast to the histones which remain constant in both these instances.

At the present time, there is quite a body of information on the organization of the histones along the DNA, and this will be described in the following section. The number and arrangement of nonhistones on the DNA have not been defined in as much detail.

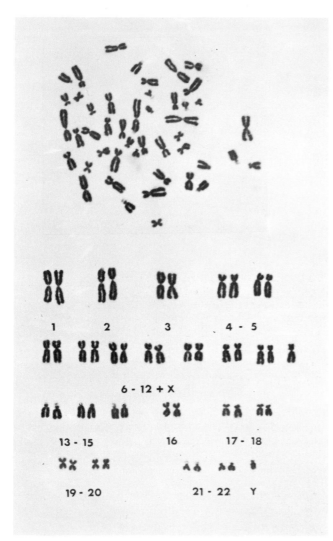

Fig. 2.7 Top: mitotic chromosomes of a normal human male (46, XY). Bottom: Unbanded karyotype: the same chromosomes arranged in homologous pairs. (Courtesy of A. T. Sumner, with permission from C. J. Bostock and A. T. Sumner, *The Eukaryotic Chromosome*. Copyright © 1978, North Holland Publishing Co., Amsterdam, The Netherlands.)

than 100 nonhistones and these include the replication and transcription enzymes, DNA polymerase and RNA polymerase, and molecules involved in the control of DNA and RNA synthesis. Thus, as would be expected in view of this, the complement

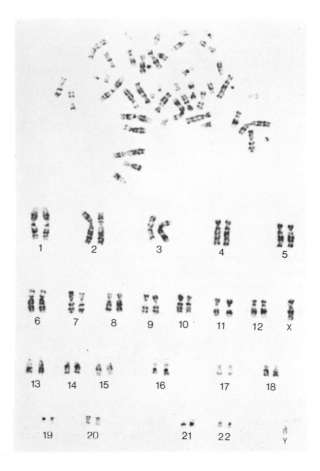

Fig. 2.8 Top: Human male mitotic chromosomes stained with acetic acid-saline-Giemsa (ASG). Bottom: The same ASG-banded chromosomes arranged in homologous pairs and numbered. (Courtesy of A. T. Sumner, with permission from A. T. Sumner *et al*, *Nature New Biology*, **232**:31–32. Copyright © 1971, Macmillan Journals Ltd., England.)

Nucleosomes and chromatin

As we have said, chromosomes are tight complexes between DNA and protein. It has long been known that the DNA in chromosomes is much longer than the chromosome length and thus some means of compacting the genetic material must be available. Indeed, in chromatin the DNA is in a tightly packed state representing at least a 100-fold contraction. This is apparently achieved by several orders of folding, but only the first order folding, the arrangement of DNA with histones in nucleosomes, has been described well to date. The existence of nucleosomes (also called nu-bodies) was first proposed by R. Kornberg; they may be visualised quite readily under the electron microscope (Fig. 2.9). A diagrammatic representation of a nucleosome, showing the proposed path of the DNA around the histone core is shown in Fig. 2.10. Biochemical analysis has revealed the following main features of the nucleosome repeat units and their components:

1. The nucleosomes occur about every 200 base pairs along the DNA. Thus, the interactions between DNA and histone are not DNA-sequence specific, from which we can conclude that the assembly of chromatin must be the direct result of histone–histone and DNA–histone interactions.
2. Reconstitution experiments have shown that each nucleosome contains eight core histones, that

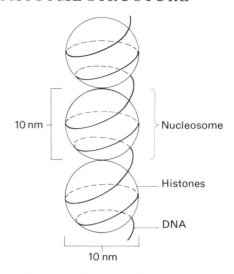

Fig. 2.10 General schematic diagram of chromatin showing the nucleosome with the DNA following a path around them. (After R. D. Kornberg, 1977, *Annu. Rev. Biochem.* **46**:931–954.)

is two molecules each of H2A, H2B, H3, and H4.
3. The core of histones is associated with approximately 140 base pairs of the DNA in the repeat unit. This is an average figure since the length of DNA per nucleosome varies between higher and lower eukaryotes and there are also reported instances of variation in the DNA content of nucleosomes within the same type of cells.

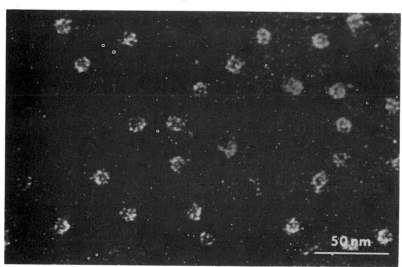

Fig. 2.9 An electron micrograph of nucleosome core particles from the chicken showing the highly compact globular nature of these structures. (Courtesy D. Olins. From D. Olins and A. Olins, 1978, *Nucleosomes: the structural quantum in chromosomes.* American Scientist **66**: 704–711. Reprinted by permission of American Scientist, Journal of Sigma Xi, the Scientific Research Society.)

4. The path of the DNA with respect to the nucleosome has been studied and it has been established that the DNA is on the outside of the structure. More specifically, the DNA is coiled almost two times around the nucleosome in a 90 base pair/turn superhelix around two symmetrically paired histone tetramers. This coiling arrangement introduces approximately one superhelical turn into the double helix per nucleosome.

5. Histone H1 is present in about one copy per nucleosome and is associated with the variable linker region of DNA between adjacent nucleosomes. The current thought is that H1 stabilizes the interaction of adjacent nucleosomes.

6. Recent work has included studies of the three-dimensional structure of the nucleosomes and their modes of interaction. Results of these studies suggest that the nucleosome core is a short, wedge-shaped cylinder measuring approximately 11 × 11 × 6 nm. The cores apparently interact primarily between the top and bottom surfaces of these cylinders.

7. The chromatin structure does vary as a result of changes in higher-order folding. These changes are probably brought about by chemical modification (for example acetylation or phosphorylation) of the histones and perhaps by interactions with specific nonhistones. These events occur as the processes of DNA replication and RNA synthesis are carried out and more will be said about them in later Topics.

Euchromatin and heterochromatin

When interphase chromosomes of metabolically active cells are chemically stained and examined under the microscope, it becomes apparent that there are two distinct types of organisation of the chromatin material. One type is lightly staining and is called euchromatin and the other type is darkly staining and is called constitutive heterochromatin. The former involves chromatin that is in a relatively uncoiled and unclumped state, whereas the latter exhibits higher-order folding of the chromatin to produce a relatively coiled and clumped state. (Indeed, when the cell enters the mitotic phase of the cell cycle, in general all of the chromosomes become more highly condensed as a result of such higher order folding.)

The distribution of constitutive heterochromatin varies from organism to organism and examples are known where parts of chromosomes or whole chromosomes are heterochromatinised. In general, constitutive heterochromatin is found interspersed in short segments among euchromatin and is located around the centromeres. Functionally, euchromatin contains DNA in an active, or potentially active state (that is, capable of being transcribed) whereas heterochromatin contains DNA in a transcriptionally inactive configuration. Characteristically heterochromatin replicates later than euchromatin in the cell cycle.

Repetitive DNA sequences

If the DNA of any organism is isolated, sheared to pieces several hundred nucleotide pairs long, denatured to single strands, and then the complementary strands are allowed to reassociate, then the rate of reassociation for a particular nucleotide sequence will be related to the number of copies of that sequence in the genome.

From this type of DNA–DNA reassociation experiment, it was discovered that prokaryotic chromosomes consist almost entirely of unique DNA sequences. By contrast, eukaryotes were found to have three general (and somewhat arbitrarily defined) frequency classes of DNA sequences in their chromosomes: (1) highly repetitive sequences (also called satellite DNA), that is, more than 100 000 copies of 5–300 nucleotides; (2) middle-repetitive sequences, that is, 10 to 100 000 copies; and (3) unique sequences. Different eukaryotic organisms contain different relative amounts of these three frequency classes of DNA sequences.

The highly repetitive, satellite DNA sequences are not transcribed and tend to be clustered in the heterochromatin regions around centromeres. The function of these sequences is unknown. More is known about the middle-repetitive sequences, however. They represent a broad, heterogeneous class of DNA sequences, some of which are definitely transcribed. The multiple copies of the genes for ribosomal RNA (rRNA), transfer RNA (tRNA), and histones are examples of middle-repetitive DNA sequences. Finally, the unique sequence DNA constitutes the major fraction of DNA in a eukaryotic organism. It seems clear at this time that most of the cellular proteins are coded by unique sequence DNA.

REFERENCES

Britten R.J. & D.E. Kohne, 1968. Repeated sequences in DNA. *Science,* **161**: 529–540.

Chambon P. 1978. Summary: The molecular biology of the eukaryotic genome is coming of age. *Cold Spring Harbor Symp. Quant. Biol.* **42**: 1209–1234.

Cold Spring Harbor Symposia on Quantitative Biology, vol. 38, 1973, *Chromosome Structure and Function,* Cold Spring Harbor Laboratory, New York.

Cold Spring Harbor Symposia on Quantitative Biology, vol. 42, 1977, *Chromatin.* Cold Spring Harbor Laboratory, New York.

Comings D.E., B.W. Kovacs, B.E. Avelino & D.G. Harris, 1975. Mechanisms of chromosome banding V. Quinacrine banding. *Chromosoma,* **50**: 111–145.

Crick F.H.C. 1976. Linking numbers and nucleosomes. *Proc. Natl. Acad. Sci. USA,* **73**: 2639–2642.

Dubochet J. & M. Noll, 1978. Nucleosome arcs and helices. *Science,* **202**: 280–286.

DuPraw E.J. 1970. *DNA and Chromosomes.* Holt, Rinehart and Winston.

Dutrillaux B. & J. Lejeune, 1975. New techniques in the study of human chromosomes: methods and applications. *Adv. Hum. Genet.* **5**: 119–156.

Elgin S.C.R. & H. Weintraub, 1975. Chromosomal proteins and chromatin structure. *Annu. Rev. Biochem.* **44**: 725–774.

Finch J.T., L.C. Lutter, D. Rhodes, R.S. Brown, B. Rushton, M. Levitt & A. Klug, 1977. Structure of the nucleosome core particles of chromatin. *Nature,* **269**: 29–36.

Gottesfeld J.M. & D.A. Melton, 1978. The length of nucleosome-associated DNA is the same in both transcribed and nontranscribed regions of chromatin. *Nature,* **273**: 317–319.

Kennell D. E. 1971. Principles and practices of nucleic acid hybridization. *Progr. Nucl. Acid Res.* **11**: 259–302.

Kornberg R.D. 1974. Chromatin structure: a repeating unit of histones and DNA. *Science,* **184**: 868–871.

Kornberg R.D. 1977. Structure of chromatin. *Annu. Rev. Biochem.* **46**: 931–954.

Lima-de-Faria A. (ed.) 1969. *Handbook of Cytology.* North-Holland, Amsterdam.

Noll M. & R.D. Kornberg, 1977. Action of micrococcal nuclease on chromatin and the location of histone H1. *J. Mol. Biol.* **109**: 393–404.

Prunell A. & R.D. Kornberg, 1977. Relation of nucleosomes to DNA sequences. *Cold Spring Harbor Symp. Quant. Biol.* **42**: 103–108.

Ris H. & D.F. Kubai, 1970. Chromosome structure. *Annu. Rev. Genet.* **4**: 263–294.

Southern E.M. 1974. Eukaryotic DNA. In *Biochemistry of Nucleic Acids,* K. Burton (ed.), vol. 6 of MTP International Review of Science, pp. 103–139. Butterworths, London.

Straus N.A. 1976. Repeated DNA in eukaryotes. In *Handbook in Genetics,* vol. 5, R.C. King (ed.), pp. 3–30. Plenum Press, New York.

Streisinger G., J. Emrich & M.M. Stahl, 1967. Chromosome structure in bacteriophage T4. III. Terminal redundancy and length determination. *Proc. Natl. Acad. Sci. USA,* **57**: 292–295.

Tartof K.D. 1975. Redundant genes. *Annu. Rev. Genet.* **9**: 355–385.

Thomas C.A. 1971. The genetic organization of chromosomes. *Annu. Rev. Genet.* **5**: 237–256.

Thomas C.A. & L.A. MacHattie, 1967. The anatomy of viral DNA molecules. *Annu. Rev. Biochem.* **36**: 485–518.

Weintraub H., S.J. Flint, I.M. Leffak, M. Groudine & R.M. Grainger. 1977. The generation and propagation of variegated chromosome structures. *Cold Spring Harbor Symp. Quant. Biol.* **42**: 401–407.

Worcel A. 1977. Molecular architecture of the chromosome fiber. *Cold Spring Harbor Symp. Quant. Biol.* **42**: 313–324.

Worcel A. & E. Burgi, 1972. On the structure of the folded chromosome of *Escherichia coli. J. Mol. Biol.* **71**: 127–147.

Wu R. & E. Taylor, 1971. Nucleotide sequence analysis of DNA II. Complete nucleotide sequence of the cohesive ends of bacteriophage lambda DNA. *J. Mol. Biol.* **57**: 491–511.

Wu R. 1978. DNA sequence analysis. *Annu. Rev. Biochem.* **47**: 607–634.

Yunis J.J. 1976. High resolution analysis of human chromosomes. *Science,* **191**: 1268–1270.

Topic 3
DNA Replication: Prokaryotes

OUTLINE
Nucleotide synthesis — *de novo* and salvage
  pathways
Purine biosynthesis
Pyrimidine biosynthesis
DNA synthesis *in vitro*
DNA replication *in vivo*
  Meselson and Stahl experiment
  discontinuous DNA replication
  enzymes involved in DNA replication
  DNA chains initiated by RNA primer.

NUCLEOTIDE SYNTHESIS

In order to be efficient, virtually all cells use two different kinds of pathway for the synthesis of nucleotides. One is a de-novo pathway, in which a sugar (ribose 5'-phosphate), certain amino acids, carbon dioxide and NH_3 are combined in a series of reactions to form the nucleotides directly. In this case neither the free purines or pyrimidines nor the nucleosides are intermediates in the pathway.

The second kind of pathway is called a salvage pathway. Here the purines, pyrimidines and nucleosides released by the breakdown of nucleic acids can, by a variety of routes, be converted back to the nucleotides needed for nucleic acid synthesis. Both pathways are important to the cell and their relative importance depends on the circumstances in which the cell finds itself. In fact, the existence of the two pathways makes it difficult to study the control of nucleotide biosynthesis.

PURINE BIOSYNTHESIS

The purine biosynthetic pathway follows essentially the same sequence of reactions in a wide range of organisms, for example *E. coli*, yeast and man.

The purine ring structure is assembled on the sugar, ribose 5'-phosphate (Fig. 3.1). A number of components is used to produce the purine ring, including some amino acids. Fig. 3.2 summarises the origin of the purine ring atoms and Fig. 3.3 presents the structures of the amino acids involved.

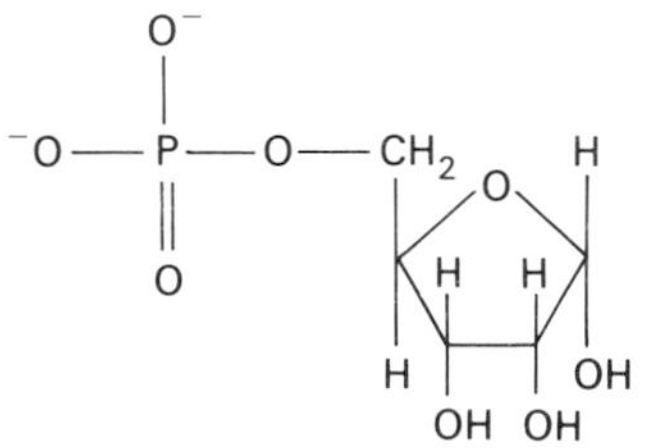

Fig. 3.1 The structure of ribose 5'phosphate.

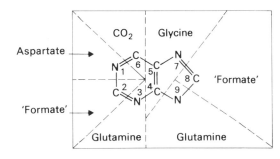

Fig. 3.2 Origin of the carbon and nitrogen atoms of the purine ring.

The primary product of the purine biosynthesis pathway is the ribonucleotide inosine 5'-monophosphate (IMP) and it is from this compound that the adenine and guanine ribonucleotides are derived (Fig. 3.4). As will be described later, these latter two compounds are then phosphorylated further to convert them to the immediate RNA and DNA precursors.

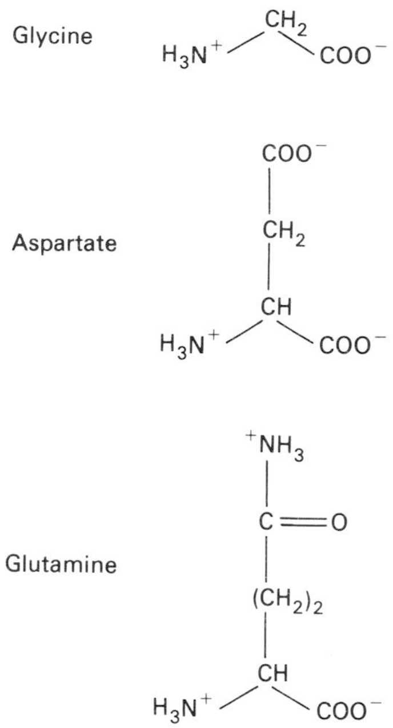

Glycine

Aspartate

Glutamine

Fig. 3.3 Structures of amino acids used in the formation of the purine (and pyrimidine) rings.

PYRIMIDINE BIOSYNTHESIS

The biosynthesis of pyrimidine nucleotides differs in general from that of the purine nucleotides in that the pyrimidine ring is assembled first and then is attached to the ribose 5'-phosphate. As was the case with the purine ring, a number of compounds are used in the construction of the pyrimidine ring and these are shown in Fig. 3.5. The pyrimidine biosynthesis pathway itself is depicted in Fig. 3.6 and the key pyrimidine intermediate in this pathway is orotate. It is the orotate molecule which becomes attached to an activated form of ribose 5'-phosphate, phosphoribosyl pyrophosphate (PRPP) to form a nucleotide, orotidine 5'-monophosphate (OMP). Removal of a carbon dioxide molecule generates uridine 5'-monophosphate (UMP). The cytosine nucleotide then is produced from UMP by the addition of an amino (NH₂) group (Fig. 3.6).

The nucleoside monophosphates generated by the de-novo purine and pyrimidine biosynthesis pathways do not participate directly in nucleic acid biosynthesis. Rather, they are first converted to the triphosphate derivatives via diphosphate intermediates. The phosphate donor for all of these reactions is the ribonucleotide, adenosine 5'-triphosphate (ATP), and the reactions themselves are catalysed by enzymes called kinases. These enzymes are obvious sites for regulation of the production of DNA and RNA precursors. The conversion of nucleoside monophosphates to diphosphates involves kinases that are specific for each base but nonspecific with regard to the ribose or deoxyribose sugar. Thus, they are involved in the synthesis of both RNA and DNA precursors. On the other hand, the synthesis of nucleoside triphosphates from the diphosphates is catalysed by a kinase that is nonspecific for both the bases and the sugars.

Therefore, by these kinase-catalysed reactions, the ribonucleoside 5'-triphosphates are produced from the ribonucleoside monophosphate end products of the purine and pyrimidine biosynthesis pathways. These triphosphates are the immediate precursors for RNA synthesis, a process which will be described in a later Topic.

The precursors for DNA synthesis are also derived from the ribonucleoside 5'-monophosphates. Firstly, the diphosphates are produced and then the ribose sugar is reduced to the deoxyribose sugar in a reduction catalysed by the enzyme ribonucleoside phosphate reductase. With the exception of dUDP, the resulting deoxyribonucleoside 5'-diphosphates (dADP, dGDP, dCDP) are then phosphorylated to produce 5'-triphosphates. As we discussed earlier, the base uracil is specific for RNA and this pyrimidine is replaced in DNA by thymine, which is actually 5-methyl-uracil. To produce the thymine deoxyribonucleotide, dUDP is dephosphorylated to dUMP and the 5-position carbon of the pyrimidine ring is then methylated in a reaction catalysed by thymidylate synthetase. The resulting deoxythymidine 5'-monophosphate is phosphorylated to produce the triphosphate. In this way (Fig. 3.7), the four DNA precursors, dATP, dGTP, dTTP and dCTP, are produced in reactions catalysed by the kinases discussed previously.

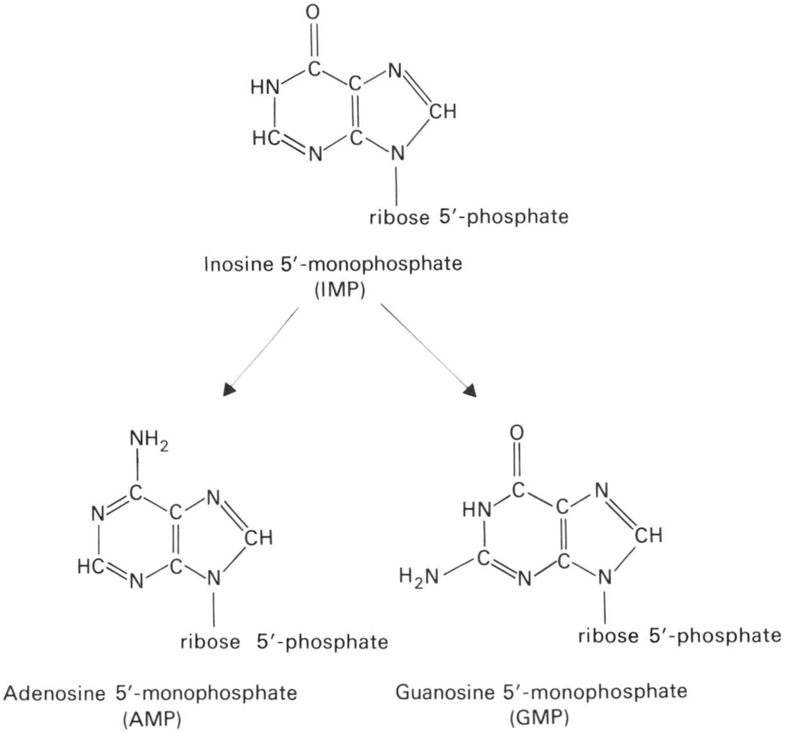

Inosine 5'-monophosphate
(IMP)

ribose 5'-phosphate

Fig. 3.4 Purine biosynthesis: production of the adenine and guanine nucleotides from inosine 5'-monophosphate.

Adenosine 5'-monophosphate
(AMP)

ribose 5'-phosphate

Guanosine 5'-monophosphate
(GMP)

ribose 5'-phosphate

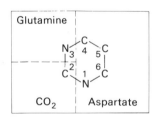

Fig. 3.5 Origin of the carbon and nitrogen atoms of the pyrimidine ring.

Since the syntheses of RNA and DNA precursors are intimately related, one might expect that the regulatory elements of the system would be complex and indeed that is the case. For example, one of the pivotal enzymes is ribonucleoside diphosphate reductase, since cells must distribute their resources appropriately between RNA and DNA synthesis. This enzyme receives complex inhibitory and stimulatory signals by a variety of deoxyribo-nucleotides so as to provide a balanced supply of precursors for DNA synthesis. The synthesis of the enzyme can also be repressed by the deoxyribo-nucleotides if their pool sizes become large.

DNA SYNTHESIS *IN VITRO*

Most of what we know of the biochemical processes of DNA synthesis has come from the work of A. Kornberg and his colleagues, which has included the in-vitro synthesis of DNA under defined conditions.

In Topic 1, it was shown that the basic building blocks for DNA synthesis are the deoxyribonucleoside 5'-triphosphates, and these precursors are polymerised into DNA as shown in Fig. 3.8. In DNA synthesis, the polynucleotide chain grows from one end by the step-wise addition of deoxyribonucleoside triphosphates. In this, the direction of synthesis is determined by the fact that the

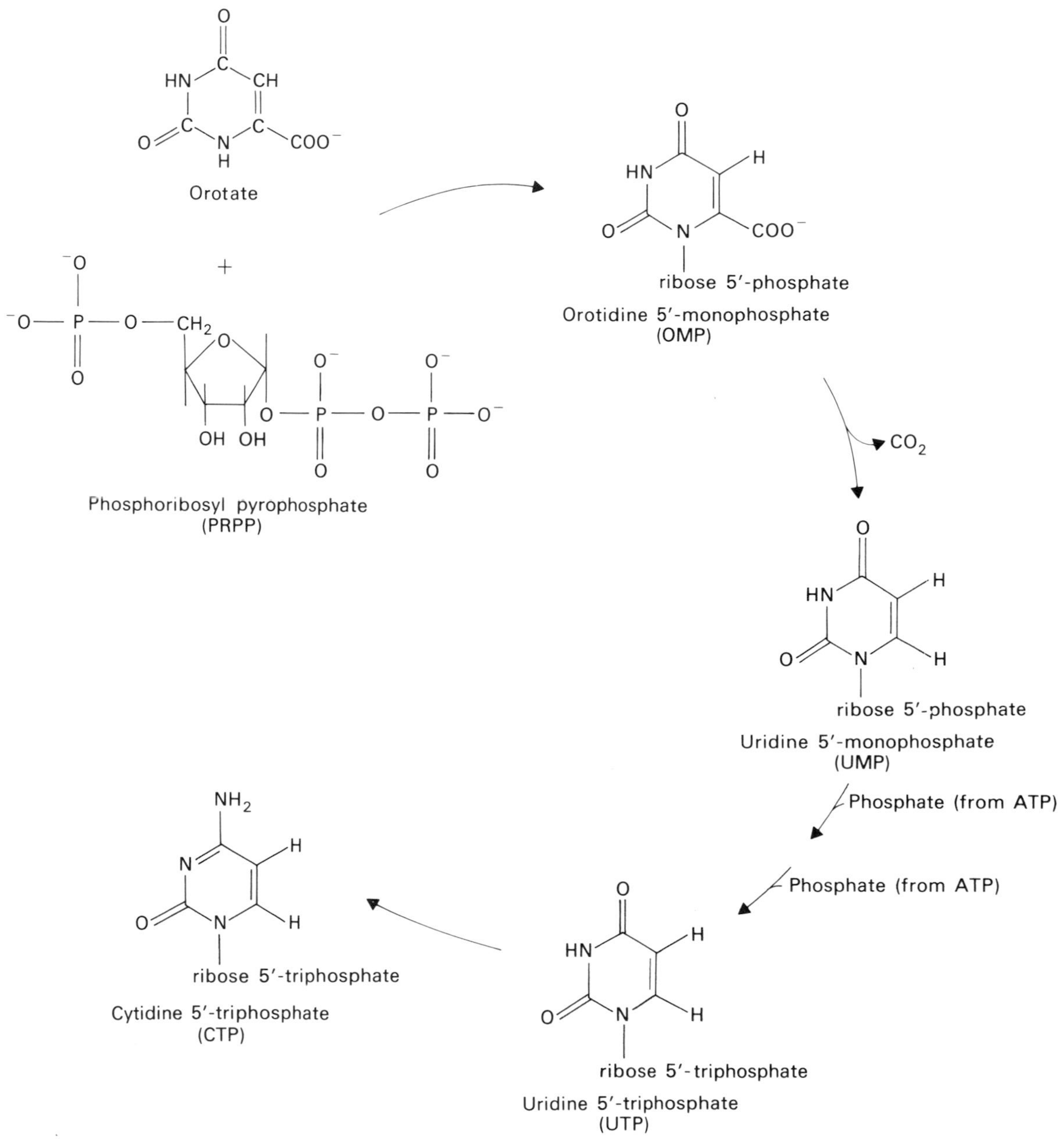

Fig. 3.6 General outline of the pyrimidine biosynthesis pathway.

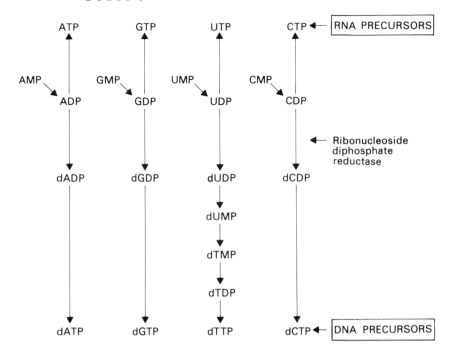

Fig. 3.7 Formation of ribonucleotide precursors for RNA and deoxyribonucleotide precursors for DNA.

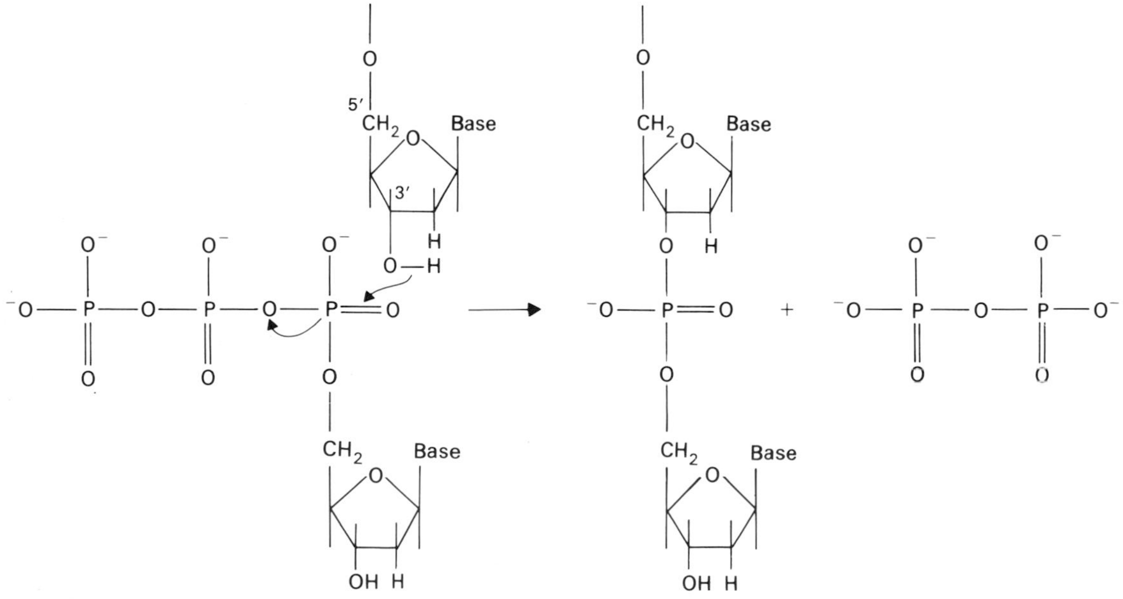

Fig. 3.8 Mechanism of DNA polymerisation.

5'-triphosphate precursors must bond to the 3'-OH of the preceding deoxyribose. Before discussing recent information about DNA synthesis, we shall review Kornberg's historical work.

In his experiments of the 1950s, Kornberg determined that the following ingredients were required for the in-vitro synthesis of DNA:

1. A mixture of all four deoxyribonucleoside 5'-triphosphates.
2. Magnesium ions (Mg^{2+}).
3. A purified enzyme, DNA polymerase, obtained from cell-free extracts of *E. coli*.
4. High-molecular-weight DNA.

In his initial experiments with the above ingredients, Kornberg obtained an approximately twenty-fold increase in the amount of DNA over that added. The reaction itself continued until one of the 5'-triphosphate precursors ran out.

That all of the four ingredients are required for DNA synthesis *in vitro* was shown by experiments in which one ingredient at a time was omitted from the reaction mixture. In each case no DNA synthesis occurred. Similar omission experiments showed that all four deoxyribonucleoside 5'-triphosphates must be present in order for DNA synthesis to proceed. The key to the reaction, though, is the DNA polymerase which promotes synthesis by effecting the 3'-5' phosphodiester linkages. This enzyme will be discussed in more detail later.

Role of DNA in the reaction
Two possibilities were proposed originally for the role of high-molecular-weight DNA in the in-vitro synthesis of DNA:

1. The DNA acts as a *primer*, that is as a growing point for the terminal random addition of nucleotides.
2. The DNA acts as a *template,* that is the DNA product is identical to that used to start the reaction in terms of the base sequence. The requirement that all four DNA precursors are present for DNA synthesis to occur indeed suggests that DNA has a template role.

All the available evidence indicates that the second possibility is correct, that is DNA acts as a template. The first indication that this is the case came from studies which showed that the A + T/G + C ratio of in-vitro synthesised DNA was the same as the DNA used to start the reaction. However, an obvious objection to this is that the results could be a coincidence since base ratios do not indicate the arrangement of bases along the DNA. A more exacting comparison of newly synthesised and template DNA can be made by *nearest neighbour analysis* which shows the relative frequency with which each of the four bases is located next to each of the four bases. This does not give information about the sequence of bases along the DNA, but it does give some notion of the general arrangement of the bases in the DNA.

Nearest neighbour analysis is a relatively straightforward procedure. DNA is isolated from the organism of interest and then is used as the template in Kornberg's in-vitro synthesising system. In this case, one of the 4 deoxyribonucleoside 5'-triphosphates (dATP in the figure) is labelled with ^{32}P (a radioactive isotope of phosphorus) and the other three are not labelled. The resulting DNA is treated with a mixture of micrococcal deoxyribonuclease and spleen diesterase which breaks the backbone of DNA between the 5'-carbon and the phosphoryl group. The digestion products, then, are deoxyribonucleoside 3'-monophosphates which can be analysed by paper electrophoresis to determine the amount of radioactivity each contains (Fig. 3.9).

In the example shown, the ^{32}P is now found in the form of dG-3'MP (deoxyguanosine 3'-monophosphate) which was the nearest neighbour to the adenine nucleotide in the DNA. Extrapolating this discussion to the entire DNA, the method provides data concerning the relative amount of radioactive label transferred from ^{32}P-labelled dATP to the four possible neighbours. This then gives information about the relative frequency with which A is next to A, G, C, and T in the DNA. These four nearest neighbours are usually written as 5'-ApA-3', GpA, CpA, and TpA. The experiment is then repeated with each of the other three DNA precursors so that, in the end, a matrix of 16 nearest neighbour

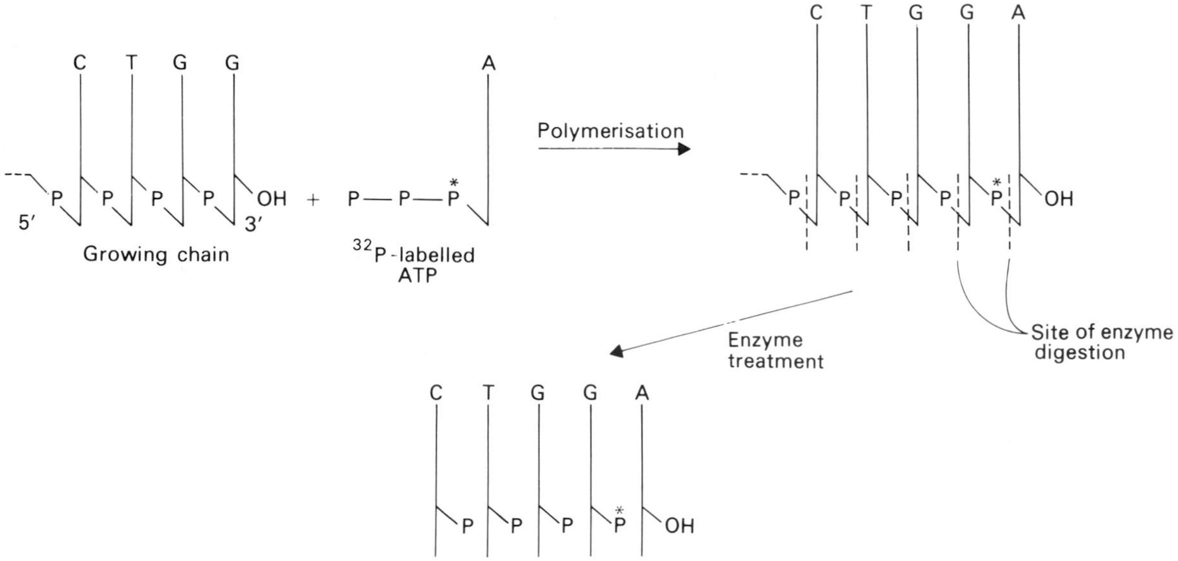

Fig. 3.9 The procedure for nearest neighbour analysis.

values is produced. The 16 frequencies found are generally unique to the DNA in question. An example of the type of data that is obtained from nearest neighbour analysis is shown in Table 3.1.

The data presented illustrate several points:

1. The sixteen possible nearest neighbour frequencies occur with a large number of frequencies.
2. The sums of the vertical columns show that the amount of A is equal to the amount of T, and that G equals C, thus indicating that the DNA was probably replicated correctly.

Table 3.1 Nearest neighbour frequencies of *Mycobacterium phlei* DNA. (With permission from J. Josse *et al.*, 1961, *J. Biol. Chem.* **236**: 804, © 1961, Journal of Biological Chemistry.)

Labelled triphosphate	Deoxyribonucleoside 3'-monophosphate isolated			
	Tp	Ap	Cp	Gp
dATP	TpA 0.012	ApA 0.024	CpA 0 063	⁀ ． ᴗ.ᴗ᧒᧒
dTTP	TpT 0.026	ApT 0.031	CpT 0.045	GpT 0.060
dGTP	TpG 0.063	ApG 0.045	CpG 0.139	GpG 0.090
dCTP	TpC 0.061	ApC 0.064	CpC 0.090	GpC 0.122
Sum	0.162	0.164	0.337	0.337

3. The two DNA strands are of opposite polarity. This is borne out by the frequency equivalence of the pairs of sequences: CpT and ApG, GpT and ApC, GpA and TpC, and CpA and TpG, as predicted by antiparallel DNA strands. Were the two strands of the same polarity, different matching sequences would have been predicted, for example TpA and ApT, GpA and CpT, CpA and GpT etc.

To show that newly synthesised DNA is made using a DNA template requires two rounds of nearest neighbour analysis. In the first round the originally isolated DNA is used in the reaction and a set of 16 nearest neighbour frequencies are obtained as described. In the second round the enzymatically synthesized DNA is used in the reaction and a second set of frequencies is produced. The results show good agreement between the two sets of frequencies, thus indicating that DNA plays a template role in DNA replication *in vitro*.

Since the early, comparatively crude, in-vitro DNA synthesis experiments, a large number of refinements have been made. For example, the reaction mixtures now duplicate the conditions that are present in growing cells such that it is possible to synthesise DNA *in vitro* that is identical in all respects with DNA produced *in vivo*. The reaction mixtures required for efficient DNA synthesis vary depending on the DNA used as the template. In general, there is a basic set of enzymes and proteins required for the synthesis of all DNA sources. Beyond that, each DNA requires a number of specific proteins for new synthesis to occur and the actual set of proteins needed depends on the DNA in question. Later discussion in this Topic will involve more details of the enzymes and proteins involved in DNA replication.

DNA REPLICATION *IN VIVO*

The Meselson and Stahl experiment
This experiment showed that when DNA replicates, the two strands separate and each serves as a template for the synthesis of a complementary strand. The analytical method used by Meselson and Stahl was caesium chloride (CsCl) density gradient centrifugation. When a concentrated solution of CsCl is centrifuged in an ultracentrifuge, the opposing forces of sedimentation and diffusion produce a stable concentration gradient of the CsCl. The concentration gradient results in a continuous increase of density along the direction of the centrifugal force. If DNA, for example, is present in the gradient, it will come to equilibrium in the region of the gradient where the solution density is equal to its buoyant density. Thus, if two species of DNA with different buoyant densities are analysed in such a gradient, at the completion of centrifugation they will have formed two distinct bands in the centrifuge tubes.

Meselson and Stahl grew a culture of *E. coli* in a medium in which the sole nitrogen source contained the 'heavy' isotope ^{15}N. The DNA synthesised during this time therefore had a higher density than normal since the ^{15}N becomes incorporated into the purine and pyrimidine rings. Then, at time zero, the cells were collected and resuspended in fresh medium where only the normal nitrogen isotope (^{14}N) was present. Several rounds of replication were allowed to proceed in the normal medium. During this time samples of the cultures were taken, the DNA was extracted and then analysed in CsCl gradients to compare its density with the known densities of heavy and normal DNA. In an exaggerated, diagrammatic form, the results are shown in Fig. 3.10.

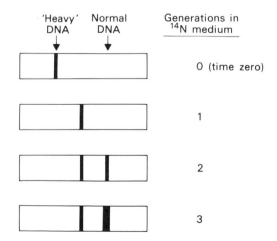

Fig. 3.10 Diagrammatic representation of the results of the Meselson and Stahl experiment.

When these experiments were done, there were two main hypotheses for DNA replication. The first said that DNA replicated *semi-conservatively*. In other words the two strands separate and act as templates for the synthesis of the new complementary strands. Thus each progeny double helix contains one parental strand and one new strand. The second hypothesis stated that DNA replicated *conservatively*, that is the double helix remains intact while serving as a template for production of a new double helix. In this case both strands of the progeny DNA would be newly synthesised. The predicted densities of DNA for the two models in the Meselson and Stahl experiment are shown in Fig. 3.11.

For the conservative replication hypothesis to be correct, some heavy DNA must be present at each generation and all new DNA must have normal density. Thus at generation number one, half of the DNA double helices should be heavy and half should be normal. On the other hand, in that generation, the semi-conservative replication model predicts that *all* DNA molecules will have one heavy and one normal strand. This would cause a hybrid DNA to form a band at a position intermediate between those for heavy and normal DNA. By contrast, no hybrid DNA will ever be formed by conservative DNA replication. Inspection of the banding data allowed Meselson and Stahl to conclude that the semi-conservative model is correct.

The discontinuous model of DNA replication

DNA replication proceeds in one overall direction such that both strands are replicated simultaneously. However, this unidirectional movement of the replication fork poses some problems since the two strands are of opposite polarity and the known DNA polymerases can only catalyse DNA synthesis in the 5′ to 3′ direction. As the DNA helix unwinds to provide the template for new synthesis, then, the DNA cannot be continuously polymerised on at least one of the strands. Some evidence for how cells get around this problem came from the work of R. Okazaki and his colleagues. They added

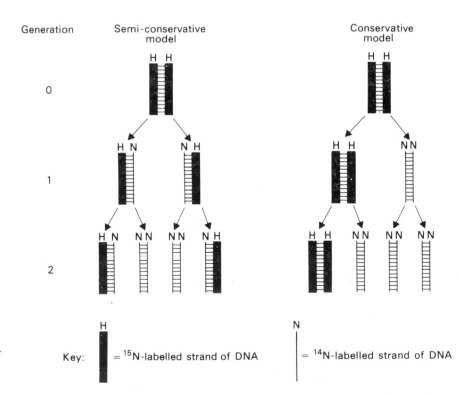

Fig. 3.11 Predicted densities for daughter DNAs produced by semi-conservative and conservative methods of DNA replication.

radioactive DNA precursors to cultures of *E. coli* for very short time intervals (0.5% of a generation time) and determined the size of the newly labelled DNA by sedimentation analysis (Fig. 3.12). They found that most of the incorporated label was present in relatively low-molecular-weight DNA. Then, as the labelling time increased, a significant fraction of the label sedimented as high-molecular-weight DNA. The conclusions they drew was that DNA synthesis is discontinuous, that is short segments (Okazaki fragments) are synthesised initially and later these become covalently bonded to the new high-molecular-weight DNA.

A model for discontinuous DNA replication which incorporates the current information about the enzymes and proteins required for this process is presented in Fig. 3.13. The evidence for this has come from biochemical, genetical, and physiological experiments.

The first step is the unwinding of the parental strands. This is aided by DNA unwinding proteins which remain bound to the single-stranded DNAs

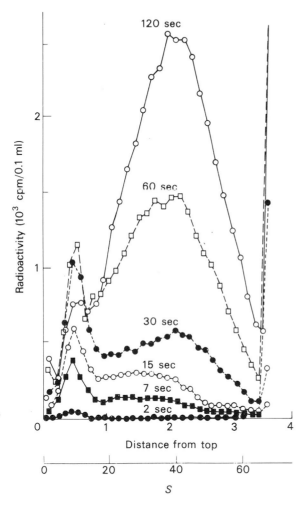

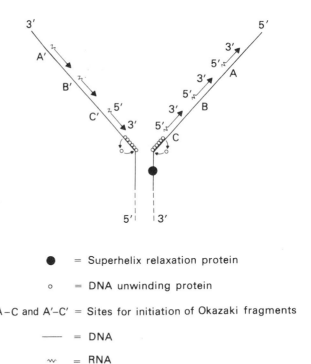

= Superhelix relaxation protein

o = DNA unwinding protein

A–C and A′–C′ = Sites for initiation of Okazaki fragments

—— = DNA

∿ = RNA

Fig. 3.12 Demonstration of discontinuous replication of DNA. Phage T4-infected *E. coli* cells were labelled with a radioactive DNA precursor. At various times samples were taken and the DNA was analysed on sucrose gradients. At early times most of the label is in low-molecular-weight DNA (near top of gradient) whereas at later time label is found in high-molecular-weight DNA. (Reproduced with permission from R. Okazaki *et al*, 1968, *Proc. Natl. Acad. Sci. USA* **59**:598.)

Fig. 3.13 A model for discontinuous DNA synthesis that includes the events proposed to occur at the replication fork. (After M. L. Gefter, 1975, *Annu. Rev. Biochem.* **44**:45.) (Reproduced, with permission, from the Annual Review of Biochemistry, Volume 44. © 1975 by Annual Reviews, Inc.)

to keep them linear, and to protect them from cleavage by nucleolytic enzymes. As the helix unwinds in this manner, a strain is imposed on the DNA. This strain is relieved by the action of superhelix relaxation proteins which cause single-strand breaks which in turn allow one strand to rotate about the other strand. One revolution occurs every ten base pairs synthesised. The single strand 'nick' is resealed by enzyme action and the replication fork moves on.

When the DNA unwinds, the single-stranded DNAs serve as templates for new DNA synthesis. In the diagram, the initiation of the Okazaki fragments occurs at A-C and A'-C' on the two complementary strands. Considering the direction of overall DNA synthesis, then, the fragments are initiated in the order A, B, C. The result is adjacent, relatively short segments of DNA that later become covalently bonded into high-molecular-weight DNA by the action of the enzyme polynucleotide ligase which seals the gaps between the segments. This cycle of events repeats itself as the DNA continues to unwind. Thus the mode of replication is discontinuous and this is certainly compatible with the 5' to 3' direction of synthesis dictated by the activity of the DNA polymerases.

Enzymes involved in DNA replication

In *E. coli* there are three DNA polymerases. Each catalyses the DNA template-directed condensation of deoxyribonucleoside 5'-triphosphates. There are significant differences in their activity in DNA synthesis and in the nuclease (DNA or RNA breakdown) activities associated with them (Table 3.2).

All three enzymes catalyse new DNA synthesis in the 5' to 3' direction. The rate of DNA synthesis catalysed by the three enzymes varies considerably with Pol III being the most active and Pol II being the least active. Exonuclease activities are associated with each of these enzymes. (An exonuclease will degrade a strand of nucleic acid from a free end whereas an endonuclease will make cuts within a strand.) Each enzyme has 3' to 5' exonuclease activity suggesting that each can trim away newly synthesized polynucleotide chains. It seems that this functions in a manner analogous to a correcting typewriter: incorrectly-paired bases are detected and excised to allow accurate replication to continue. This function of the enzymes is responsible for the high degree of accuracy found in DNA synthesis.

DNA chains are initiated by RNA primers

Careful studies of the properties of the three DNA polymerases showed that each can only add nucleotides to the free 3'-OH of pre-existing DNA chains. None can initiate new DNA chains (i.e. Okazaki fragments). The explanation is that the initiation of DNA synthesis involves the synthesis of a short RNA primer to which deoxyribonucleotides are added by the action of DNA polymerases. Thus each Okazaki fragment is initiated by an RNA segment whose synthesis is catalysed by the enzyme RNA polymerase. The role of the RNA and DNA polymerases is summarised in Fig. 3.14.

It is important to note that, in these models, the requirements for 5' to 3' chain growth and semiconservative replication are both met.

Table 3.2 Properties of *E. coli* DNA polymerases.

	Pol I	Pol II	Pol III
Molecular weight	109 000	120 000	180 000
Number of polypeptides	1	1	2
			(140 000 & 40 000)
Molecules per cell	400	100	10
5' to 3' exonuclease activity	yes	no	yes
3' to 5' exonuclease activity	yes	yes	yes

Model for DNA synthesis

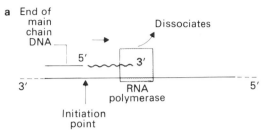

a End of main chain DNA

Dissociates

5′ 3′

3′ 5′

RNA polymerase

Initiation point

RNA polymerase synthesises short RNA primer and dissociates from DNA

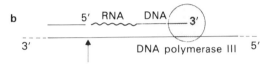

b 5′ RNA DNA 3′

3′ 5′

DNA polymerase III

DNA polymerase III catalyses addition of DNA precursors to RNA chain

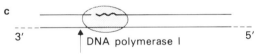

c

3′ 5′

DNA polymerase I

DNA polymerase 1 extends DNA main chain, simultaneously excising the RNA primer

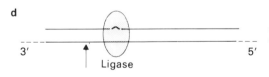

d

3′ 5′

Ligase

Polynucleotide ligase seals gap between extended main chain DNA (where RNA primer degraded) and the polymerase III-synthesised DNA

Fig. 3.14 Roles of DNA polymerases and RNA polymerases in DNA replication in *E. coli*. (After J. D. Watson, 1976, *Molecular Biology of the Gene*. W. A. Benjamin, Menlo Park.)

REFERENCES

Cold Spring Harbor Symposia for Quantitative Biology, vol. 33, 1968, *Replication of DNA in microorganisms,* Cold Spring Harbor Laboratory, New York.

Davidson J.N. 1972. *The Biochemistry of the Nucleic Acids,* 7th edn. Chapman and Hall, London.

De Lucia P. & J. Cairns, 1969. Isolation of an *E. coli* strain with a mutation affecting DNA polymerase. *Nature,* **224:** 1164–1166.

Dressler D. 1975. The recent excitement in the DNA growing point problem. *Annu. Rev. Microbiol.* **29:** 525–559.

Emmerson P.T. 1974. DNA replication in *Escherichia coli.* In *Biochemistry of Nucleic Acids,* K. Burton (ed.), vol. 6, pp. 141–164. MTP International Review of Science, Butterworths, London.

Gefter M.L. 1975. DNA replication. *Annu. Rev. Biochem.* **44:** 45–78.

Gefter M.L., Y. Hirota, T. Kornberg, J.A. Wechsler & C. Barnoux, 1971. Analysis of DNA polymerase II and III in mutants of *E. coli* thermosensitive for DNA synthesis. *Proc. Natl. Acad. Sci. USA,* **68:** 3150–3153.

Gilbert W. & D. Dressler, 1968. DNA replication: the rolling circle model. *Cold Spring Harbor Symp. Quant. Biol.* **33:** 473–484.

Gottesman M.M., M.L. Hicks & M. Gellert, 1973. Genetics and function of DNA ligase in *E. coli. J. Mol. Biol.* **77**: 531–547.

Goulian M. 1971. Biosynthesis of DNA. *Annu. Rev. Biochem.* **40**: 855–898.

Goulian M., P. Hanawalt & M. Fox, 1976. *DNA Synthesis and its Regulation.* Benjamin, Menlo Park, California.

Gudas L.J., R. James & A.B. Pardee, 1976. Evidence for the involvement of an outer membrane protein in DNA initiation. *J. Biol. Chem.* **251**: 3470–3479.

Klein A. & F. Bonhoeffer, 1972. DNA replication. *Annu. Rev. Biochem.* **41**: 301–332.

Kornberg A. 1960. Biologic synthesis of deoxyribonucleic acid. *Science,* **131**: 1503–1508.

Kornberg A. 1974. *DNA Synthesis.* W.H. Freeman, San Francisco.

Kornberg A., I.R. Lehman, M.J. Bessman & E.S. Simms, 1956. Enzymic synthesis of deoxyribonucleic acid. *Biochim. Biophys. Acta,* **21**: 197–198.

McPherson A., I. Molineux & A. Rich, 1976. Crystallization of a DNA-unwinding protein: Preliminary X-ray analysis of *fd* bacteriophage gene 5 product. *J. Mol. Biol.* **106**: 1077–1081.

Masters M. & P. Broda, 1971. Evidence for the bidirectional replication of the *E. coli* chromosome. *Nature New Biol.* **232**: 137–140.

Meselson M. & F.W. Stahl, 1958. The replication of DNA in *Escherichia coli. Proc. Natl. Acad. Sci. USA,* **44**: 671–682.

Okazaki R.T., K. Okazaki, K. Sakobe, K. Sugimoto & A. Sugino, 1968. Mechanism of DNA chain growth. I. Possible discontinuity and unusual secondary structure of newly synthesized chains. *Proc. Natl. Acad. Sci. USA,* **59**: 598–605.

Sugino A., S. Hirose & R. Okazaki, 1972. RNA-linked nascent DNA fragments in *Escherichia coli. Proc. Natl. Acad. Sci. USA,* **69**: 1863–1867.

Wickner, S.H. 1978. DNA replication proteins of *Escherichia coli. Annu. Rev. Biochem.* **47**: 1163–1191.

Topic 4
DNA Replication and the Cell Cycle
in Eukaryotes

OUTLINE
Outline of the eukaryotic cell cycle
The G1 phase
The S phase
   discontinuous DNA synthesis
   DNA polymerases
   replication units
   chromatin replication
The G2 phase
Molecular aspects of mitosis.

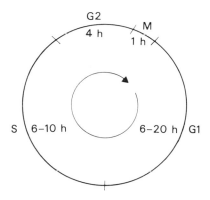

Fig. 4.1 Typical eukaryotic cell cycle.

OUTLINE OF THE EUKARYOTIC CELL CYCLE

In prokaryotes the process of DNA replication is well defined. Bacteria growing in a nutrient medium, for example, synthesise DNA throughout the cell cycle and then two daughter cells are produced by the laying down of a new cell wall midway along the cell. By contrast the cell cycle in eukaryotes is a lot more complicated. In this Topic we will discuss the biochemical aspects of the cell cycle, and in the subsequent Topic the cell division events of mitosis and meiosis will be discussed at the morphological level.

In most somatic cells of higher plants and animals, the cell cycle may be represented as shown in Fig. 4.1, that is there is a distinct phasing of events between divisions. The sequence of phases are G1 (gap 1), S (synthesis), G2 (gap 2) and M (mitosis), the first three collectively describing the interphase stage of the cell cycle. Following G2 the cell divides to produce two daughter cells by mitosis and each daughter cell then begins a new cell cycle. In mammalian cells in culture, the cell cycle takes approximately 24 hours, and indeed most of the information we have to date about the molecular biology of the cell cycle has come from experiments done with such cultures. For euka-

ryotic cell cultures in general, though, the relative time spent in each of the four phases of the cell cycle varies, as does the cell cycle time itself (Fig. 4.2). Clearly cells *in vivo* are under different control conditions and this is responsible for the wide range of cell-cycle times. At the present, our knowledge of the S and M stages at the molecular level is incomplete and much of what goes on in G1 and G2 is not known. What is known for each of the four phases will now be discussed in turn.

THE G1 PHASE

The G1 phase commences after mitosis has been completed. At the general level it is characterised by a change in the chromosomes from the condensed mitotic state to the more extended interphase state, and by a series of events leading to the initiation of DNA replication.

In a homogeneous population of cultured cells there is variability in the cell-cycle time and this is a major problem in attempts to work with synchronously dividing cells. The G1 phase of the cell cycle is far more variable in length than the other three phases and it is this that is responsible for

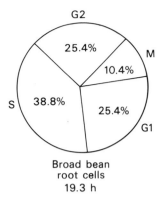

Broad bean
root cells
19.3 h

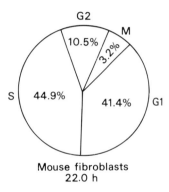

Mouse fibroblasts
22.0 h

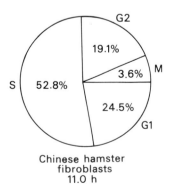

Chinese hamster
fibroblasts
11.0 h

Fig. 4.2 Relative time in the four phases of the cell cycle for three cell types. (After B. Kihlman *et al*, 1966, *Hereditas* **55**:386.)

the variability in generation times within a cell population and for such variability in the different cell types of an organism. The cause of the variability of G1 is not known although there is some evidence to suggest that the length of this phase is related to cell mass and hence to protein synthesis.

Apparently G1 is a very significant phase in the cell cycle since cells that stop dividing normally arrest in this phase. One hypothesis for which there is some evidence is that there is a restriction point within G1 and, once the biochemical event(s) associated with this point has occurred, the cell is then irreversibly committed to initiate DNA synthesis and proceed through cell division. Whether a cell then continues to proliferate or whether it becomes specialised and non-dividing would presumably depend upon molecular regulatory signals acting at the restriction point. One example of regulation at this level is that changes or differences in the rates of cell reproduction for cells of the same genetic constitution are brought about principally by alteration of the G1 phase. Thus, there may be a number of regulatory genes operating in G1 but we know very little about this area at this time.

The same may be said of the specific molecular events occurring in G1. With the de-condensation of the chromosomes after mitosis, we can presume that a large number of potential sites in the chromatin are made available for RNA transcription and that a series of molecules required for the initiation of DNA synthesis are produced as a result of this. However, some organisms do not have a defined G1 phase, for example the slime mould *Physarum polycephalum,* the fission yeast *Schizosaccharomyces pombe,* the cleavage stages of sea-urchin embryos, *Xenopus* embryos and mouse embryos. This indicates that the events leading to DNA synthesis may precede mitosis in the G2 phase in some systems rather than being in G1.

THE S PHASE

Discontinuous DNA synthesis

During the S phase the chromatin material is duplicated. In cell cultures the S period lasts for 6–10 hours whereas *in vivo* it may be as short as

10 min, as in *Schizosaccharomyces pombe,* or as long as 35 hours, as in mouse ear skin cells. J.H. Taylor in 1957 showed by autoradiographic studies of root tip cells of the broad bean pulse-labelled with tritium-labelled thymidine, that DNA replication was semi-conservative (Fig. 4.3). He labelled cells during the DNA synthesis phase and showed by autoradiography that all chromatids were labelled. Then he allowed the cells to go through one more division cycle in the absence of label and found that at any one point along each chromosome, only one chromatid was labelled. There is also recent evidence from a number of workers that, as in prokaryotes, DNA replication in eukaryotes is discontinuous. For example, when the animal virus SV40 (simian virus 40) replicates its DNA *in vitro,* newly synthesised fragments can be isolated that have the ability to self-anneal, that is, form short stretches of double helix. This is the result one would expect if Okazaki fragments were being generated on the two complementary template strands. Also, in mammalian cells there is evidence that nucleotides are polymerised into 100-nucleotide short-lived strands that are then joined into longer ones.

As we discussed in the preceding Topic, all prokaryotic DNA polymerases are unable to initiate DNA synthesis; rather they polymerise deoxyribonucleotides on an RNA primer. This has led to a search for RNA primers during DNA replication in eukaryotes and a variety of results have been obtained to date. Thus, in some systems there appears to be good evidence for RNA primers whereas in others there is no evidence at all. At the moment, then, one cannot generalise regarding RNA primers and more work needs to be done in this area.

DNA polymerases

The enzymes required for DNA replication have been studied in a number of eukaryotes. In higher eukaryotes three DNA polymerases, alpha (α), beta (β), and gamma (γ), have been distinguished based on molecular weight, chromatographic properties, sensitivity to the inhibitor N-ethylmaleimide (NEM), sensitivity to salts, and the ability to copy various templates. These properties are summarised in Table 4.1. All of the alpha and beta poly-

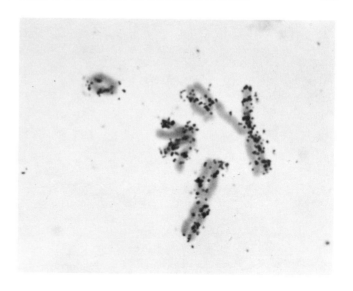

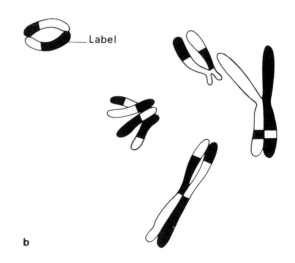

Fig. 4.3   (a) Autoradiogram of broad bean (*Vicia faba*) chromosomes showing semi-conservative DNA replication. Cells were incubated with ³H-thymidine to label all the DNA and then they were 'chased' with cold thymidine for one generation. (× 1875) (Courtesy of J. H. Taylor.)
(b) Interpretative sketch of the autoradiogram showing that at any one point, only one chromatid is labelled (shaded area) thus indicating semi-conservative DNA replication and, coincidentally, showing sister chromatid exchange. (Courtesy of J. H. Taylor.)

Table 4.1 Properties of mammalian DNA polymerases. (After A. Weissbach, 1977. *Annu. Rev. Biochem.* **46**, 25.)

Polymerase type	Molecular weight	Inhibition by N-ethylmaleimide	Salt effect
α	120 000–300 000	+	Inhibited by >25 mM NaCl
β	30 000–50 000	−	Stimulated by 100–200 mM NaCl; inhibited by 50 mM phosphate
γ	150 000–300 000	+	Stimulated by 100–250 mM KCl and 50 mM phosphate

merase molecules are located in the nucleus, whereas the gamma enzyme is apparently localized in the mitochondria. The alpha and beta enzymes are usually found with other proteins in a replication complex. It is impossible to generalise about these enzymes, since there is much variability among the eukaryotic organisms that have been studied. For example, in some organisms there are multiple forms of the enzymes whereas there is a single major enzyme in others. Plants, protozoa and fungi lack the beta polymerase, and eukaryotic microorganisms in general have an enzyme called alpha although it is very different from the mammalian alpha polymerase. In many instances the enzymes have not been purified to homogeneity and thus it

is difficult to make comparisons or even to estimate the number of enzyme types or subtypes that a cell possesses.

Replication units

In *E. coli* there is a single origin of replication in the chromosome and DNA replication proceeds bidirectionally from that point. Electron microscopy studies of replicating chromosomes of eukaryotes show that each chromosome has a number of replication units (RUs) or *replicons,* each of which has a specific origin and two termini for the replication process. This arrangement of chromosomes into replication units is necessary so that the enormous amount of DNA relative to a bacterial

Fig. 4.4 Autoradiographs of replicating eukaryotic chromosomes showing distinct replication units. Both are from *Xenopus* tissue culture cells labelled with ^{3}H-thymidine. (a) shows one replicating unit which had been initiated some hours before ^{3}H-thymidine was supplied, and the divergent 'V'-tracks give direct evidence that replication is bi-directional. (b) also shows evidence of distinct replication units and the middle unit shows sister strand separation. (Both photographs courtesy of H. G. Callan. Photograph (a) is with permission from the *Proceedings of the Royal Society of London B* **181**: 19–41, 1972. The Royal Society, Copyright © 1972.)

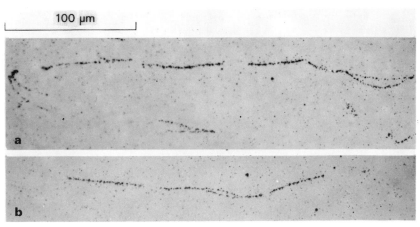

cell can be replicated within a reasonable time.

EM studies show that DNA replication begins by the opening of a 'bubble' in the DNA representing two replication forks (Fig. 4.4). These forks migrate bidirectionally as the DNA is synthesised until they reach specific termination points. Very little is known about the origin and termination signals, but it is presumed that specific nucleotide sequences provide recognition sites for the DNA polymerase-replication complex. The simplest model here is that a specific initiator protein binds with a genetically-determined origin sequence and this facilitates the binding of the DNA replication complex and hence the initiation of DNA replication.

Careful examination of replicating chromosomes under the electron microscope shows that the spacing between origins of replication varies between 7 and 100 μm (3×10^4–3×10^5 base pairs) in a wide range of organisms such as yeast, eukaryotic microorganisms in general, plants, birds, and mammalian cells. Thus, for example, there are perhaps as many as 100 RUs per chromosome in HeLa cells. The rate of movement of the replication complex in RUs is very similar in most eukaryotes that have been examined, ranging from 1 to 15×10^3 nucleotides of DNA per minute at 37° C. This rate of replication fork movement does vary within a single system, however, as a result of genetic controls, temporal controls or in response to environmental changes.

Turning to DNA replication for all the chromosomes as a whole, there is a temporal ordering of replication through the S phase. Specifically, there seems to be a highly ordered and complex pattern of replication of RUs in the nucleus. A diagrammatic representation of this is shown in Fig. 4.5. Here initiation of replication of RUs begins at the o regions. Replication proceeds bidirectionally producing the 'bubbles' in the DNA. Termination of

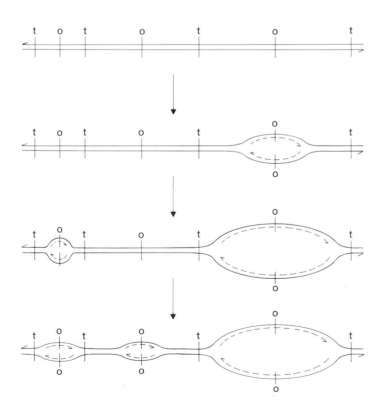

o = Origin of replication in replication unit
t = Terminus
——— Template DNA
– – – New DNA

Fig. 4.5 Temporal sequencing of DNA initiation in replication units of eukaryotic chromosomes. (After J. A. Huberman and A. D. Riggs, 1968, *J. Mol. Biol.* **32**:327.)

the process involves fusion of adjacent replication forks. Apparently the sequence of movement of replication complexes is species-specific and is repeated generation after generation. How the signals for these processes are coordinated is not known, but it is clear that protein synthesis is required for continued initiation of replication of sets of RUs. There is also some evidence that the more GC-rich DNA is replicated first and then the AT-rich DNA regions are replicated later in the S phase.

The DNA replication itself is dependent upon protein synthesis. This has been shown by the use of inhibitors of protein synthesis such as cycloheximide and puromycin. The addition of either of these to mammalian cells in cultures causes an immediate, rapid decline in the initiation of DNA synthesis, but not in the elongation of DNA chains already begun. Since DNA replication is discontinuous, the overall replication of chromosomes cannot be completed in the absence of new protein synthesis. DNA synthesis is also dependent on RNA synthesis but the inhibition of the latter process does not have an immediate effect on the former process. Thus the addition of the RNA synthesis inhibitor actinomycin D to cells early in the S phase blocks DNA replication late in S. Further, ribosomal RNA synthesis must continue to within an hour or so of the transition from G1 to S if DNA synthesis is to be initiated. This is correlated with the genetic, physiological and biochemical evidence which suggests that progression of cells into the S phase is determined by a regulatory protein synthesised during the G1 phase. This protein apparently acts as a positive effector to de-repress DNA replication. Indeed a large number of DNA binding proteins have been identified in eukaryotes and one or more of these may serve to facilitate initiation of RNA replication.

Chromatin replication
DNA is only one component of chromatin and the replication of DNA in the S phase must therefore be tightly coupled with replication of histones and probably of nonhistones. DNA replication in eukaryotes, then, should be considered a component of chromatin replication.

Newly synthesised DNA in mammalian cells is found in chromatin material which differs from non-replicating chromatin in a number of ways, for example, protein and enzyme content, and sensitivity to nuclease digestion. One interpretation of this is that the nucleosomes are less densely packed along the newly made, as compared to the mature, chromosomal DNA. A maturation time of 2–15 min is required to convert the former to the mature state.

In eukaryotes, then, DNA replication is initiated within nucleosomal units which do not dissociate from the DNA through the replication process. When new DNA is made, histones rapidly become associated with it and together they form nucleosomes.

THE G2 PHASE

In the G2 phase, the chromosomes condense in preparation for mitosis. The condensation process involves higher-order folding of the chromatin fibril but the mechanism for it is poorly understood. Inhibitor studies have shown that both RNA and protein synthesis are required for the completion of G2. The end of the G2 phase is delineated by the onset of mitosis. This transition is very hard to define by light microscopy.

MOLECULAR ASPECTS OF MITOSIS

Mitosis is the division of a cell into two genetically identical daughter cells. By the beginning of mitosis the chromosomes have duplicated and the chromosomes are then distributed to the progeny cells by the division process. The resulting cells are then in the G1 phase. The behaviour of the chromosomes in mitosis will be described in the next Topic; here we shall discuss the general morphological and biochemical changes of mitosis.

Mitosis is characterised by large changes in cell structure and function but the molecular bases of these events are not clearly understood. As mitosis commences, RNA synthesis decreases, then stops by the metaphase stage, and resumes again by late telophase. It is possible that this is related to the unavailability of transcription initiation sites owing to the highly condensed state of the chromosomes. As a result of the cessation of RNA synthesis, the rate of protein synthesis drops drastically, and then resumes in late telophase concomitant with the rise in the rate of RNA synthesis. The decrease in protein synthesis is not due to the lack of messenger RNA but rather a change in competency of the ribosomes to carry out the process. Along with the inhibition of RNA and protein synthesis, the nuclear membrane and nucleolus breaks down, and these structures are reformed in daughter nuclei in late telophase. The molecular signals for these events are not known.

In summary the cell cycle of a eukaryotic cell may be depicted as shown in Fig. 4.6. The sequence of events in the cycle is presumably dependent on the transcription and translation of cell-cycle genes in a particular temporal order. Some progress has been made in identifying cell-cycle genes in a number of organisms (for example yeast and mammalian cells) by the selection for genetic mutants with conditional blocks in the cell cycle. Study of these mutants has provided valuable information about the ordering and regulation of cell-cycle events at the biochemical level.

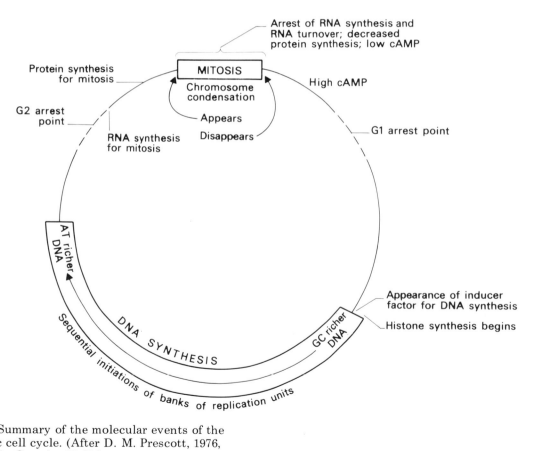

Fig. 4.6 Summary of the molecular events of the eukaryotic cell cycle. (After D. M. Prescott, 1976, *Advances in Genetics* **18**:99.)

REFERENCES

Bollum F.J. 1975. Mammalian DNA polymerases. *Prog. Nucl. Acid Res. Mol. Biol.* **15**: 109–144.

Callan H.G. 1973. DNA replication in the chromosomes of eukaryotes. *Cold Spring Harbor Symp. Quant. Biol.* **38**: 195–203.

Cold Spring Harbor Symposia for Quantitative Biology, 1977, vol. 42 'Chromatin.' Cold Spring Harbor Laboratory.

Edenberg H.J. & J.A. Huberman. 1975. Eukaryotic chromosome replication. *Annu. Rev. Genetics* **9**: 245–284.

Gefter M.L. 1975. DNA replication. *Annu. Rev. Biochem.* **44**: 45–78.

Hartwell L.H. 1974. *Saccharomyces cerevisiae* cell cycle. *Bacteriol. Rev.* **38**: 164–198.

Huberman J.A. & H. Horwitz. 1973. Discontinuous DNA synthesis in mammalian cells. *Cold Spring Harbor Symp. Quant. Biol.* **38**: 233–238.

Huberman J.A. & A.D. Riggs. 1968. On the mechanism of DNA replication in mammalian chromosomes. *J. Mol. Biol.* **32**: 327–341.

Klein A. & F. Bonhoeffer. 1972. DNA replication. *Annu. Rev. Biochem.* **41**: 302–322.

Loeb L.A. 1974. Eucaryotic DNA polymerases, In *The Enzymes,* vol. X, P.D. Boyer (ed.). pp. 174–210. Academic Press, New York.

Pardee A.B., R. Dubrow, J.L. Hamlin & R.K. Kletzien. 1978. Animal cell cycle. *Annu. Rev. Biochem.* **47**: 715–750.

Petes T.D., C.S. Newlon, B. Byers & W.L. Fangman. 1973. Yeast chromosomal DNA: size, structure, and replication. *Cold Spring Harbor Symp. Quant. Biol.* **38**: 9–16.

Prescott D.M. 1970. The structure and replication of eukaryotic chromosomes. *Adv. Cell Biol.* **1**: 57–117.

Prescott D.M. 1976. The cell cycle and the control of cellular reproduction. *Advances in Genetics* **18**: 99–177.

Seale R.L. 1977. Persistence of nucleosomes on DNA during chromosome replication. *Cold Spring Harbor Symp. Quant. Biol.* **42**: 433–438.

Sheinin R., J. Humbert & R.E. Pearlman, 1978. Some aspects of eukaryotic DNA replication. *Annu. Rev. Biochem.* **47**: 277–316.

Simchen G. 1978. Cell cycle mutants. *Annu. Rev. Genetics* **12**: 161–191.

Taylor J.H., P.S. Woods & W.L. Hughes. 1957. The organization and duplication of chromosomes as revealed by autoradiographic studies using tritium-labeled thymidine. *Proc. Natl. Acad. Sci. USA,* **43**: 122–128.

Watson J.D. 1971. The regulation of DNA synthesis in eukaryotes. *Adv. Cell Biol.* **2**: 1–46.

Weintraub H., A. Worcel & B. Alberts. 1976. A model for chromatin based upon symmetrically paired half-nucleosomes. *Cell,* **9**: 409–417.

Weissbach A. 1977. Eukaryotic DNA polymerases. *Annu. Rev. Biochem.* **46**: 25–47.

Topic 5
Mitosis and Meiosis

OUTLINE
Mitosis
Meiosis
   first meiotic division
   second meiotic division.

MITOSIS

As has already been discussed, DNA synthesis is continuous through the bacterial cell cycle. As the DNA content doubles, the cell enlarges and then begins to synthesise a dividing wall in the central region of the cell. This serves to segregate the two daughter chromosomes to different cell compartments and, when the wall is completed, the bacterium separates into two progeny cells. This process is repeated as long as the bacteria keep growing. In Topic 4 we described the eukaryotic cell cycle in detail and clearly it differs markedly from that of a bacterium. Thus dividing somatic (non-gamete producing) cells go through four distinct but interrelated phases of a cell cycle described as G1, S, G2 and M. The chromosomes are replicated during the S phase of the cycle and then later, in a process called *mitosis,* (the M phase), the duplicated chromosomes are segregated into two daughter cells that have the same genetic content as the parent cell. (An exception will be discussed in a later Topic.)

Before describing mitosis, we must define some terms that concern the chromosome content of eukaryotic organisms. The number of chromosomes per nucleus is generally constant for all individuals of a species, and varies from one species to another. Thus, for example, man has 46, rat has 42, and the pea has 14 chromosomes. For the somatic cells of these organisms and of other eukaryotes (except the lower eukaryotes) the chromosomes are present in pairs, that is there are 23

pairs in man, and so on. Conventionally, the somatic cells of these organisms are described as being *diploid* in chromosome number. By contrast, the mature germ cells (gametes — produced by meiosis) of a sexually-reproducing individual contain only half the somatic number of chromosomes, that is one member of each pair. The gametes are described as being *haploid* in chromosome number. The symbol N is used to signify the haploid chromosome number and the term 2N refers to the diploid chromosome number. For later consideration it should be mentioned that the cells of the lower eukaryotes such as yeast and *Neurospora* have a haploid chromosome number.

To return now to the process of mitosis. Mitosis is the mechanism by which the chromosome content of a somatic cell (haploid or diploid) is kept constant through successive cell divisions. When an interphase cell is stained with basic dyes, the nucleus becomes visible under the light microscope and is seen to be surrounded by a membrane. Within the nucleus one or two RNA-rich regions, the nucleoli, are apparent (Fig. 5.1). During the S phase of the cell cycle (which is part of interphase) the chromosomes divide but remain attached at one constricted region which is called the centromere. The two daughter strands of a duplicated chromosome that are held together in this way by a centromere are called chromatids.

Mitosis itself is a continuous process but for descriptive purposes it is broken down into four stages called *prophase, metaphase, anaphase,* and *telophase*. A series of photographs showing the typical appearances of the various phases of mitosis are shown in Fig. 5.2. For the purposes of simplifying the discussion, in the following the events of mitosis are illustrated for a hypothetical, diploid cell with two pairs of chromosomes and the main features of each phase are listed. The same events occur for haploid cells undergoing mitosis.

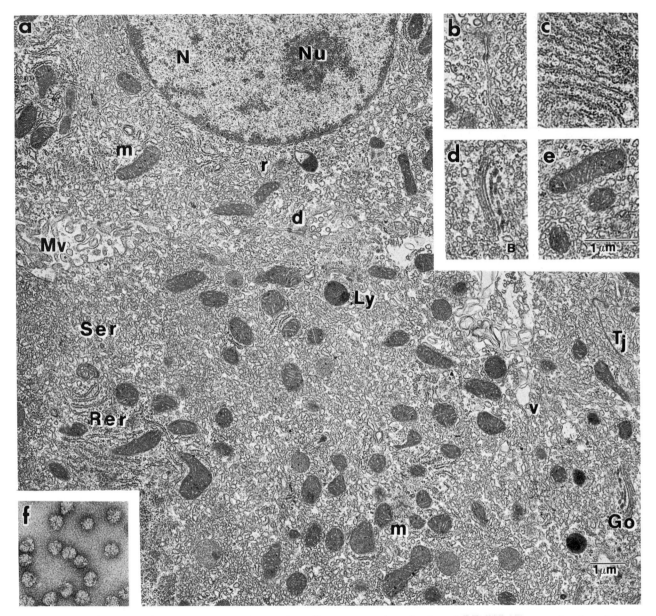

Fig. 5.1 An electron micrograph of a eukaryotic cell (from rat liver) with 'close ups' of some of the major cell parts. (Courtesy of M. Boublik.)

(a) General ultrastructure of rat liver cells (× 9600). Key to parts: N, nucleus; Nu, nucleolus; Tj, tight junction (where the cell membranes of adjacent cells have fused); d, desmosome (a portion of the cell membrane specialized for adhesion to a neighbouring cell); Mv, microvilli (invaginations of the cell surface); r, ribosomes; Rer, rough endoplasmic reticulum; Ser, smooth endoplasmic reticulum; m, mitochondria; v, vacuole; Ly, lysosome (small vesicles containing digestive enzymes); Go, Golgi apparatus (a stack of flattened vesicles).

(b) Cell membrane with desmosome (14 400).
(c) Rough endoplasmic reticulum (× 14 400).
(d) Golgi apparatus (× 14 400).
(e) Mitochondria (× 14 400).
(f) Ribosomes (× 160 000).

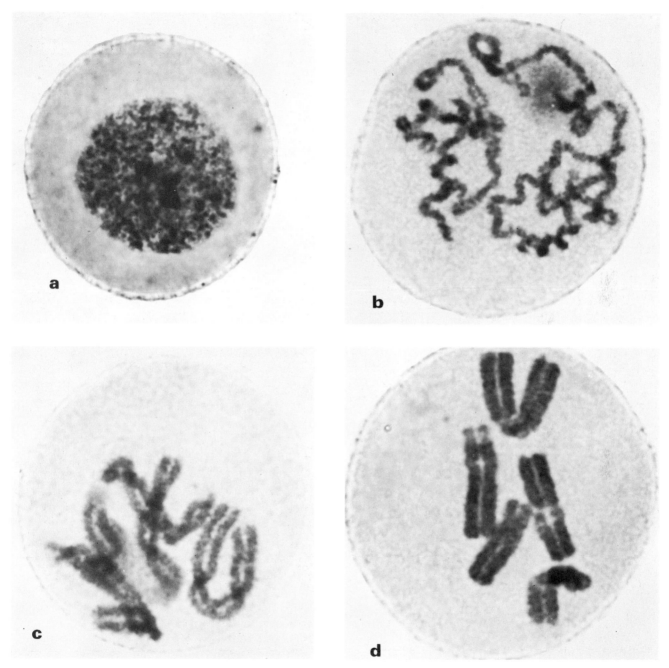

Fig. 5.2 The stages of mitosis in *Trillium erectum*. Since this is a plant, there is no centriole and spindle fibres are present but not easy to see in photomicrographs. The mitosis shown is occurring in pollen and the cell in this case is haploid. (a) Interphase. (b) Pro-phase. (c) Late prophase. (d) Metaphase. (e) Anaphase. (f) Telophase. (All photomicrographs × 2000). (Courtesy of A. H. Sparrow and R. F. Smith, Brookhaven National Laboratory.)

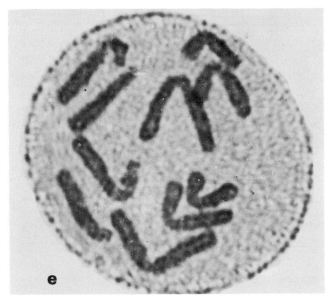

Fig. 5.2 (*continued*)

Prophase (Fig. 5.3a)
a. The chromosomes become visible as a result of coiling events.
b. Each chromosome is seen to consist of two sister chromatids.
c. Near the end of prophase, the nucleolus and the nuclear membrane disappear.

Metaphase (Fig. 5.3b)
a. Spindle fibres appear and radiate from the opposite poles of the cells. In animals the fibres are attached to structures called centrioles which are at the two poles, whereas no such structures are found in plant cells.
b. Some spindle fibres become attached to the centromere regions of the chromosomes.
c. The sister chromatids become aligned in one plane in the middle of the cell in a region called the metaphase plate.

Anaphase (Fig. 5.3c)
a. This stage is begun by division of the centromere of each chromosome.

b. The two sister chromatids of each chromosome then separate. Each resulting daughter chromosome remains attached to a daughter centromere.
c. The two daughter centromeres migrate towards the opposite poles of the cell as a result of the properties of the spindle fibres. This serves to segregate the two identical copies of each chromosome to the two poles.

Telophase (Fig. 5.3d)
a. The migration of the daughter chromosomes to the two poles is completed. Thus, the two identical replicates of each chromosome are separated and reformed into two groups in the cell.
b. A nuclear membrane forms around each set of chromosomes.
c. The nucleolus or nucleoli reform.
d. The spindle fibres disappear.
e. The chromosomes uncoil and become 'invisible' under the light microscope. Two typical interphase nuclei are now apparent.
f. In most cases telophase is followed by cell division (cytokinesis).

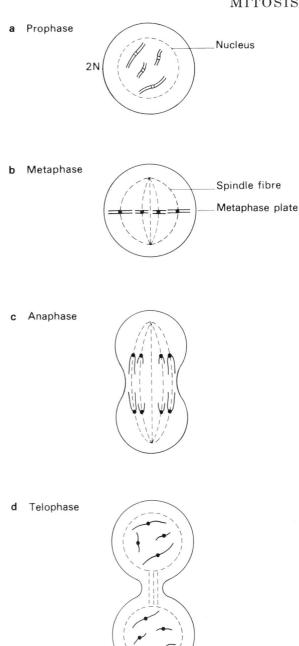

a Prophase

Nucleus

2N

b Metaphase

Spindle fibre

Metaphase plate

c Anaphase

d Telophase

Fig. 5.3 Diagrammatic representation of mitosis in an hypothetical diploid animal cell with a haploid chromosome number of two. (a) Prophase. (b) Metaphase. (c) Anaphase. (d) Telophase.

For studies of genetics, the key points of mitosis are:

1. Homologous chromosomes divide to give two chromatids each in the S phase of the cell cycle.
2. The homologous chromosomes (two sister chromatids each) align *independently* at the metaphase plate.

MEIOSIS (Sexual reproduction)

The sexual cycle of a diploid organism involves the alternation of haploid and diploid states (Fig. 5.4).

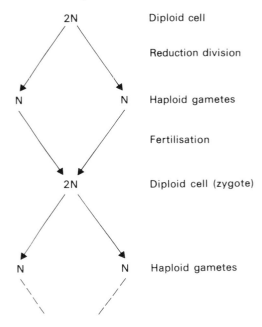

2N   Diploid cell

Reduction division

N   N   Haploid gametes

Fertilisation

2N   Diploid cell (zygote)

N   N   Haploid gametes

Fig. 5.4 The alternation of haploid (N) and diploid (2N) states from generation to generation.

In meiosis, a diploid cell containing two sets of chromosomes undergoes two consecutive divisions the first of which is preceded by DNA replication. In general meiosis resembles mitosis but close examination reveals the fact that the two processes are quite different. From each diploid cell, meiosis results in four haploid cells which are gametes in most eukaryotes and progeny in some lower eukaryotes. A cytological description of meiosis is shown in Fig. 5.5.

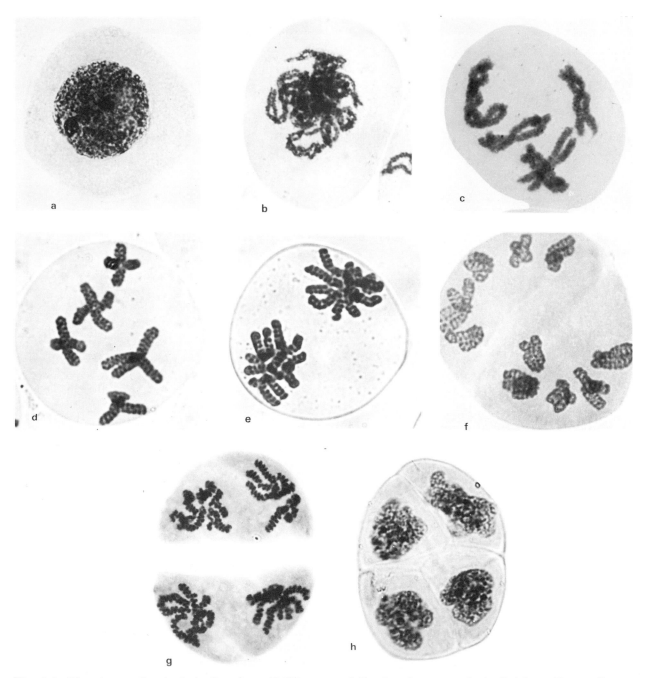

Fig. 5.5 The stages of meiosis in the plant, *Trillium erectum.* (a) Prophase I (early). (b) Prophase I (middle). (c) Prophase I (late). (d) Metaphase I. (e) Anaphase I. (f) Metaphase II. (g) Anaphase II. (h) Early interphase following the two meiotic divisions. Four cells are apparent. (All photomicrographs × 1000 approx. Courtesy of A. H. Sparrow and R. F. Smith, Brookhaven National Laboratory.)

First Meiotic division

Prophase I (Fig. 5.6a, b)
a. The diploid number of chromosomes become
visible as they condense by coiling. Unlike the
prophase stage in mitosis, each chromosome appears to be single.
b. Homologous chromosomes become paired.
c. Each chromosome later is seen to have divided
to produce two chromatids (Fig. 5.6b).
d. It is at this stage of meiosis (the tetrad stage)
that genetic exchange occurs between paternally-
and maternally-derived homologues by a process
called crossing-over. This event is correlated with
the appearance of *genetic recombination* of paternal
and maternal characters among the progeny as
determined by genetic analysis. The site of crossing-over is called a chiasma (plural = chiasmata).

Metaphase I (Fig. 5.6c)
a. At the onset of metaphase I the nucleolus and
nuclear membrane disappear.
b. The undivided centromeres become aligned on
the spindle fibres and the associated chromatids
become oriented on the metaphase plate.

Anaphase I (Fig. 5.6d)
In a process very similar to mitotic anaphase the
homologous centromeres migrate to the opposite
poles of the spindle apparatus.

Telophase I/Interphase II (Fig. 5.6e)
a. Cell division occurs to produce two progeny
cells. Each of these cells has only one complete
haploid set of chromosomes (two chromatids each)
and there is an equal probability that a particular
chromosome will be paternal or maternal in origin.
b. In a brief interphase, the chromosomes elongate
and a nuclear membrane reforms.

Second meiotic division

The second meiotic division is almost identical to a
mitotic division. In prophase II (Fig. 5.7a) the
chromosomes condense and the centromeres divide
and in metaphase II (Fig. 5.7b) the chromosomes
become oriented at the metaphase plate.

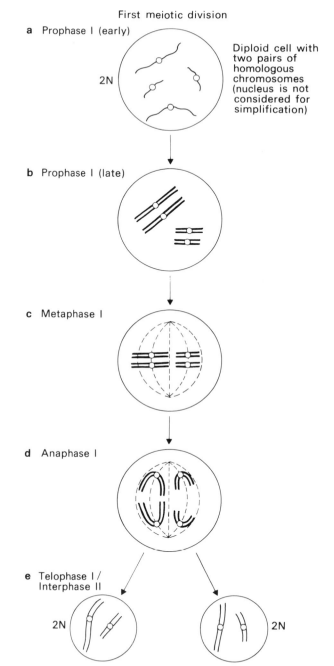

First meiotic division

a Prophase I (early)

2N

Diploid cell with
two pairs of
homologous
chromosomes
(nucleus is not
considered for
simplification)

b Prophase I (late)

c Metaphase I

d Anaphase I

e Telophase I /
Interphase II

2N   2N

Fig. 5.6 Diagrammatic representation of the first
meiotic division in an hypothetical animal cell with a
haploid chromosome number of two. (a) Prophase I
(early). (b) Prophase II (late). (c) Metaphase I.
(d) Anaphase I. (e) Telophase I/Interphase II.

Anaphase II (Fig. 5.7c)
The centromeres migrate to the opposite poles of the spindle, pulling the chromatids with them.

Telophase II (Fig. 5.7d)
a. Each of the two cells produced by the first division now divides. Thus four haploid cells are produced for each diploid cell that goes through meiosis.
b. The chromosomes become less condensed and a nuclear membrane forms.

Thus each of the four haploid cells produced contains half the number of chromosomes of a normal diploid cell, that is one of each homologous pair. The chromosomes present in the haploid cell are randomly distributed with respect to paternal and maternal origin.

The behaviour of chromosomes in the meiotic division is directly relevant to the segregation of genes and this relationship will be developed in detail in later Topics. Of particular relevance to genetics is the tetrad stage of the first meiotic division which is the point where crossing-over occurs, and the random segregation of each of the four chromatids of a homologous pair of chromosomes to the four haploid cells independently of the four chromatids of every other chromosome pair.

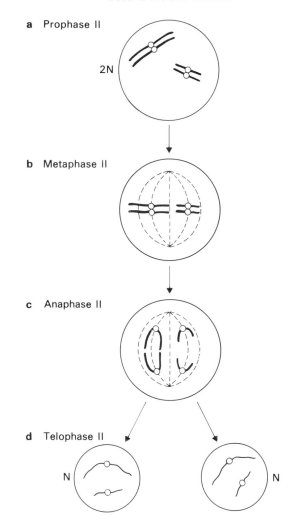

Second meiotic division

a Prophase II

2N

b Metaphase II

c Anaphase II

d Telophase II

N   N

Fig. 5.7 Second meiotic division—a continuation from Fig. 5.6. (a) Prophase II. (b) Metaphase II. (c) Anaphase II. (d) Telophase II.

REFERENCES

Brachet J. & A.E. Mirsky (eds.). 1961. *The Cell*, vol. 3: *Meiosis and Mitosis*. Academic Press, New York.
Brinkley B.R. & E. Stubblefield. 1970. Ultrastructure and interaction of the kinetochore and centriole in mitosis and meiosis. *Adv. Cell Cycle* **1**: 119–186.
Henderson S.A. 1970. The time and place of meiotic crossing over. *Annu. Rev. Genet.* **4**: 295–324.

John B. & K.R. Lewis. 1965. *The Meiotic System*. Springer-Verlag, New York.
Moses M.J. 1968. Synaptinemal complex. *Annu. Rev. Genet.* **2**: 363–412.
Stern H. & Y. Hotta. 1969. *Biochemistry of Meiosis*. In *Handbook of Cytology*, C.A. Lima-de-Faria (ed.). North Holland, Amsterdam.
Westergaard M. & D. von Wettstein. 1972. The synaptinemal complex. *Annu. Rev. Genet.* **6**: 74–110.

Topic 6
Mutation, Mutagenesis and Selection

OUTLINE
Glossary of terms
Mutagenesis
Spontaneous mutations
Induced mutations
   X-ray, gamma-ray
   5-BU, 2-AP
   nitrous acid
   hydroxylamine
   acridines
Repair of mutational damage
   photoreactivation
   excision repair
Mutant isolation
   visible mutants
   nutritional mutants
   conditional mutants
   enrichment procedures.

GLOSSARY OF TERMS

Phenotype — observable physical, biochemical etc. properties of an organism.

Genotype — genetic constitution of an organism.

Allele — one of the alternative forms of a gene.

Homozygous — in a diploid organism, the presence of identical alleles of a gene on the two homologous chromosomes.

Heterozygous — in a diploid organism, the presence of different alleles of a gene on the two homologous chromosomes.

Dominant allele — one which exhibits its phenotypic effect either when homozygous or heterozygous.

Recessive allele — one which exhibits its phenotypic effect only when homozygous.

Spontaneous mutation — one which occurred without a known cause.

Induced mutation — one which apparently is caused by the application of physical or chemical perturbations.

Transition mutation — where a purine-pyrimidine base pair is changed to the other purine-pyrimidine base pair, that is A-T to G-C or vice versa.

Transversion mutation — where a purine-pyrimidine base pair is changed to a pyrimidine-purine base pair, for example A-T to T-A.

MUTAGENESIS

In a previous Topic we discussed the structure of DNA. A gene is a specific sequence of nucleotides in the DNA and different genes have different sequences of nucleotides. Mutations are changes in the sequence of base pairs of the DNA such as transitions and transversions, and also insertions or deletions of base pairs. Experiments have shown that most single base-pair changes are reversible. The consequences of a mutation depend on its location within a gene and thus not all mutations result in an altered (mutant) phenotype of the organism under study. The induction of mutations is called mutagenesis and the agent involved is called a *mutagen*.

SPONTANEOUS MUTATIONS

No mutagens are involved in the production of spontaneous mutations. Both base-pair changes and chromosome aberrations can occur spontaneously. For example, the adenine molecule can exist in two forms called tautomers. In its more stable configuration, adenine forms two hydrogen bonds with thymine in DNA, but cannot hydrogen

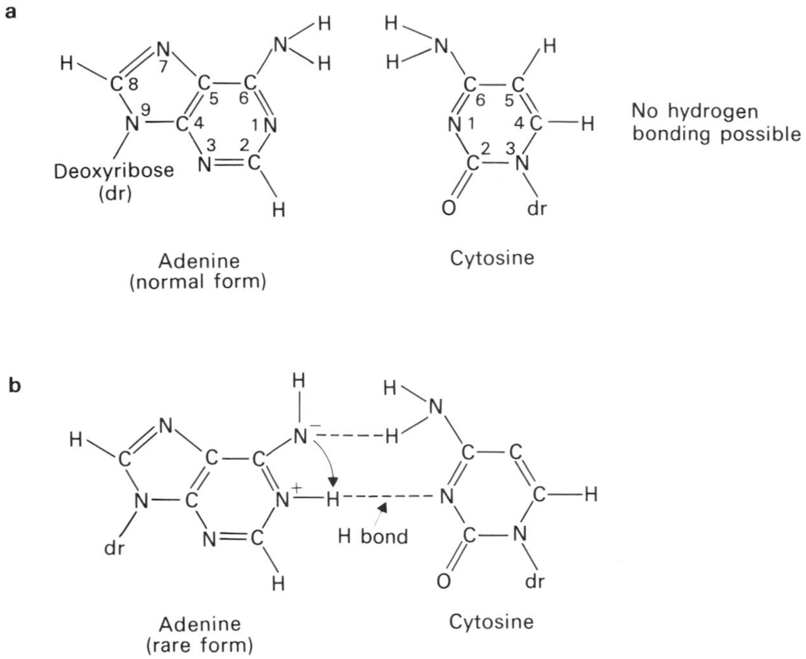

a

Adenine
(normal form)

Cytosine

No hydrogen
bonding possible

Deoxyribose
(dr)

b

Adenine
(rare form)

Cytosine

H bond

Fig. 6.1 Shift of adenine to rare
form results in formation of
unusual adenine-cytosine base
pairing.

bond with cytosine. However, if adenine undergoes
a tautomeric shift such that a hydrogen atom moves
from the 6-amino group to the 1-N position, then
hydrogen bonding with cytosine can occur at two
positions (Fig. 6.1).

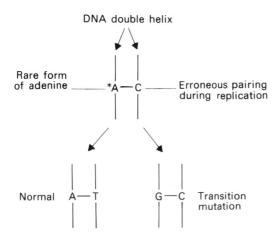

Fig. 6.2 Spontaneous mutation: erroneous pairing of
adenine with cytosine during replication leads to
transition mutation.

If the A-C pairing occurs while DNA is replicating, then at the ensuing round of replication one
of the two daughter DNA helices will have a G-C
pair instead of an A-T at that position. This is an
example of a *transition* mutation (Fig. 6.2). The
consequences of this mutation to the organism
depend on its location within the gene.

INDUCED MUTATIONS

Mutations can be induced by either physical or
chemical means. Irradiation is an example of the
former with X-rays, gamma-rays, and ultraviolet
light being the most commonly used mutagens. The
mode of action of X- or gamma rays is the breakage
of chromosomes which may result in chromosomal
rearrangements, or the events may be lethal to the
cell.

On the other hand, chemical mutagens can act
in a variety of ways depending on the properties of
the chemical and its reactions with the bases of the
DNA. Some examples of chemical mutagens and
their mode of action will now be given.

5-Bromouracil

5-Bromouracil (5-BU) is a base analogue; that is to say its structure closely resembles one of the bases normally found in DNA. 5-BU can exist in two states. In its normal keto state it exhibits properties similar to those of thymine and thus will pair with adenine in DNA. Rarely it switches to the enol state and in this form it will pair specifically with guanine (Fig. 6.3).

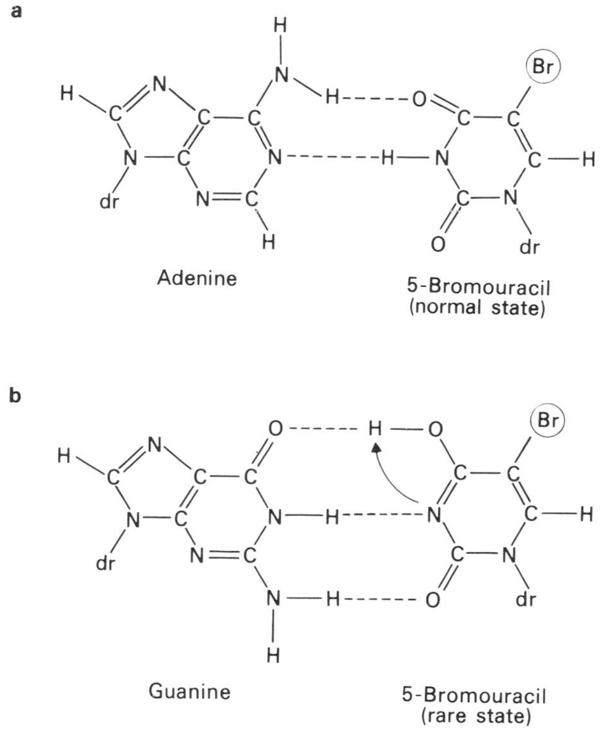

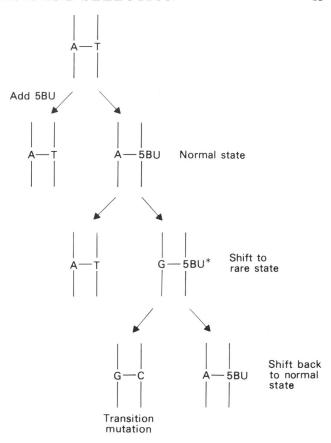

Fig. 6.4 Mutagenic action of 5-BU when it incorporates into DNA in the normal state and then shifts to the rare state during the next round of replication.

Fig. 6.3 Pairing properties of 5-bromouracil (5-BU): (a) In its normal keto state, 5-BU pairs with adenine; (b) In its rare enol state, 5-BU pairs with guanine.

Mutations can, therefore, be induced by 5-BU (and in general by base analogue mutagens) in two ways. The first involves the incorporation of the normal form of 5-BU into DNA during replication. If 5-BU shifts to its rare enol state during the next round of replication, then the result will be a transition mutation from A-T to G-C (Fig. 6.4).

The second way that 5-BU can induce mutations occurs if the base analogue is incorporated into the DNA while it is in the rare enol state. This would dictate insertion opposite a G on the complementary strand. Subsequent replication with a shift of the 5-BU to the normal keto state will result in a transition mutation from G-C to A-T (Fig. 6.5).

Thus 5-BU can induce either AT to GC or GC to AT transition mutations. (In the jargon of this area, 5-BU is said to induce two-way transition mutations.) Therefore it is possible to correct a 5-BU-induced transition mutation by treating with 5-BU for a second time. This is called reversion of the mutation.

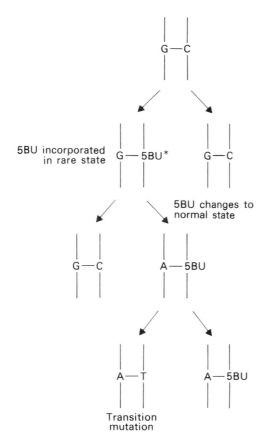

Fig. 6.5 Mutagenic action of 5-BU when it incorporates into DNA in the rare state and then shifts to the normal state during the next round of replication.

2-Aminopurine (2-AP)

2-AP is also a base analogue and, like 5-BU, it can exist in two states (Fig. 6.6). In its normal state it behaves like adenine and will form two hydrogen bonds with thymine. In its rare imino state 2-AP behaves like guanine and forms two hydrogen bonds with cytosine. Thus 2-AP can induce transition mutations both from AT to GC and from GC to AT. 2-AP-induced mutations can therefore be reverted by 2-AP treatment.

Nitrous acid (NA)

Nitrous acid (HNO_2) is a deaminating agent, that is it acts by removing amino groups (NH_2) from the bases. In some instances this alters their base-pairing abilities and hence induces mutations. Three of the bases have amino groups; namely, adenine, guanine, and cytosine. When adenine is treated with NA, it is changed to hypoxanthine which will pair with cytosine. This results in an AT to GC transition mutation (Fig. 6.7a).

Treatment of guanine with NA removes the amino group from the 2-carbon position and produces xanthine. However, since both guanine and xanthine pair with cytosine, no base pair mutation results (Fig. 6.7b).

A mutation does result when NA acts on cytosine, however. Deamination of cytosine produces uracil which, of course, pairs with adenine. This results in a GC to AT transition mutation, that is the opposite of NA's effect on adenine (Fig. 6.7c).

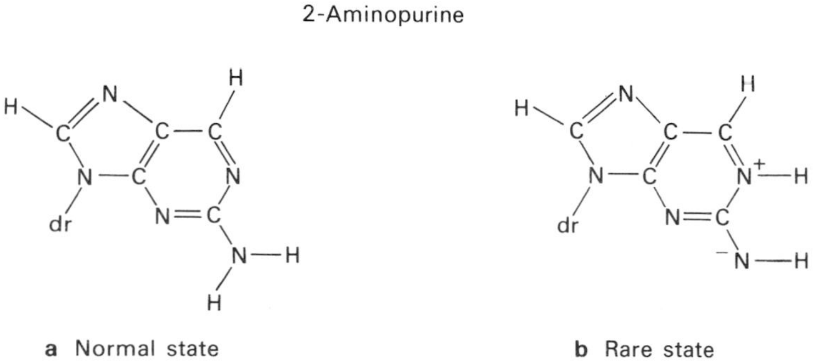

2-Aminopurine

Fig. 6.6 Structure of 2-amino-purine: (a) In its normal state (pairs with thymine) and (b) in its rare imino state (pairs with cytosine).

a Normal state   b Rare state

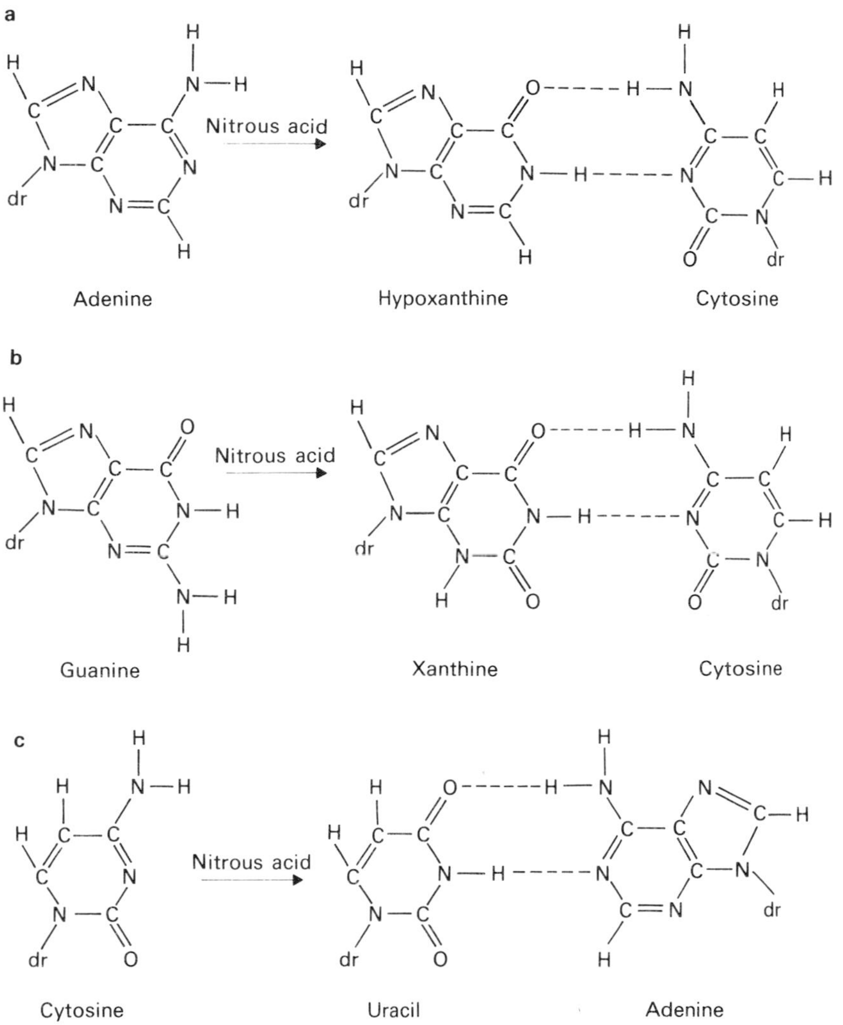

Fig. 6.7 Mutagenic action of nitrous acid. (a) Deamination of adenine by nitrous acid treatment produces hypoxanthine which pairs with cytosine (transition mutation); (b) Deamination of guanine produces xanthine which pairs with cytosine (no base-pair substitution); (c) Deamination of cytosine produces uracil which pairs with adenine (transition mutation).

Thus, mutations induced by nitrous acid can be reverted by nitrous acid treatment. Or, in other words, nitrous acid induces two-way transition mutations.

Hydroxylamine

Hydroxylamine (NH_2OH) induces mutations in a specific way in that it only reacts with cytosine, hydroxylating it so that it can then only pair with adenine (Fig. 6.8). Thus hydroxylamine induces one-way transition mutations from GC to AT. Because of this, mutations induced by hydroxylamine cannot be reverted by treatment with this

same mutagen, although they could be induced to revert by 5-BU, 2-AP or NA treatment since these can bring about a GC to AT transition.

Acridines

Acridine treatment results in the addition or deletion of one base pair in the DNA. This has serious consequences since the amino acid sequence of a protein coded for by a stretch of DNA altered in this fashion will be changed drastically. This will become more apparent in later discussions of messenger RNA translation.

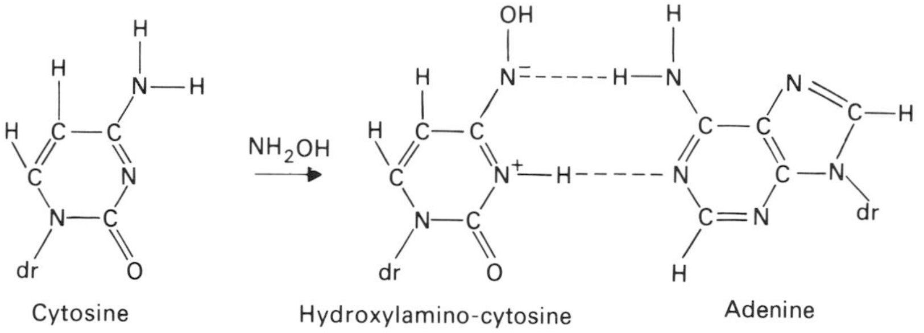

Fig. 6.8 Mutagenic action of hydroxylamine.

Cytosine   Hydroxylamino-cytosine   Adenine

When present at relatively low concentrations, acridines act by becoming inserted between adjacent base pairs in the DNA. When this occurs, it 'stretches' the distance between adjacent base pairs to 0.68 nm, which is precisely double the normal distance. The consequences of this depend on whether the acridine molecule is inserted into the template strand (the one being copied) or into the strand being synthesised. In the former case, a randomly chosen base is inserted opposite the acridine molecule when the replication fork passes by. At the next round of replication, the correct complementary base is paired with the inserted base with the result that one base pair is added to the DNA in that region. This is called an *insertion* mutation. These events are summarised in Fig. 6.9a.

Alternatively, if the acridine becomes inserted into the newly-synthesised strand, it blocks one of the bases on the template strand from having a complementary base. Then, if the acridine is lost before the next round of replication, the result will be a deletion of a base pair (Fig. 6.9b). Therefore, it is possible to revert an acridine-induced mutation by a second treatment with acridine.

Table 6.1 Summary of the modes of action of various chemical mutagens. (After W. Hayes, 1968, *The Genetics of Bacteria and Their Viruses*. Blackwell Scientific Publications, Oxford.)

Mutagen	Base-pair changes
5-bromouracil	AT ↔ GC two-way transitions
2-aminopurine	AT ↔ GC two-way transitions
Nitrous acid	AT ↔ GC two-way transitions
Hydroxylamine	GC → AT one-way transition
Acridines	+1 or −1 insertion or deletion

The mutagens described in this section, then, have different modes of action and cause different mutational changes. These are summarised in Table 6.1.

The chemical mutagens that have been described in this section are some that are commonly used in the laboratory. A large variety of other chemicals appear to cause mutations and there is increasing public awareness of this as industrial effluents, cosmetic ingredients, food additives, etc. are examined carefully for any mutagenic activity in test organisms.

REPAIR OF MUTATIONAL DAMAGE

Throughout the lifetime of an organism, the cells are exposed to a number of agents that have the potential to damage the DNA and hence induce mutations. Examples of such agents are ultraviolet light (from sunlight) and chemicals in the environment. Accumulated damage to the DNA over a period of time is considered by some scientists to be a cause of transformation of cells to the neoplastic (cancerous) state. Thus to enhance a cell's chance of survival a variety of inherent repair mechanisms have evolved that serve to reverse the effects of some spontaneous and induced mutations. A model system that has proved useful for the elaboration of two such mechanisms is pyrimidine dimers in which adjacent thymines or cytosines on the same DNA strand become bonded together when cells are irradiated with ultraviolet light (Fig. 6.10). The dimers distort the DNA so no pairing occurs with the purines of the opposite strand. Failure to remove the dimers may be a lethal event to the cell or if the wrong nucleotides are inserted opposite the dimers during replication

a

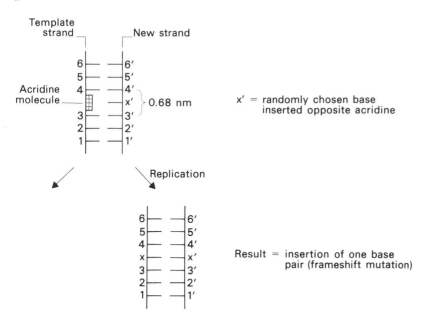

b

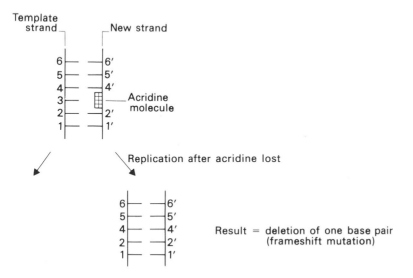

Fig. 6.9 Mutagenic action of acridines by intercalation into DNA. (a) Generation of addition mutation when acridine inserts in template strand. (b) Generation of deletion mutation when acridine inserts into newly-synthesising strand. (After W. Hayes, 1968, *The Genetics of Bacteria and their Viruses*. Blackwell Scientific Publications, Oxford.)

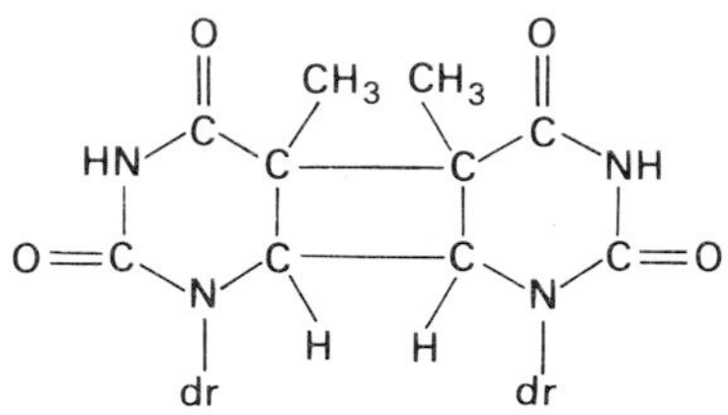

Fig. 6.10 Structure of thymine dimer induced by ultraviolet light.

a mutation may result. In this section two repair mechanisms, *photoreactivation* and *excision repair*, will be considered.

Photoreactivation
In this form of repair the UV-induced pyrimidine dimers are reversed directly to the original form.

This reaction is dependent upon exposing the cells to visible light following irradiation, hence the name 'photoreactivation' (Fig. 6.11). The reversal process itself is catalysed by an enzyme called a *photolyase* which monomerizes the dimers when activated by a photon of light in the wavelength range 320–370 nm. Examination of a large number

Fig. 6.11 Photoreactivation of thymine dimer induced by UV light. (a) Segment of DNA double helix distorted by a thymine dimer. (b) Attachment of photoreactivating enzyme (photolyase) to the thymine dimer region. (c) Absorption of photon of light in the blue end of the spectrum causes the enzyme to split the thymine dimer, allowing the A-T base pairs to reform. The enzyme then dissociates from the DNA. (After M. W. Strickberger, 1976, *Genetics*. Macmillan, New York.)

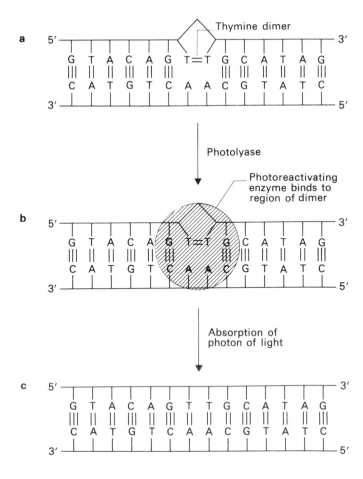

of organisms leads us to the conclusion that the photolyase repair enzymes are probably ubiquitous.

Excision repair

The second repair mechanism was discovered separately by P. Boyce and P. Howard-Flanders, and by R. Setlow and W. Carrier. In this case the dimers are excised from the DNA by the action of nucleases and the single-stranded gap is filled in by the action of polymerase and ligase enzymes (Fig. 6.12). Since this reaction does not require activation by light, it is also called 'dark repair'. In fact the excision repair mechanism in *E. coli*

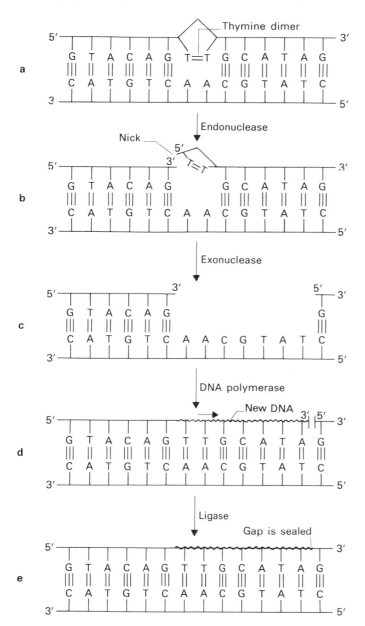

Fig. 6.12 Excision repair of thymine dimers in DNA. (a) Segment of DNA distorted by a UV-light induced thymine dimer. (b) Endonuclease activity 'nicks' the DNA on the 5′ side of the dimer. (c) Exonuclease action removes the dimer and other nucleotides on the same strand in the 5′ to 3′ direction. (d) DNA polymerase I fills in the single-stranded gap catalysing the synthesis of new DNA in the 5′ to 3′ direction. (e) The gap between old and new DNA is sealed by the action of polynucleotide ligase. (After M. W. Strickberger, 1976, *Genetics*. Macmillan, New York.)

is very complex and the details will not be given here. Generally, there are specific correcting nucleases involved with repair. The first step involves recognition of the dimer in the DNA and this is followed by the formation of a single-strand nick by the action of a correcting endonuclease. These enzymes are normally small proteins with a molecular weight of about 30 000 daltons. All of the ones that have been characterised make their incision close to the damage in the DNA strand. The result of the enzyme activity is a free 5′-end that is the substrate for the 5′ to 3′ exonuclease activity of DNA polymerase I. Concomitant with the removal of part of one strand of DNA, including the pyrimidine dimer, a new segment of DNA is polymerised into the gap in the 5′ to 3′ direction by the action of DNA polymerase I. This leaves a single-stranded nick in the DNA strand which is sealed by polynucleotide ligase. The dark repair process has also been shown to occur in other organisms and a number of the enzymes involved have been identified in several systems.

The genetic basis of excision repair has been studied in several organisms, for example by selecting for UV-light sensitivity. Five genes, *uvrA–E*, have been shown to mutate in *E. coli* to give this phenotype and the biochemical bases for the sensitivity have been examined in some cases. Thus *uvrA*, *uvrB* and *uvrC* strains have very similar sensitivity to UV light and, while the latter has normal levels of the dimer-specific endonuclease, *uvrA* and *uvrB* lack the enzyme activity and hence excision repair cannot proceed. The biochemical defect in the *uvrC* strain is not known. The *uvrD* mutant is less sensitive to UV irradiation than the three strains discussed above and, in addition, for reasons not yet known, it shows extensive degradation of the DNA after UV treatment. The *uvrE* strain is UV-sensitive and shows an increased rate of spontaneous mutation compared with the wild type.

A number of recombination-deficient (rec) mutations have also been found in *E. coli* and these are also UV-sensitive. These strains are not defective in excision repair, however, their primary role being in the recombination process.

Finally, as would be expected, mutations in the genes for polynucleotide ligase (lig) or for DNA polymerase I (polA) result in decreased levels of excision repair. A genetic map showing the scattered locations of genes involved with dimer repair in *E. coli* is shown in Fig. 6.13. It is likely that similar enzymes play similar roles in other organisms also.

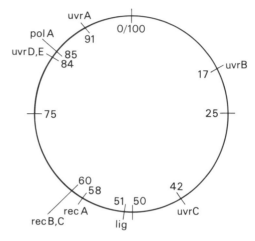

Fig. 6.13 Map location of the genes involved in the repair of DNA in *E. coli*. (After U. Goodenough, 1978, *Genetics*. Holt, Rinehart and Winston, New York.)

MUTANT ISOLATION

Spontaneous mutations occur very rarely among populations of any organism. Therefore, scientists use mutagens to increase the frequency of occurrence of mutations. Even so, as is apparent from the previous discussions, the mutagens generally used to induce mutations act at the base-pair level and therefore they do not induce mutations in specific genes. Geneticists and biochemists usually want to study mutants that carry mutations affecting specific functions. Therefore a number of screening and enrichment procedures have been developed to help isolate specific mutants from among a heterogeneous mutagenised population of cells or organisms. These procedures will be described in the context of the types of mutants that are useful for genetical and/or biochemical types of analysis.

The first class of mutants to consider is the visible mutants. As their name suggests, these have phenotypes that are different from the normal (wild type) organism at the gross macroscopic or microscopic level. In diploid organisms, mutations that give rise to a visible mutant phenotype are often recessive so the mutant phenotype is only seen when the mutation is homozygous. Thus, visible mutants in a diploid organism may only become apparent after further breeding of a mutagenised population of the organism. By contrast, visible mutants are immediately apparent in haploid organisms.

There are numerous examples of visible mutants. In diploid organisms, eye colour mutants, body colour mutants, wing shape mutants in *Drosophila*, coat colour mutants in animals, and petal colour mutants in plants are some examples of this class of mutants. In haploid organisms, examples are mutants of *Pneumococcus* bacteria which produce rough instead of smooth colonies when grown on solid media, mutants of yeast which produce colonies which are much smaller than the wild-type strain (*petite* mutants), and mutants of the fungus *Neurospora* which grow with a colonial morphology instead of with the web-like growth habit which is characteristic of the wild-type strain (Fig. 6.14).

The second class of mutants affects the ability of

Fig. 6.14 (a, below) Morphology of wild-type *Neurospora crassa* growing on solid medium. (b) Morphology of a colonial mutant of *Neurospora*. (Photographs by P. J. Russell.)

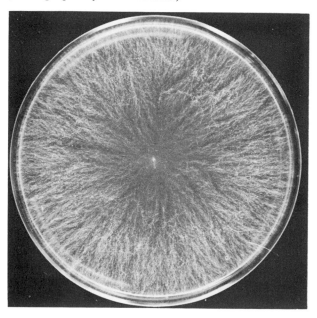

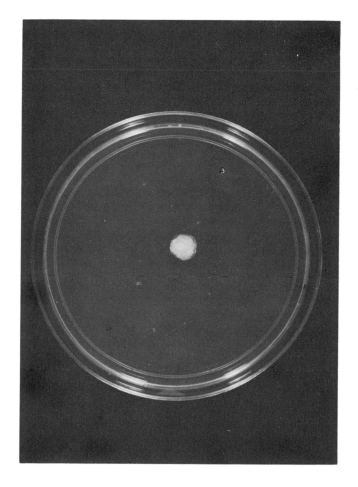

the organism to grow. These types of mutants are most often isolated and studied in haploid organisms such as *E. coli*, yeast, and *Neurospora*. The reason for this is that these organisms can grow on a well-defined simple minimal medium which provides a carbon source, salts and trace elements (and sometimes vitamins) which they use to make all of the cell molecules such as amino acids, purines, pyrimidines, vitamins, etc. The standard laboratory strain of an organism that can grow on the minimal medium is called the wild-type or prototrophic strain. By mutagenesis of the wild-type strain, it is possible to isolate strains that cannot grow on the minimal medium alone but which can grow on a supplemented minimal medium. These *auxotrophic* mutants (also called nutritional mutants) are unable to make an ingredient essential for cell growth. Thus the mutation involved may affect the synthesis of an amino acid, purine, pyrimidine, vitamin, or other essential molecule. That ingredient must be present as a supplement to the minimal medium so that the mutant strain can grow.

For many organisms that will grow to discrete colonies on solid medium in a petri dish, auxotrophic mutants can be isolated using the *replica plating* technique developed by E. and J. Lederberg (Fig. 6.15). In outline, samples of a cell culture (that has or has not been treated with a mutagen) are spread on the surface of a completely supplemented medium which is incubated to allow colonies (clones of the cell that landed on the medium) to grow. On this medium both prototrophic and auxotrophic cells will grow. The pattern of colonies can then be transferred onto velveteen cloth which acts like thousands of tiny inoculating needles. By gently pressing a plate containing minimal medium onto the velveteen cloth, the pattern of colonies can be replicated onto that medium. Only prototrophic strains will grow on the minimal medium and thus, by comparing the patterns of colonies on the two media, auxotrophic strains can be detected and isolated from the original master plate. These auxotrophs can be tested as to the exact nature of their nutritional requirements by setting up a new master plate and replica plating

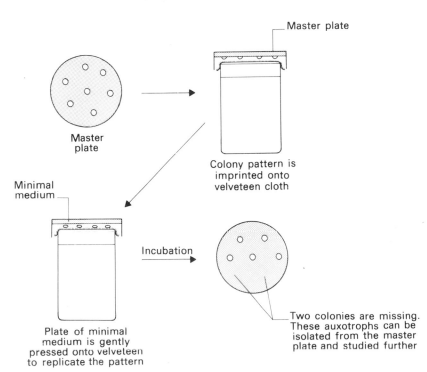

Fig. 6.15 The isolation of auxotrophic mutants of a colony-forming organism by replica plating.

to a number of different media, for example minimal plus amino acids, minimal plus vitamins, and so on.

Some degree of mutant enrichment can be achieved with replica plating by the choice of the media onto which the colony pattern is replicated. For example, one can specifically isolate mutants that require adenine for growth by replica plating onto a medium containing all possible supplements except adenine.

It is possible to enrich for auxotrophic mutants by using some other procedures:

1. Antibiotic selection. This procedure takes advantage of the fact that growing, but not non-growing, cells of certain organisms can be killed by treating them with antibiotics. For example, if a mutagenised population of *E. coli* cells is placed in a medium that is restrictive (nonpermissive) for the type of auxotroph desired, then only the strains not requiring the omitted supplement will be able to grow. If penicillin is added to the culture, the growing cells will be killed, leaving the auxotrophs surviving. The same principle applies for nystatin (an antibiotic named after or for New York state) enrichment in yeast.

2. Filtration enrichment. This procedure is used with filamentous fungi such as *Neurospora*. If a mutagenised population of cells is incubated in a selective medium, the mutant strains desired will not grow, whereas everything else will produce filamentous hyphae. The latter may be removed by passing the culture through cheesecloth. The mutant, ungerminated cells pass through this material. When this is repeated over several days, there is significant enrichment for the auxotrophic mutant type desired.

Another class of mutants that is very valuable in studies of macromolecular synthesis, cell function and regulation, is the conditional mutants. The most common examples are the temperature-sensitive mutants which, unlike the wild type, do not grow well or at all at either high or low temperatures. These heat-sensitive (hs) and cold-sensitive (cs) mutants can be isolated easily following mutagenesis and the application of a suitable enrichment procedure. For example the replica-

plating method can be adapted by incubating the replicated colonies at either high or low temperature. Note that it is possible to isolate conditional auxotrophs in this way also. These can be avoided if the medium in *all* plates is fully supplemented.

An enrichment procedure that has been applied to the isolation of heat-sensitive and cold-sensitive mutants is tritium suicide. Tritium suicide is defined as the death of cells caused by the decay of the radioactive isotope of hydrogen, tritium (^{3}H), incorporated into their macromolecules, such as proteins and nucleic acids. This technique has been used with a number of organisms including *E. coli*, yeast, *Neurospora*, and cultured mammalian cells, to enrich for particular mutant types in a heterogeneous mutagenised population of cells. Selection by tritium suicide can be made relatively specific by the choice of the tritiated precursor and the culture conditions used. For example, in some recent experiments in the author's laboratory,

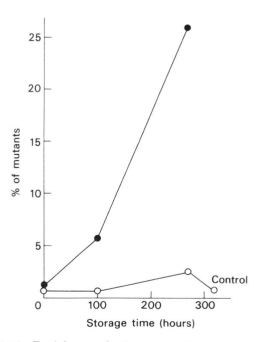

Fig. 6.16 Enrichment for heat-sensitive mutants of *Neurospora crassa* as a function of time after the incorporation of tritiated amino acids. (●) ^{3}H-labelled culture; (○) unlabelled, control culture. (After P. J. Russell and M. P. Cohen, 1976, *Mut. Res.* **34**:359.)

tritium suicide was used to enrich for heat-sensitive mutants of *Neurospora*. Mutagenised asexual spores (conidia) were incubated in minimal medium at 35°C (the chosen nonpermissive temperature) in the presence of relatively high amounts of ^{3}H-amino acid mixture (protein precursor). After two hours the conidia were washed free of radioactivity and stored in a refrigerator. At various times thereafter, samples were taken to test for viability and for the presence of heat-sensitive (hs) mutants. A culture that had not been incubated in the presence of the radioactive precursor was used as a control (Fig. 6.16).

The data indicated clearly that significant death of cells occurs when tritium is incorporated into their macromolecules. In addition, as time of storage increased, there was significant enrichment for heat-sensitive mutants. The reason for this is that at 35°C the heat-sensitive mutants are unable to grow, they are metabolically quiescent and therefore much less of the tritiated precursor is incorporated into their macromolecules than is the case with non-mutants. The high level of radioactivity incorporated into the latter cells results in their death, whereas the heat-sensitive mutants have a better chance of survival.

REFERENCES

Auerbach C. & B.J. Kilbey. 1971. Mutation in eukaryotes. *Annu. Rev. Genet.* **5**: 163–218.

Beadle G.W. & E.L. Tatum. 1945. Neurospora. II. Methods of producing and detecting mutations concerned with nutritional requirements. *Am. J. Bot.* **32**: 678–686.

Boyce R.P. & P. Howard-Flanders. 1964. Release of ultraviolet light-induced thymine dimers from DNA in *E. coli* K12. *Proc. Natl. Acad. Sci. USA*, **51**: 293–300.

Drake J.W. 1969. Mutagenic mechanisms. *Annu. Rev. Genet.* **3**: 247–268.

Freese E. 1959. The specific mutagenic effect of base analogues on phage T4. *J. Mol. Biol.* **1**: 87–105.

Hanawalt P.C. 1972. Repair of genetic material in living cells. *Endeavour* **31**: 83–87.

Howard-Flanders P. 1968. DNA repair. *Annu. Rev. Biochem.* **37**: 175–200.

Kelley R.B., M.R. Atkinson, J.A. Huberman & A. Kornberg. 1969. Excision of thymine dimers and other mismatched sequences by DNA polymerase of *E. coli. Nature* **224**: 495–501.

Lederberg J. & E.M. Lederberg. 1952. Replica plating and indirect selection of bacterial mutants. *J. Bacteriol.* **63**: 399–406.

Lester H.E. & S.R. Gross. 1959. Efficient method for selection of auxotrophic mutants of Neurospora. *Science* **129**: 572.

Littlewood B.S. & J.R. Davies. 1973. Enrichment for temperature-sensitive and auxotrophic mutants in *Saccharomyces cerevisiae* by tritium suicide. *Mut. Res.* **17**: 315–322.

Moat A.G., N. Peters & A.M. Srb. 1959. Selection and isolation of auxotrophic yeast mutants with the aid of antibiotics, *J. Bacteriol.* **77**: 673–681.

Russell P.J. & M.P. Cohen. 1976. Enrichment for auxotrophic and heat-sensitive mutants of *Neurospora crassa* by tritium suicide. *Mut. Res.* **34**: 359–366.

Setlow R.B. & W.L. Carrier, 1964. The disappearance of thymine dimers from DNA: an error-correcting mechanism. *Proc. Natl. Acad. Sci. USA*, **51**: 226–231.

Tatum E.L., R.W. Barratt, N. Fries & D. Bonner. 1950. Biochemical mutant strains of Neurospora produced by physical and chemical treatment. *Am. J. Bot.* **37**: 38–46.

Woodward V.W., J.R. De Zeeuw & A.M. Srb. 1954. The separation and isolation of particular biochemical mutants of Neurospora by differential germination of conidia, followed by filtration and selective plating. *Proc. Natl. Acad. Sci. USA*, **40**: 192–200.

Topic 7
Transcription

OUTLINE
The central dogma
RNA synthesis (transcription)
Only one strand of DNA is transcribed
Prokaryotic RNA polymerases
Classes of RNA
Eukaryotic RNA polymerases
Messenger RNA
Transfer RNA
Ribosomes and ribosomal RNA
Biosynthesis of prokaryotic ribosomes
Biosynthesis of eukaryotic ribosomes.

THE CENTRAL DOGMA

The genetic material of a cell has a major function to direct protein synthesis. The genome contains all of the information for the structure and function of an organism, but not all of the genes are active at any one time. The regulatory elements for this will be discussed in later Topics.

The DNA itself is not a direct template for protein synthesis. Rather, the genetic information of DNA is first transferred to molecules of RNA in a process called *transcription*. The RNA template is then used to produce the sequence of amino acids of proteins in a process called *translation*. The relationship of DNA to protein is summarised by the central dogma (Fig. 7.1).

As was discussed in Topic 1, RNA differs from DNA in two respects; namely, it has a ribose sugar instead of deoxyribose, and the base, uracil, is used instead of thymine. The RNA chain is usually linear and the ribonucleotide constituents are linked by 3',5'-phosphodiester bonds as is the case with DNA chains.

RNA SYNTHESIS (Transcription)

RNA synthesis occurs by a similar process to DNA replication. The DNA double helix unwinds at the start points for transcription (this is controlled by regulatory signals within the cell) to allow RNA synthesis to commence. The precursors for the RNA transcripts are ATP, GTP, CTP and UTP and these are polymerised into an RNA chain in the 5' to 3' direction using one of the DNA chains as a template (Fig. 7.2). Thus, the RNA product has a nucleotide sequence complementary to that of the template DNA strand, and in fact is a faithful RNA copy of the non-template DNA strand. Considering the large number of start points in the genome for RNA synthesis, the RNA chains produced are relatively short.

In transcription of RNA from DNA, the enzyme DNA-dependent RNA polymerase (usually called RNA polymerase) catalyses the reaction. The enzyme 'obeys' the complementary base-pairing rules

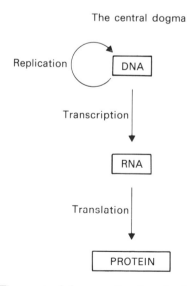

The central dogma

Replication — DNA

Transcription

RNA

Translation

PROTEIN

Fig. 7.1   The central dogma of molecular biology.

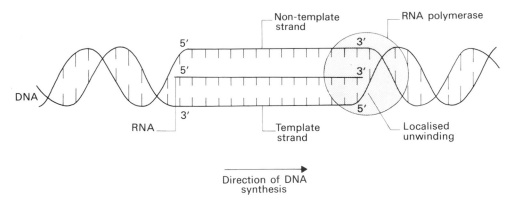

Fig. 7.2 Schematic of the transcription of RNA from a DNA double helix.

in the same way as does DNA polymerase. The RNA chain, then, is a faithful copy of DNA with the exception that uracil is put into RNA where thymine is found in the non-template strand of DNA. The enzyme itself can only polymerise the RNA chain in the 5′ to 3′ direction and, as was apparent in the discussions of DNA synthesis, it can initiate chains. This latter property is obviously essential considering the fact that different RNA molecules are required in different cells or at different times in the same cells.

In the diagram for transcription (Fig. 7.2) it was implied that only one of the two strands of DNA is copied into RNA in a given region. Some evidence for this came from the work of J. Marmur and his colleagues in their studies of the RNA species produced when bacteriophage SP8 infects the bacterium *Bacillus subtilis*. The SP8 phage has a double-stranded DNA chromosome which has a density of 1.743 gm/ml. When this DNA was separated into the two polynucleotide chains, it was found that their densities differed significantly (1.756 and 1.764) such that they could be separated easily by CsCl equilibrium density gradient centrifugation (Fig. 7.3).

Marmur and his colleagues then allowed SP8 to infect *Bacillus subtilis* growing in a medium containing ^{32}P such that all RNA species transcribed from the phage genome during the life cycle became radioactively labelled. They reasoned that the RNAs would be complementary to the DNA strand

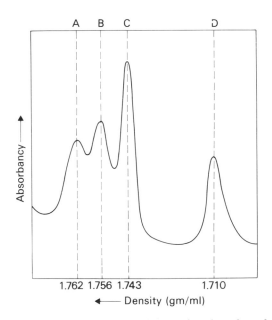

Fig. 7.3 Results of an experiment showing that the two strands of DNA in phage SP8 have different buoyant densities. (D) Reference DNA, (C) Native, double-stranded SP8 DNA, (A,B) Denatured 'heavy' and 'light' SP8 DNA strands, respectively. (After J. Marmur, 1963). *Cold Spring Harbor Symp. Quant. Biol.* **28**: 191.)

from which they were transcribed. Since a number of phage-specific RNA molecules are produced during SP8 infection, prima facie it was possible that all were transcribed from one of the strands, or that some were transcribed from each. This was examined by determining whether the single-stranded RNA molecules formed stable hybrids with one or both denatured DNA strands. fortunately it was found that stable DNA-RNA hybrids were only formed with the heavy DNA strand, from which it was concluded that only one DNA strand is transcribed.

This conclusion is now generally accepted for all organisms. However, further experiments have shown that, in most cases, the particular DNA strand used as the template strand for RNA synthesis varies throughout the genome and appears to be gene-specific.

PROKARYOTIC RNA POLYMERASES

Among prokaryotic RNA polymerases, the enzyme from *E. coli* has been studied most extensively. This enzyme has a sedimentation coefficient of 11–13S and a molecular weight of approximately 500 000 daltons. Relatively gentle treatment of the soluble enzyme dissociates a 95 000-dalton polypeptide called the sigma factor (σ), leaving the so-called core polymerase. The latter can be further dissociated into four polypeptide chains. These are the beta-prime (β') subunit (165 000 daltons), beta (β) subunit (155 000 daltons) and two copies of the alpha (α) subunit (41 000 daltons each). A fifth subunit called omega (ω) (12 000 daltons) has been reported by some researchers but its presence in the core polymerase is controversial.

The core enzyme alone can make an RNA copy of DNA. The roles of the subunits in this process have been studied extensively. Isolated beta-prime subunits, for example, have been shown to bind to DNA *in vitro*, whereas the alpha and beta subunits do not. Thus, apparently, beta-prime is the DNA-binding subunit. Our knowledge of the function of the beta subunit has come from studies of anti-biotic-resistant mutants. Rifampicin inhibits the initiation of transcription and streptolydigin inhibits the elongation of the RNA chain. These antibiotics bind to the beta subunit of wild-type cells but do not bind to the beta subunit isolated from mutants resistant to these drugs. Thus, the beta

subunit may be involved in the catalysis of phosphodiester bond formation. The roles of the alpha and omega subunits are not known.

The dissociable polypeptide of RNA polymerase, sigma, plays a very important role in RNA synthesis. This 95 000-dalton subunit is necessary for the core enzyme to initiate RNA synthesis at specific sites along the DNA. These sites are called promoter regions. If sigma is not bound to the core enzyme, the latter initiates RNA synthesis at random sites along the DNA and, moreover, both strands are copied instead of one.

After RNA synthesis has begun the sigma factor dissociates from the core enzyme, which then continues the transcription process. The sigma factor may then associate with another core enzyme to initiate another RNA transcription at the same or a different promoter. Thus the sigma factor acts catalytically in terms of the initiation of transcription at specific sites.

Evidence for this role of the sigma factor in RNA synthesis came from experiments performed by A. Travers and R. Burgess. In one experiment they measured the in-vitro incorporation of gamma-^{32}P-ATP and gamma-^{32}P-GTP into RNA in an assay mixture containing varying amounts of the sigma factor. The radioactive label in the precursors for these experiments was located in the third (gamma) phosphate group (Fig. 7.4).

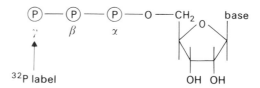

(P) = Phosphate group

Fig. 7.4 γ-^{32}P-labelled ribonucleoside triphosphate.

In RNA synthesis the nucleoside triphosphate at the 5'-end of the chain may retain all three phosphate groups. During the polymerisation of the remainder of the chain, however, both the beta- and gamma-phosphate groups are removed. Therefore the amount of ^{32}P found in the RNA synthesised during the experiment directly reflects the

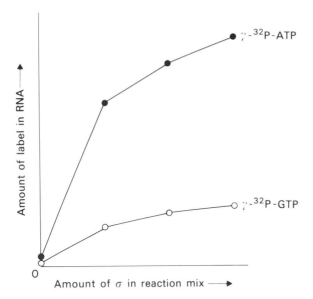

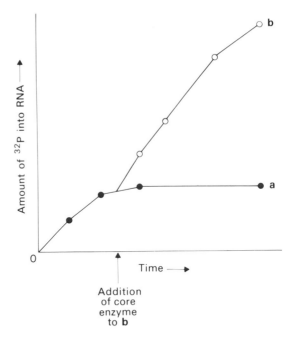

Fig. 7.5 Demonstration that the sigma factor of RNA polymerase stimulates RNA synthesis by increasing the number of new RNA chains initiated. Here the incorporation of label from γ-^{32}P-labelled ATP or GTP into RNA was used as a measure of the number of RNA chains initiated while the amount of sigma factor in the reaction mix was varied. (After A. A. Travers and R. R. Burgess, 1969. *Nature,* **222**:354.)

Fig. 7.6 Demonstration that the sigma factor is a catalyst in the initiation of RNA synthesis. Curve (a): The kinetics of RNA chain initiation with core enzyme and sigma under ionic conditions where the former cannot be re-used. Curve (b): When large amounts of fresh core enzyme are added to the reaction mix, renewed initiation occurs showing that sigma is recycled. (After A. A. Travers and R. R. Burgess, 1969. *Nature* **222**:354.)

number of RNA chains initiated. The results were as shown in Fig. 7.5.

The data indicate clearly that the principal effect of sigma is to increase the number of RNA chains initiated.

In a second experiment, Travers and Burgess provided evidence that sigma acts as a catalyst in RNA chain initiation; in other words sigma can be re-used (Fig. 7.6). Here, an in-vitro assay system was used in which the ionic strength was kept low. Under these conditions initiation of RNA chains will occur and the enzyme will remain attached to the DNA in an enzyme–DNA–RNA complex. Thus, each enzyme can initiate only once. In their experiment the kinetics of initiation again were followed by the incorporation of gamma-^{32}P-ATP and gamma-^{32}P-GTP into RNA when complexes of core enzyme with sigma were added to the reaction mixture. After several minutes no further initiation occurred (curve **a** in the diagram). At this point the

amount of initiation corresponded to approximately one initiation per complex molecule.

It was reasoned that if sigma is involved only with initiation, then free sigma factors should be present in the reaction mixture at this time. If that was the case then addition of a large amount of core enzyme (no sigma) to the reaction mixture should stimulate new RNA chain initiation, since new core enzyme-sigma factor initiation complexes could form. Indeed this resulted in renewed initiation (curve **b**), thus supporting the hypothesis that sigma is recycled and is involved only in RNA chain initiation.

There is evidence that the core RNA polymerase undergoes conformational changes. Specifically, its sensitivity to salt, proteases (protein degrading enzymes) and inhibitors changes when it becomes bound to DNA and when RNA synthesis is being catalysed. And, as was noted before, the enzyme is no longer sensitive to the RNA synthesis initiation inhibitor, rifampicin, once chain initiation has begun.

Just as there are start signals for the initiation of RNA synthesis, so there are specific nucleotide pair sequences in the DNA whose function is to terminate RNA synthesis. Two types of stop signal are known. One is recognised by a multimeric form of a 50 000-dalton protein called the rho factor. It is not yet clear whether rho interacts with RNA polymerase or whether it binds to DNA *in vivo*. The second type of stop signal involves a relatively long sequence of AT base pairs which seem to be read by the RNA polymerase enzyme itself.

CLASSES OF RNA

Before discussing the RNA polymerases of eukaryotic organisms, we must first briefly describe the major classes of transcribed RNA found in cells of both prokaryotes and eukaryotes. These will be discussed in much more detail later.

1. Messenger RNA (mRNA). This RNA serves as a template for protein synthesis. The size of a mRNA molecule is a function of the length of the polypeptide chain for which it codes. Thus mRNAs of a cell represent a heterogeneous population of molecules.

2. Transfer RNA (tRNA). Each tRNA molecule is able to bind covalently with a specific amino acid and to form hydrogen bonds with a three-nucleotide sequence (codon or triplet) on an mRNA. The latter event occurs on the ribosome which is the site of protein synthesis. All tRNA molecules have a molecular weight of about 25 000–30 000 daltons and a sedimentation coefficient of 4S (the S stands for Svedberg units and, as a rule, the larger the S value, the larger is the molecule or particle).

3. Ribosomal RNA (rRNA). Ribosomes are made up of rRNA molecules and proteins. The rRNA molecules have discrete sizes which are described with generalised S values for prokaryotes and

eukaryotes. Thus, prokaryotic ribosomes are considered to contain 23S, 16S, and 5S rRNA molecules and the cytoplasmic (non-organellar) ribosomes of eukaryotes (here usually referring to the 'higher' eukaryotes) contain 28S, 5.8S, 18S, and 5S rRNA molecules. We will use these S values in general discussions of eukaryotic ribosomes but we will refer to specific values when dealing with 'lower' eukaryotes in later sections.

EUKARYOTIC RNA POLYMERASES

Three types of nuclear RNA polymerases (I, II, and III) have been found in a number of eukaryotic cells. These have been shown to have distinct specialised roles in RNA transcription, and they vary in their sensitivity to the RNA synthesis inhibitor alpha-amanitin. None is sensitive to the prokaryotic RNA synthesis inhibitor rifampicin.

RNA polymerase I. This enzyme is not inhibited at all by alpha-amanitin. It is found within the nucleolus and is responsible for the transcription of the genes for the 28S, 5.8S, and 18S rRNA molecules.
RNA polymerase II. This polymerase is inhibited at low concentrations of alpha-amanitin and is required for the synthesis of most other RNA species, notably mRNA.
RNA polymerase III. This enzyme is inhibited at high concentrations of alpha-amanitin and functions in the transcription of tRNA (4S RNA) and 5S ribosomal RNA.

MESSENGER RNA

General properties
Genes coding for proteins are transcribed into mRNA molecules which become associated with ribosomes where protein synthesis occurs. In prokaryotic organisms where no nuclear membrane exists it is possible for mRNA to associate with ribosomes before the mRNA chain itself has been completed. In eukaryotes, owing to the cellular compartmentation, the mRNAs must migrate out

of the nucleus to the cytoplasm where the ribosomes active in protein synthesis are found.

Most of the genome of prokaryotic and eukaryotic organisms codes for proteins, and thus most of the DNA can be transcribed into mRNA. However, at any one time, less than 10–20% of the total cellular RNA is present as mRNA.

When the mRNA population of a cell is examined it is found to be heterogeneous in size, with a fairly wide range of S values. This results from the fact that the number of nucleotide pairs in a gene is directly related to the size of the protein for which they code. Thus, since proteins vary considerably in the number of amino acids they contain, their mRNAs vary in the same way.

Prokaryotic mRNA

In prokaryotic organisms, mRNAs have relatively short lifetimes, that is they are quite labile. By this we mean that a newly synthesised mRNA molecule may only function for about two minutes in organisms such as *E. coli*. After this time, which is approximately 10% of a cell cycle, the RNA will be destroyed. Thus to continue to synthesise a particular protein, the gene for that protein must be transcribed continuously. As will be more apparent when we discuss the regulation of gene expression, this affords the cell a very rapid means of conserving resources. Specifically, if a protein is no longer needed for a cell to function, the gene can be blocked from being transcribed and any extant mRNA will soon be destroyed. This allows the organism, then, to respond rapidly to changing needs when its external environment changes by turning off the transcription of some genes and turning on the transcription of a different set of genes.

Before considering eukaryotic mRNA, it must be noted that, while most prokaryotic mRNAs are short lived, there are some mRNA species known in prokaryotes that have relatively long lives. These appear to be resistant to the nucleases that are responsible for the rapid destruction of the short-lived mRNAs. In *E. coli* for example, at least five ribonuclease (RNase) activities have been demonstrated and some of these, if not all, may be involved in mRNA breakdown.

Eukaryotic mRNA

Longevity of mRNAs

Eukaryotic mRNAs are usually considered to be long lived. Compared with the prokaryotic mRNAs in terms of actual times, this is certainly true. However, there are many mRNAs in eukaryotic organisms that are functional for only an hour or so, which is about 5–10% of the cell cycle. One might certainly consider these to be short lived in these terms, in much the same way as we did for the short-lived prokaryotic mRNAs.

It is fair to say, though, that the lifetimes of eukaryotic mRNAs, when expressed as a fraction of the normal cell-cycle time, are highly varied. This is a practical consequence of the functional characteristics of a eukaryotic cell as compared with a prokaryotic cell. In particular, eukaryotic organisms are differentiated into various cell types which have different functions. For example, some cell types are specialised to produce one type of protein almost entirely. Characteristically, the mRNA for that protein will be very stable.

Chromatin structure and transcription

As was discussed in Topic 2, the DNA of eukaryotic nuclear chromosomes is organised into nucleosome structures. This raises the obvious question: what happens to the nucleosomes when transcription occurs? A number of studies have shown that transcriptionally-active chromatin is more sensitive to deoxyribonuclease attack than is nontranscribing chromatin. This result indicates that the transcribed segments of the chromatin must have a different structural conformation than the inactive chromatin. However, this is not due to the absence of any of the four core histones, H2A, H2B, H3 and H4 which are still found in the nucleosome structures. It is not clear at this time whether H1 is still present in active chromatin or whether the core histones are chemically modified. The latter is a distinct possibility since, for example, acetylation of the histones results in an increase in template activity in an in-vitro system. In conclusion, then, the transcription machinery does not appear to alter the nucleosome structure in any major way, although clearly some configurational changes do occur, resulting in the histones becoming more loosely bound to the DNA.

5'- and 3'-end modifications

Unlike prokaryotic mRNAs, eukaryotic mRNAs are modified at both the 5' and 3' ends. The modifications occur post-transcriptionally and involve the action of specific enzymes, only some of which have been characterised to date.

The 5' ends of most eukaryotic mRNAs do not have a free 5'-nucleoside triphosphate, but instead they are *capped*. In other words, after transcription of the mRNA, a modification occurs in which a guanine nucleotide is added to the terminal 5'-nucleotide by a 5'-5' linkage. In addition, methyl groups are added to the now terminal guanine and to the 2'-hydroxyl group of the adjacent nucleotide (Fig. 7.7). The capping phenomenon is ubiquitous among eukaryotes although there are slight variations of the 'cap' structure itself. At the functional level, the cap structure appears to be essential for the formation of the mRNA–ribosome complex and therefore for the initiation of protein synthesis.

The 3' ends of most eukaryotic mRNAs (histone mRNA is a general exception) are modified by the addition of 50 to 200 adenine nucleotides. These so-called poly(A) segments are added after transcription has been completed with the aid of the enzyme poly(A)-polymerase. There appear to be multiple forms of the enzyme in cells and they all catalyse the reaction shown in Fig. 7.8.

The exact role of the poly(A) segment is not clear although it is suggested that it stabilises the mRNA against nuclease attack. Some evidence to support this comes from studies in which the translatability of a mRNA from which the poly(A) had been removed by enzymatic means was tested in an in-vitro system. Compared with the control, the mRNA in this instance was much more labile.

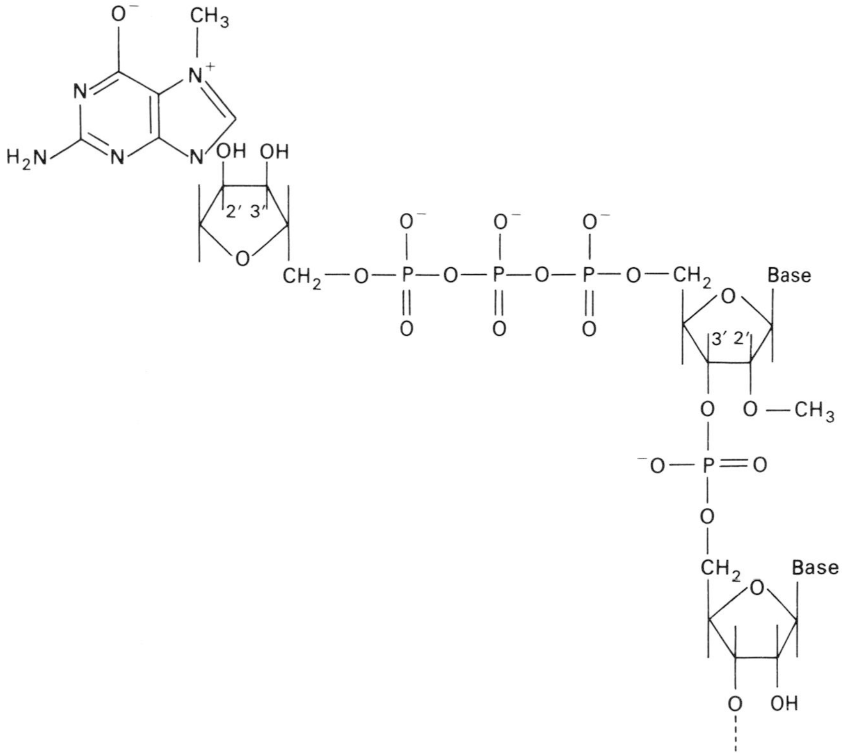

Fig. 7.7 The 'cap' structure at the 5' end of eukaryotic mRNAs.

$$\text{RNA} + n\text{ATP} \xrightarrow[\text{Mg}^{2+}]{\substack{\text{Polyadenylate} \\ \text{polymerase}}} \text{RNA} - (\text{A})n + n\text{PPi}$$

Fig. 7.8 The synthesis of poly(A) sequences at the 3′ end of many eukaryotic mRNAs catalysed by polyadenylate polymerase.

Non-coding sequences

Not all of the nucleotide sequence between the 5′-cap and the 3′-poly(A) segment in eukaryotic mRNAs is used to code for the amino acid sequence in proteins. At the 5′ end there is most probably a short, untranslated segment that is involved with ribosome recognition for the initiation of protein synthesis. This is also the case for prokaryotic mRNAs. Comparisons of the amino acid sequences of proteins and the nucleotide sequences of some eukaryotic mRNAs has shown that there is a segment of the latter at the 3′ end that is not translated. The number of nucleotides in the non-coding sequence (this excludes the poly(A)) may be as many as one-third to one-half of those found in the coding region itself. Thus mature mRNAs may be significantly longer than is required for their coding capacity. The function of the non-translated segment is unknown at present but it has been suggested that it may contain binding sites that recognise the proteins that must be involved in the synthesis, processing, transport, association with ribosomes, and degradation of mRNAs. Since these properties are common to all mRNAs, it would be reasonable to expect to find a sequence of nucleotides that is identical or very similar within the non-coding segments of various mRNA molecules. Indeed, for a number of mRNAs, including those for chicken ovalbumin, mouse immunoglobulin light chain, and rabbit alpha- and beta-globin, there is a AAUAAA sequence located in nearly identical positions, that is about 20 residues internal to the poly(A) segment.

Messenger RNA precursors

In the nucleus of eukaryotic cells, a rapidly-labelled RNA species can be isolated that is distinct from ribosomal and transfer RNA or their precursors. This RNA is called heterogeneous nuclear RNA (hnRNA) and in general it has a relatively short lifetime in the nucleus. The hnRNA is heterodisperse in length and is much longer than cytoplasmic mRNA. In addition the hnRNA has both the 5′-cap and 3′-poly(A) modifications that are characteristic of the cytoplasmic mRNAs. This has lent credence to the hypothesis that the hnRNA molecules are precursors to the functional mRNAs, the former being 'processed' to the latter in the nucleus. However, in most eukaryotes, a large fraction (approximately 90%) of the hnRNA that is made is rapidly degraded and thus only a small portion of the synthesised material is transported to the cytoplasm in the form of mRNA. The reason for this 'wastage' is not known.

There is now quite a body of evidence to support the precursor-product relationship of hnRNA and mRNA. For example, when the animal virus Adenovirus 2 (Ad2) infects a mammalian cell in tissue culture, late in infection the viral DNA codes for a series of 13 distinct mRNA molecules. One of these codes for a protein subunit of the viral capsule (the hexon polypeptide). At the 5′-end of this mRNA, there are three sequences that are transcribed from three non-contiguous regions of the viral DNA and these 'leader' segments are attached to the main part of the hexon mRNA coding for the polypeptide. The order and polarity of the three segments is the same as their arrangement on the viral chromosome. Examination of the nuclei of the infected cells showed that there are RNA molecules present that are much longer than the mRNAs engaged in protein synthesis. The hypothesis has been put forward, therefore, that late in infection the Ad2 DNA codes for a large primary transcript which undergoes maturation and processing to produce the mature hexon mRNA (Fig. 7.9). The same primary transcript is apparently processed in a number of ways to generate different mRNAs coding for different proteins. In all cases the final product is a 'mosaic' mRNA made up of sequences that were not contiguous on the viral chromosome.

The most favoured mechanism for the processing of the long precursor RNA involves looping-out and excision of the unwanted intervening sequences and then RNA–RNA 'splicing' or 'ligation' of the

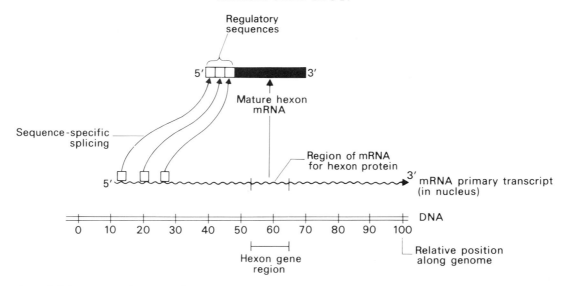

Fig. 7.9 A model for the production of mature Adenovirus-2 hexon mRNA by RNA·RNA splicing of parts of a long transcription units during late stages of cell injection. (After J. M. Berget *et al*, 1977. *Cold Spring Harbor Symp. Quant. Biol.* **42**:523.)

conserved segments into the final molecule. The function of the intervening sequences is not known but it is speculated that they play a role in the regulation of processing of the precursor-RNA. This phenomenon has also been reported for other viral-induced mRNA molecules, for haemoglobin mRNA, for immunoglobulin mRNA, for ovalbumin mRNA, and for other cellular mRNAs. It appears, therefore, that this is a general occurrence in eukaryotic organisms and the mechanism proposed certainly serves to explain the identity of the 5′ and 3′ modifications in hnRNA and mRNA. The molecular mechanisms for the RNA–RNA splicing reactions are currently under study. All of these results, of course, now make the definition of a 'gene' in eukaryotes very complex.

TRANSFER RNA

Transfer RNA molecules have a central role in protein synthesis and interact with a wide range of other molecules. This class of molecules is predominantly located in the soluble portion of the cytosol (hence its earlier, name, soluble RNA) and repre-

sents about 10–15% of the total cellular RNA in both prokaryotes and eukaryotes. Each molecule is able to combine specifically with one of the amino acids in a reaction catalysed by one of the set of enzymes called amino acyl synthetases. The resulting amino acyl tRNAs migrate to specific sites on the ribosome and interact with specific three-nucleotide sequences (codons) on the mRNA so that the correct amino acid can be inserted into the growing polypeptide chain. These events will be described in detail in the next Topic.

Transfer RNAs can be extracted from most cells with buffered aqueous phenol and are found to be small, with an S value of about 4, and a molecular weight of about 25 000–30 000 daltons. The nucleotide chain length is remarkably uniform for all the tRNAs of prokaryotes and eukaryotes, ranging from 76 to 85. Thus, since all tRNAs have quite similar properties and are all about the same length, it is reasonably difficult to purify single species. An early approach was to use counter current distribution and this permitted R. Holley and his colleagues to purify and then sequence a tRNA from yeast that is specific for the amino acid, alanine (tRNA.ala). The original sequence determination required large amounts of pure tRNA, whereas only a small amount of highly radioactive ^{32}P-labelled tRNA is needed for the rapid sequencing method developed by F. Sanger and his colleagues that is now widely used. This so-called 'fingerprinting' technique involves the controlled degradation of the RNA with enzymes and separation of the oligonucleotide products by two-dimensional electrophoresis.

All the tRNA molecules sequenced to date (more than 40) appear to conform to the clover-leaf model with a stem, three large arms consisting of a stem and a loop, and occasionally an extra arm. One general nomenclature for this is shown in Fig. 7.10. Some of the modified bases are shown in Fig. 7.11.

Certain features of the primary sequence of tRNAs appear to be constant. For example, the 3'-terminal sequence –CCA.OH and the sequence T–ψ–C (where ψ is pseudouridine) in loop IV appear universal. The anticodon is a sequence of three nucleotides that must bond with the codon of mRNA during polypeptide elongation, and hence this region varies according to the tRNA in question. However, the nucleotide to the 5' side of the

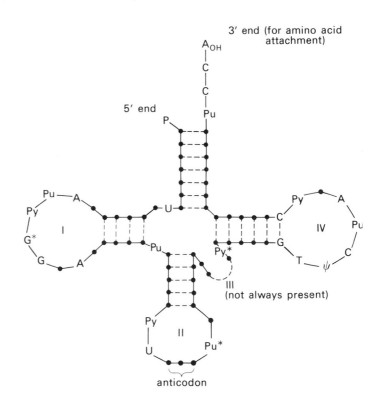

Fig. 7-10 A schematic of a 'generalised' tRNA molecule shown in the cloverleaf configuration. (After A. L. Lehninger, 1975, *Biochemistry*. Worth Publishers, New York.)

Pu = purine
Py = pyrimidine
P = phosphate
A,U,G,C = normal RNA bases
* = modified base
T = ribothymidine
ψ = pseudouridine
---- = hydrogen bonding

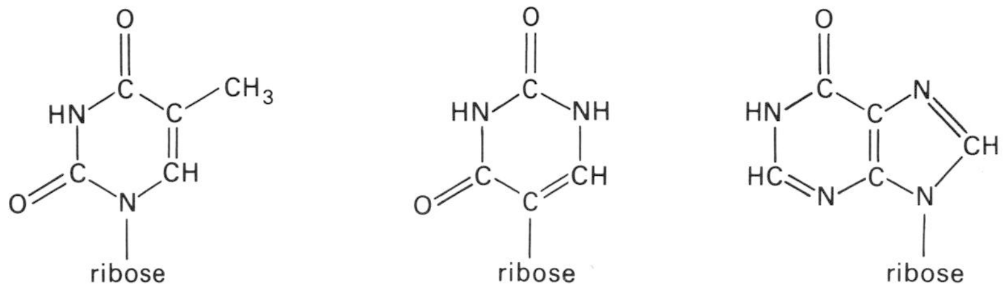

Ribothymidine (T) Pseudouridine (ψ) Inosine (I)

Fig. 7.11 Structures of some of the modified bases found in tRNA.

anticodon is always U and that to the 3′ side is always a modified purine. The nucleotides present in the stem regions are very variable but the number of base pairs in a particular stem is fairly constant.

In spite of having more than 40 tRNA nucleotide sequences to compare, no general conclusions can be drawn regarding the functions of the various parts of the molecules. One exception to this is the sequence T–ψ–C–G which probably represents the common ribosome binding site of tRNA. Thus, it is likely that the functions of the tRNA are dependent upon the tertiary structure of the molecules.

It is apparent then that, unlike mRNAs, the tRNA molecules are extensively modified. These modifications occur post-transcriptionally. Indeed, the mature tRNAs not only have modified bases, but also are considerably shorter (up to 40 nucleotides or so) than the primary gene transcripts. Thus, in both prokaryotes and eukaryotes there is evidence for precursor-tRNAs (pre-tRNAs) which must be clipped and trimmed to produce the mature molecules. In prokaryotes, the precursors either contain only one tRNA molecule with extra 'leader' and 'trailer' sequences at the 5′ and 3′ ends, or they contain several species of tRNA molecules within a long precursor. The latter precursors reflect the tandem tRNA arrangement from which they were transcribed, and they may be described as:

5′-leader – (tRNA-spacer)n – tRNA – trailer-3′

At least two enzymes are necessary for the processing of these pre-tRNAs. One of these, RNase P, catalyses the removal of the 5′ leader sequence and the other, RNase Q, catalyses the removal of the 3′ trailer sequence. Evidence for this has come from studies of mutant strains of *E. coli* with temperature-sensitive defects in the enzymes. At the non-permissive temperature, partially processed pre-tRNAs accumulate and can be isolated for sequence comparison with mature tRNAs.

In eukaryotes, there is also good evidence for the existence of pre-tRNAs although studies have generally been done with mixtures of molecules rather than with purified individual species as was done in prokaryotes. In any event, pre-tRNAs from several eukaryotes sediment at about 4.8S compared with 3.8S for mature tRNAs. These molecules are transcribed by RNA polymerase III and the data obtained from studies with denaturing agents indicate that there are an additional 15–35 nucleotides on the pre-tRNAs. Processing of the pre-tRNA molecules presumably occurs in the nucleus although, as yet, the enzymes involved and their sites of action have not been defined in any detail.

Tertiary structure of tRNAs

As was indicated earlier, all of the sequenced tRNA molecules can be arranged in a similar clover-leaf structure. However, this model is derived entirely from analysis of the primary nucleotide sequence and the maximization of hydrogen-bonding. The application of X-ray crystallography to stable crystals of tRNA (yeast phenylalanine tRNA was the first) has permitted the elucidation of its tertiary structure at the 0.3 nm resolution level, including the location of the major groups (Fig. 7.12). From the data, it was possible to conclude that all of the double-helical stems predicted by the clover-leaf model do exist. In addition, there are other hydrogen bonds which bend the clover-leaf into a

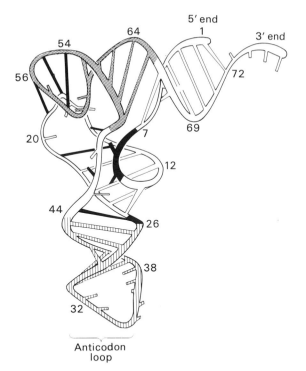

5′ end
1
3′ end
64
54
56
72
20
7
69
12
44
26
38
32
Anticodon
loop

Fig. 7.12 Schematic model of yeast phenylalanine-tRNA. The ribose phosphate backbone is drawn as a continuous cylinder with bars to indicate hydrogen-bonded base pairs. The positions of single bases are indicated by short rods. The TψC arm is heavily stippled, and the anticodon arm is marked by vertical lines. Tertiary structure interactions are illustrated by black rods. The numbers indicate the nucleotide position starting at the 5′ end. (With permission from S. H. Kim *et al, Science* **185**, 435–440. © 1974, American Association for the Advancement of Science.)

stable tertiary structure that has a rough L-shaped appearance. In this structure, the amino-acid acceptor CCA group at the 3′ end of the chain is located at the opposite end from the anticodon loop.

RIBOSOMES AND RIBOSOMAL RNA

Ribosomes are the sites of protein biosynthesis within the cell. Most of the reactions that take place on the ribosome are very similar in both pro-

karyotic and eukaryotic cells. Indeed, the major differences between bacterial and eukaryotic ribosomes is their size.

Ribosomes are usually prepared by breaking open the cells in buffered solutions containing magnesium ions. After centrifuging the cell lysate to sediment cell debris and subcellular organelles, the supernatant is centrifuged at 250 000g for 60 min. This yields a pellet of ribosomal material from which pure ribosomes can be isolated in a few more steps.

An important property of ribosomes is that under certain ionic conditions *in vitro* they dissociate into subunits. This is commonly brought about by substantially reducing the magnesium ion concentration. This potentiates studies of the complete ribosome and of the ribosomal subunits.

Ribosomes are usually characterised by their sedimentation coefficients obtained by analytical ultracentrifugation experiments. In eukaryotic cells, ribosomes sediment at about 80S whereas in bacteria the ribosomes sediment at about 70S. Both of these types of ribosomes have very similar structures. They are complexes of rRNA and proteins with molecular weights ranging from 2.7 million daltons (Mdal) for bacterial ribosomes to about 4 Mdal for mammalian ribosomes. Among eukaryotic ribosomes, the molecular weight of an undissociated (monomeric) ribosome becomes greater as the complexity of the organism increases.

Each ribosome consists of two unequally sized subunits. Both subunits must associate in order to function in protein synthesis. The relative S values for the ribosomes and their subunits, and the molecular weights of the rRNA molecules they contain are summarised in Fig. 7.13.

More than half of the mass of each ribosomal subunit consists of rRNA, the rest being protein. In both prokaryotes and eukaryotes, the small subunit contains only one species of rRNA complexed with a number of proteins. The large ribosomal subunit of prokaryotes contains two species of rRNA, 23S and 5S, complexed with proteins. In eukaryotes, the large subunit contains three rRNA species, 28S, 5.8S, and 5S, and a large number of proteins. The molecular weight of the 28S rRNA

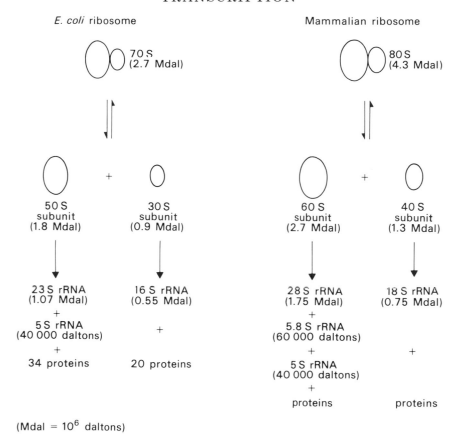

E. coli ribosome

Mammalian ribosome

70 S
(2.7 Mdal)

80 S
(4.3 Mdal)

50 S
subunit
(1.8 Mdal)

+

30 S
subunit
(0.9 Mdal)

60 S
subunit
(2.7 Mdal)

+

40 S
subunit
(1.3 Mdal)

23 S rRNA
(1.07 Mdal)
+
5 S rRNA
(40 000 daltons)
+
34 proteins

16 S rRNA
(0.55 Mdal)

+

20 proteins

28 S rRNA
(1.75 Mdal)
+
5.8 S rRNA
(60 000 daltons)
+
5 S rRNA
(40 000 daltons)
+
proteins

18 S rRNA
(0.75 Mdal)

+

proteins

(Mdal = 10^6 daltons)

Fig. 7.13 A comparison of the subunit molecular weights and composition of ribosomes from *E. coli* and a mammal (Mdal = 10^6 daltons).

varies from 1.3 Mdal in eukaryotes such as higher plants, algae, protozoa, and fungi, to 1.75 Mdal in higher eukaryotes. When the rRNAs are extracted from the large subunit, the 5.8S is found to be hydrogen-bonded to the 28S rRNA. It can be released by gentle heat treatment.

The ribosomal proteins of *E. coli* have been studied extensively. Since a number of protein factors necessary for protein biosynthesis associate transiently with ribosomes, there is a question of how one defines a ribosomal protein. One operational definition is that a ribosomal protein is a protein that remains attached to the ribosome after washing with high-salt solutions (which removes

the transiently associated proteins) provided it is present in approximately molar yield. Most of the ribosomal proteins isolated from the ribosomes are basic proteins. isolated from the ribosomes are basic proteins. In the small subunit of *E. coli* there are 20 proteins and in the large subunit there are 34 proteins. All 54 proteins are immunologically distinct with the exception of two which differ by only an acetyl group. The proteins exhibit enough differences in mass and charge that they can be displayed by two-dimensional acrylamide gel electrophoresis. The exact role of each ribosomal protein in ribosome structure and function is not known.

In eukaryotic ribosomes, there are a correspondingly larger number of ribosomal proteins and it is fair to say that the same generalisations that we have made for prokaryotic ribosomal proteins apply also to these.

BIOSYNTHESIS OF PROKARYOTIC RIBOSOMES

A lot of work has been done to investigate how the protein and RNA components of the ribosome are assembled. One approach, taken by M. Nomura and his colleagues, was to take purified ribosomal subunits of *E. coli* and to dissociate them chemically into their component rRNAs and proteins, and then to permit them to reassociate under appropriate ionic conditions. The 30S subunit was the first to be studied in this way and it was found that the 20 proteins and the 16S rRNA could interact to form a complete, functional subunit at 37°C, the normal physiological temperature for *E. coli*. Since no other factors were present in the solution, the process was called self-assembly. This experiment, reported by Nomura's group in 1969, is summarised in Fig. 7.14.

Similar results have recently been obtained with the 50S subunit of *E. coli*. Thus both subunits have the ability to self-assemble. Further reconstitution experiments by Nomura's group, in which one protein at a time is omitted from the reaction mixture, has provided information about the sequence of steps involved in the assembly reaction, and about the necessity for all proteins to be present for the subunit to be functional.

The elegant in-vitro reconstitution experiments do not completely reflect the assembly of *E. coli* ribosomal subunits *in vivo*, however. In *E. coli* the rRNA genes are arranged in tandem repeats containing the 16S, 23S, and 5S rRNA sequences. The primary transcripts of each repeat unit is a precursor 30S (p30S) RNA that is cleaved to produce the p16S, p23S, and p5S RNAs which are the immediate precursors to the mature 16S, 23S, and 5S rRNAs, respectively. Normally, the p30S molecule is not observed in wild-type cells since the

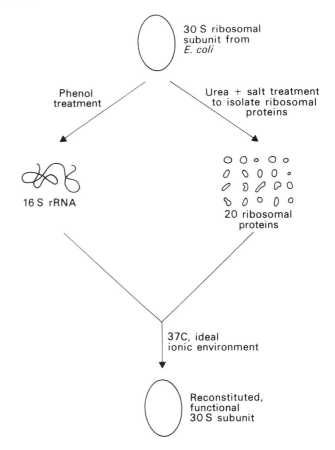

Fig. 7.14 Schematic of experiment that demonstrated the reconstitution of functional 30S ribosomal subunits of *E. coli* from the component RNA and protein molecules. (After M. Nomura, 1969. *Sci. Am.* **221**:28.)

cleavages occur while p30S is still being transcribed. However, this cleavage is partially blocked in mutant strains that are deficient in RNase III, and the p30S molecule accumulates. Analysis of this RNA revealed the presence of the 16S, 23S, and 5S sequences within it. The polarity of the mature rRNAs within the p30S RNA is:

$$5' - 16S - 23S - 5S - 3'$$

and a model for the processing steps based on in-vitro cleavage studies with RNase III is as shown in Fig. 7.15.

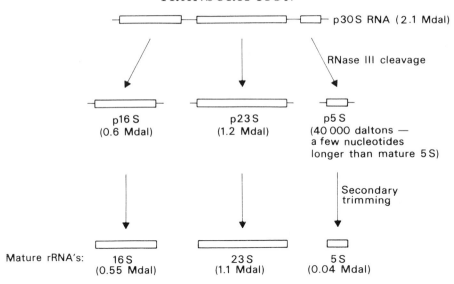

Fig. 7.15 Proposed scheme for the cleavage of a precursor RNA (p30S) to the mature rRNAs in *E. coli*.

It is proposed that an enzyme other than RNase III is required for the endonucleolytic cleavages that convert intermediate precursors to the mature RNAs. These final cleavages occur when the precursor rRNAs are associated with ribosomal proteins in preribosomal particles. In view of this, one must be cautious in extrapolating Nomura's in-vitro assembly information to the in-vivo state, since the conformations of the precursor rRNAs may facilitate different protein–protein or protein–RNA interactions than would the mature rRNAs. In other words, the end product is the same in the two, but the sequence of steps may be different.

For proper ribosome assembly to occur the 16S and 23S RNAs must be methylated. This methylation occurs post-transcriptionally and most of the methyl groups are added to the bases with only a few added to the 2′-OH of the ribose moiety. All of the methylations of the 23S rRNA occur on the p30S RNA component whereas most, if not all, of the 16S rRNA methyl groups are added when it is in a mature form. Thus methylation of 23S rRNA is an early event in RNA processing and methylation of 16S rRNA is a late event.

BIOSYNTHESIS OF EUKARYOTIC RIBOSOMES

In outline, eukaryotic ribosomes are made as follows: The rRNA genes are transcribed into a high-molecular-weight ribosomal precursor RNA (pre-rRNA), which is then modified and cleaved at specific sites to yield a number of discrete intermediate RNAs and finally the mature 18S, 5.8S and 28S rRNAs. Assembly with ribosomal proteins and with 5S rRNA (which is transcribed from separate rRNA genes) occurs in the nucleolus during pre-rRNA processing. The resultant ribosomal subunits are then released from the nucleus into the cytoplasm.

Ribosomal RNA maturation has been studied in a number of eukaryotes including mammalian cells, amphibians, insects, higher plants and fungi. The results indicate that all eukaryotes have similar pathways for the biosynthesis of ribosomes. In the following, ribosome production in higher eukaryotes and in fungi (yeast) will be described for the purpose of comparison and to point out the generalities involved.

Ribosomal DNA

In all eukaryotes studied to date, the genes for 18, 5.8, and 28S rRNA are multiple and clustered at the sites of the nucleolar organizer regions. Saturation hybridisation experiments between purified rRNA and nuclear DNA have shown the extent of this gene multiplicity, and this varies from about 100 to 1000 depending on the organism being studied. In general the more complex the eukaryote is in an evolutionary sense, the more copies there are. In most organisms, the 5S rRNA is coded by extranucleolar genes whose multiplicity is usually higher than that for the other rRNA genes. The 5S rRNA genes may be clustered in the genome as in humans or scattered throughout the genome as in *Xenopus laevis,* the South African clawed toad. At least in yeast and slime molds, the 5S rRNA genes are interspersed with the other rRNA genes.

The DNA comprising the 18S + 5.8S + 28S gene clusters consists of three basic elements:

a. sequences corresponding to the mature rRNAs,
b. sequences transcribed as part of an initial precursor molecule, but not found in the mature rRNAs ('transcribed spacer (TS) sequences'), and
c. additional 'nontranscribed spacer (NTS) sequences' interspersed among the transcribed sequences.

Molecular studies of isolated rDNA have produced a generalised picture for the repeating unit of rDNA in eukaryotes (Fig. 7.16).

Here A-G is the repeating unit. Within that unit B-C codes for 18S, D-E for 5.8S, and E-F for 28S rRNA. Regions A-B and C-D are TS sequences which are homogeneous from repeat to repeat, and F-G is a NTS sequence which is heterogeneous in length between repeats in higher eukaryotes such as *Xenopus,* is slightly heterogeneous in length in

Drosophila, and is homogeneous in length in lower eukaryotes such as yeast. Differences in the lengths of the TS sequences, and slight differences in the length of the 28S rRNA sequence, account for the different sized pre-rRNAs seen in various organisms.

Pre-rRNA transcription and processing

Ribosome formation commences with the transcription of pre-rRNA. In HeLa (mammalian) cells it is a 45S molecule with a molecular weight of about 4.5 Mdal. In yeast (a lower eukaryote) it has a sedimentation coefficient of 35S with a molecular weight of about 2.5 Mdal. Within the pre-rRNA molecule, the arrangement of the rRNA sequences has been determined to be 5' – 18S – 5.8S – 28S – 3', which is somewhat analogous to the arrangement of the rRNA sequences in the prokaryotic pre-rRNA molecule.

To produce mature ribosomes, the pre-rRNA molecule undergoes modifications (methylation and pseudouridylation), associates with ribosomal proteins and 5S rRNA, and is processed enzymatically to remove TS sequences. These events occur in the nucleolus and many of the non-ribosomal proteins involved are presumably stable, specific nucleolar proteins.

In HeLa cells over 100 methyl groups are added to 45S RNA. This methylation occurs close to the point of polymerisation of ribonucleotides by the RNA polymerase, that is, during transcription. The significance of methylation is poorly understood. but it is clear that it is essential for ribosome maturation. In HeLa cells all of the methylation sites are within the parts of the 45S RNA corresponding to the mature rRNA sequences, and all except for about six are on the 2'-OH of ribose moieties (Fig. 7.17).

Fig. 7.16 Generalised diagram for the repeating unit in the ribosomal DNA if eukaryotes. A-G is the repeating unit. B-C, D-E and E-F code for 18S, 5.8S and 28S rRNAs, respectively. A-B and C-D are transcribed spacer sequences and F-G is a non-transcribed spacer sequence.

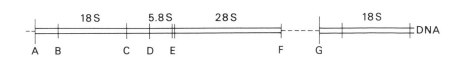

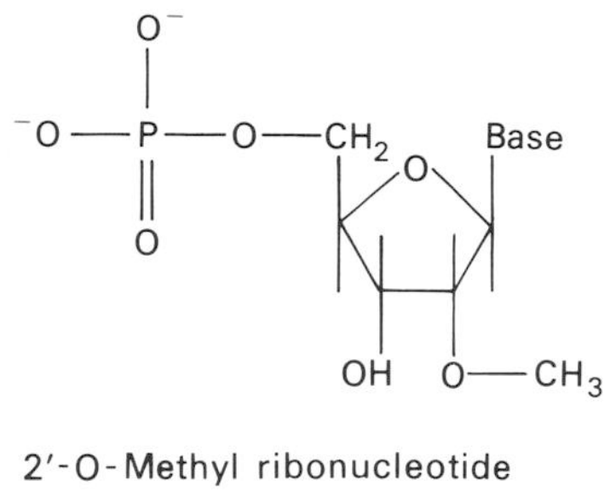

O⁻
|
⁻O—P—O—CH₂   Base
‖
O

OH   O—CH₃

2'-O-Methyl ribonucleotide

Fig. 7.17 Site of methylation on the ribose moiety in rRNA.

Mature 18S, 28S, and 5.8S rRNAs of eukaryotic cells also contain many pseudouridine residues (c.f. tRNA modification). This modification also occurs in the nucleolus on pre-rRNA but the significance of these residues is poorly understood.

The nucleolar 45S pre-rRNA molecule also associates with ribosomal proteins which have been synthesised in the cytoplasm and then transported to the nucleolus, and with 5S rRNA which, in most organisms, is transcribed elsewhere in the nucleus. Thus 'nascent ribosomal particles' can be extracted from nucleoli of actively growing cells.

Ribosome maturation, then, involves specific cleavages of the pre-rRNA resulting in the elimination of TS regions leaving only mature rRNA. These cleavages take place in the nascent ribosomal particles and are accompanied by the formation of specific protein–protein and protein–nucleic acid interactions as the mature ribosomal subunits are formed. Based on continuous-labelling and pulse-chase experiments with a radioactive RNA precursor, and on other supportive experiments, the maturation pathway for rRNA production in HeLa cells is proposed to be as shown in Fig. 7.18.

In this case the TS sequences represent about 50% of the primary transcript and these are non-conserved, being eliminated by the specific cleavages indicated.

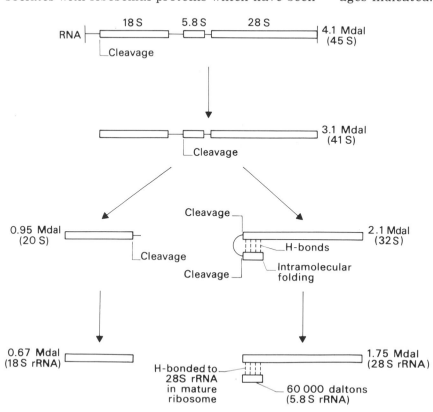

Fig. 7.18 Proposed scheme for the processing of 45S precursor-RNA in HeLa cells to produce the mature rRNA species. (After B. C. H. Maden, 1971. *Prog. Biophys. Mol. Biol.* **22**: 127.)

The maturation scheme is essentially similar in all other eukaryotes examined to date. The details of rRNA processing vary owing to differences in the sizes of the initial transcript and of the mature rRNA components, differences in the extent of the RNA modifications, and differences in the number of stable (and hence detectable) intermediates. In general the proportion of the primary transcript that is spacer decreases as one descends the evolutionary scale. In yeast, for example, the rRNA maturation scheme is as depicted in Fig. 7.19. Here about 20% of the primary transcript is nonconserved.

It should be pointed out that not all organisms exhibit an intermediate between the pre-rRNA and the 18S rRNA. Examples of those which do not are plants, *Xenopus* and *Neurospora*.

The entire process of ribosome production, in this case in HeLa cells, is summarised in Fig. 7.20.

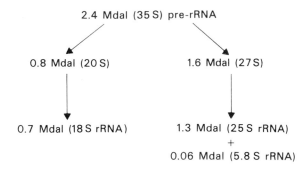

Fig. 7.19 Proposed scheme for the production of mature rRNAs from 35S pre-rRNA in the yeast, *Saccharomyces cerevisiae*. (After S. A. Udem and J. R. Warner, 1972. *J. Mol. Biol.* **65**:227.)

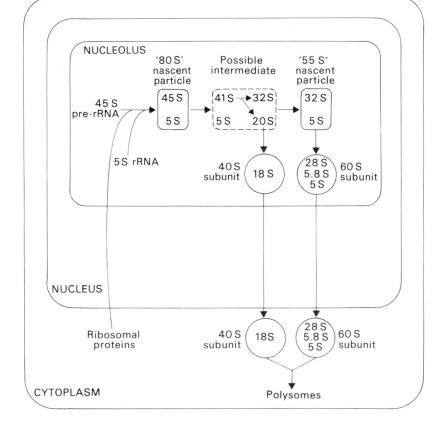

Fig. 7.20 Schematic representation of ribosome formation in HeLa cells showing the formation of precursor ribonucleoprotein particles in the nucleolus and their processing to mature ribosomal subunits. In the diagram the rectangular particles are precursors and the circular particles are mature subunits. The numbers outside the particles indicate their 'S' values, and the numbers inside indicate the 'S' values of the rRNA they contain. (After B. E. H. Maden, 1971. *Prog. Biophys. Mol. Biol.* **22**:127.)

The numbers refer to the sedimentation coefficient of the RNAs.

Very little is known in eukaryotes about the enzymes that must be present for RNA processing to occur. Recently, some potential processing enzymes have been identified, including a specific endoribonuclease and a 3′-OH-specific exonuclease.

Clearly, the synthesis of ribosomes is a highly complex process. Eukaryotic cells must possess elaborate mechanisms for regulating ribosome production according to needs. This control could be at the level of transcription of pre-rRNA and/or at post-transcriptional levels. Comparatively little specific information is available on this point.

REFERENCES

General

Adhya S. & M. Gottesman. 1978. Control of transcription termination. *Annu. Rev. Biochem.* **47**: 217–249.

Biswas B.B., R.K. Mandal, A. Stevens & W.E. Cohn (eds.). 1974. *Control of Transcription.* Plenum, New York.

Brenner S., F. Jacob & M. Meselson. 1961. An unstable intermediate carrying information from genes to ribosomes for protein synthesis. *Nature,* **190**: 576–581.

Chamberlin M.J. 1974. The selectivity of transcription. *Annu. Rev. Biochem.* **43**: 721–775.

Chambon P. 1977. Summary: The molecular biology of the eukaryotic genome is coming of age. *Cold Spring Harbor Symp. Quant. Biol.* **42**: 1209–1234.

Cold Spring Harbor Symposium for Quantitative Biology, 1970. *Transcription of Genetic Material,* vol. 35. Cold Spring Harbor Laboratory, New York.

Darnell J.E. 1968. Ribonucleic acids from animal cells. *Bacteriol. Rev.* **32**: 262–290.

Darnell J.E. 1977. Gene regulation in mammalian cells: Some problems and the prospects for their solution. In *Cell Differentiation and Neoplasia,* 30th Annual Symposium on Fundamental Cancer Research, (ed. G. Saunders). M.D. Anderson Hospital and Tumor Institute, Houston, Texas.

Davidson E.H. & R. Britten. 1973. Organization, transcription, and regulation in the animal genome. *Q. Rev. Biol.* **48**: 565–613.

Gilbert W. 1976. Starting and stopping sequences for the RNA polymerase. In *RNA Polymerase,* R. Losick and M. Chamberlin (eds.), pp. 193–206. Cold Spring Harbor Laboratory, New York.

Losick R. 1972. In vitro transcription. *Annu. Rev. Biochem.* **41**: 409–446.

Marmur J., C.M. Greenspan, E. Palecek, F.M. Kahan, J. Levine & M. Mandel. 1963. Specificity of the complementary RNA formed by *Bacillus subtilis* infected with bacteriophage SP8. *Cold Spring Harbor Symp. Quant. Biol.* **28**: 191–199.

Perry R.P. 1976. Processing of RNA. *Annu. Rev. Biochem.* **45**: 605–629.

Sanger F., G.G. Brownlee & B.G. Barrell. 1965. A two-dimensional fractionation procedure for radioactive nucleotides. *J. Mol. Biol.* **13**: 373–398.

Schweizer E., C. MacKechnie & H.O. Halvorson. 1969. The redundancy of ribosomal and transfer RNA genes in *Saccharomyces cerevisiae. J. Mol. Biol.* **40**: 261–277.

Sirlin J.L. 1972. *The Biology of RNA.* Academic Press, New York.

Stewart P.R. & D.S. Letham (eds.). 1977. *The Ribonucleic Acids,* (2nd ed.). Springer-Verlag, New York.

Travers A. 1974. Bacterial transcription. In *Biochemistry of Nucleic Acids,* K. Burton (ed.), vol. 6 in MTP International Review of Science, pp. 191–218. Butterworths, London.

Weinberg R.A. 1973. Nuclear RNA metabolism. *Annu. Rev. Biochem.* **42**: 329–354.

Weissbach H. & S. Pestka (eds.). 1977. *Molecular Mechanisms of Protein Biosynthesis,* Academic Press, New York.

RNA Polymerase

Burgess R.R. 1969. Separation and characterization of the subunits of ribonucleic acid polymerase. *J. Biol. Chem.* **244**: 6168–6176.

Burgess R.R. 1971. RNA polymerase. *Annu. Rev. Biochem.* **40**: 711–740.

Chambon P. 1975. Eukaryotic nuclear RNA polymerases. *Annu. Rev. Biochem.* **44**: 613–638.

Losick R. & M. Chamberlin (eds.). 1976. *RNA Polymerase.* Cold Spring Harbor Laboratory, New York.

Pribnow D. 1975. Nucleotide sequence of an RNA polymerase binding site at an early T7 promotor. *Proc. Natl. Acad. Sci. USA,* **72**: 784–788.

Travers A.A., R. Buckland, M. Goman, S.S.G. Le Grice & J.G. Scaife. 1978. A mutation affecting the sigma subunit of RNA polymerase changes transcriptional specificity. *Nature,* **273**: 354–358.

Travers A.A. & R.R. Burgess. 1969. Cyclic reuse of the RNA polymerase sigma factor. *Nature,* **222**: 537–540.

Messenger RNA

Berget S.M., A.J. Berk, T. Harrison & P.A. Sharp. 1977. Spliced segments at the 5′ termini of adenovirus-2 late mRNA: A role for heterogeneous nuclear RNA in mammalian cells. *Cold Spring Harbor Symp. Quant. Biol.* **42**: 523–529.

Bonner J., R.B. Wallace, T.D. Sargent, R.F. Murphy & S.K. Dube. 1977. The expressed portion of eukaryotic chromatin. *Cold Spring Harbor Symp. Quant. Biol.* **42**: 851–857.

Both G.W., A.K. Banerjee & A.J. Shatkin. 1975. Methylation-dependent translation of viral messenger RNAs *in vitro. Proc. Natl. Acad. Sci. USA,* **72**: 1189–1193.

Brawerman G. 1974. Eukaryotic messenger RNA. *Annu. Rev. Biochem.* **43**: 621–642.

Brawerman G. 1976. Characteristics and significance of the polyadenylate sequence in mammalian messenger RNA. *Progr. Nucl. Acid. Res. Mol. Biol.* **17**: 117–148.

Breathnach R., J.L. Mandel & P. Chambon. 1977. Ovalbumin gene is split in chicken DNA. *Nature,* **270**: 314–318.

Cheng C.C., G.G. Brownlee, N.H. Carey, M.T. Doel, S. Gillam, & M. Smith. 1976. The 3′-terminal sequence of chicken ovalbumin messenger RNA and its comparison with other messenger RNA molecules. *J. Mol. Biol.* **107**: 527–547.

Darnell J.E. 1978. Implications of RNA.RNA splicing in evolution of eukaryotic cells. *Science,* **202**: 1257–1260.

Darnell J.E., R. Evans, N. Fraser, S. Goldberg, J. Nevins, M. Salditt-Georgieff, H. Schwartz, J. Weber & E. Ziff. 1977. The definition of transcription units for mRNA. *Cold Spring Harbor Symp. Quant. Biol.* **42**: 515–522.

Darnell J.E., R. Wall & R.J. Tushinski. 1971. An adenylic acid-rich sequence in messenger RNA of HeLa cells and its possible relationship to reiterated sites in DNA. *Proc. Natl. Acad. Sci. USA,* **68**: 1321–1325.

Edmonds M., M.H. Vaughan & H. Nakazoto. 1971. Polyadenylic acid sequences in the heterogeneous nuclear RNA and rapidly labelled polyribosomal RNA of HeLa cells: Possible evidence for a precursor relationship. *Proc. Natl. Acad. Sci. USA,* **68**: 1336–1340.

Edmonds M. & M.A. Winters. 1976. Polyadenylate polymerases. *Prog. Nucleic Acid Res. Mol. Biol.* **17**: 149–179.

Furuichi Y., M. Morgan, S. Muthukrishnan & A.J. Shatkin. 1975. Reovirus messenger RNA contains a methylated, blocked 5′-terminal structure: m7G(5′)ppp(5′)GmpCp- *Proc. Natl. Acad. Sci. USA,* **72**: 362–366.

Furuichi Y., M. Morgan, A.J. Shatkin, W. Jelenik, M. Salditt-Georgieff & J.E. Darnell. 1975. Methylated, blocked 5′ termini in HeLa cell mRNA. *Proc. Natl. Acad. Sci. USA,* **72**: 1904–1908.

Geiduschek E.P. & R. Haselkorn. 1969. Messenger RNA. *Annu. Rev. Biochem.* **38**: 647–676.

Goodman H.M., M.V. Olson & B.D. Hall. 1977. Nucleotide sequence of mutant eukaryotic gene: The yeast tyrosine-inserting ochre suppressor SUP4-0. *Proc. Natl. Acad. Sci. USA,* **74**: 5453–5457.

Jeffreys A.J. & R.A. Flavell. 1977. The rabbit beta-globin gene contains a large insert in the coding sequence. *Cell,* **12**: 1097–1108.

McKnight S.L., M. Bustin & O.L. Miller. 1977. Electron microscopic analysis of chromosome metabolism in the *Drosophilia melanogaster* embryo. *Cold Spring Harbor Symp. Quant. Biol.* **42**: 741–754.

Nevins J.R. & J.E. Darnell. 1978. Groups of Adenovirus Type 2 mRNA's derived from a large primary transcript: Probable nuclear origin and possible common 3′ ends. *J. Virol.* **25**: 811–823.

Perry R.P., D.E. Kelley, K. Frederici & F. Rottman. 1975a. The methylated constituents of L cell messenger RNA: Evidence for an unusual cluster at the 5′ terminus. *Cell,* **4**: 387–394.

Perry R.P., D.E. Kelley, K. Frederici & F. Rottman. 1975b. Methylated constituents of heterogeneous nuclear RNA: Presence of blocked 5′ terminal structures. *Cell,* **6**: 13–19.

Proudfoot N.J. 1976. Sequence analysis of the 3′ noncoding regions of rabbit alpha- and beta-globin messenger RNAs. *J. Mol. Biol.* **107**: 491–525.

Reeves R. 1977. Structure of *Xenopus* ribosomal gene chromatin during changes in genomic transcription rates. *Cold Spring Harbor Symp. Quant. Biol.* **42**: 709–722.

Sripati C.E., Y. Groner & J.R. Warner. 1976. Methylated, blocked 5′ termini of yeast mRNA. *J. Biol. Chem.* **251**: 2898–2904.

Tilghman S.M., D.C. Tiemeir, J.G. Seidman, B.M. Peterlin, M. Sullivan, J.V. Maizel & P. Leder. 1978. Intervening sequence of DNA identified in the structural portion of a mouse beta-globin gene. *Proc. Natl. Acad. Sci. USA,* **78**: 725–729.

Tonegawa S., A.M. Maxam, R. Tizard, O. Bernhard & W. Gilbert, 1978. Sequence of a mouse germ-line gene for a variable region of an immunoglobulin. *Proc. Natl. Acad. Sci. USA,* **75**: 1485–1489.

Wei C.M., A. Gershowitz & B. Moss. 1976. 5′-terminal and internal methylated nucleotide sequences in HeLa cell mRNA. *Biochemistry,* **15**: 397–401.

Weintraub H. & M. Groudine. 1976. Chromosomal subunits in active genes have an altered configuration. *Science,* **193**: 848–856.

Winicov I. & R.P. Perry. 1976. Synthesis, methylation, and capping of nuclear RNA by a subcellular system. *Biochemistry,* **15**: 5039–5046.

Transfer RNA

Holley R.W., J. Apgar, G.A. Everett, J.T. Madison, M. Marquisee, S.H. Merrill, J.R. Penswick & A. Zamir. 1965. Structure of a ribonucleic acid. *Science,* **147:** 1462–1465.

Nishimura S. 1974. Transfer-RNA: Structure and biosynthesis. In *Biochemistry of Nucleic Acids,* K. Burton (ed.), vol. 6 in MTP International Review of Science, pp. 289–322. Butterworths, London.

Smith J.D. 1972. Genetics of transfer RNA. *Annu. Rev. Genet.* **6:** 235–256.

Smith, J.D. 1976. Transcription and processing of transfer RNA precursors. *Progr. Nucl. Acid Res. Mol. Biol.* **16:** 25–73.

Sussman J.L. & S.H. Kim. 1976. Three-dimensional structure of a transfer RNA in two crystal forms. *Science,* **192:** 853–858.

Ribosomes

Attardi G. & F. Amaldi, 1970. Structure and synthesis of ribosomal RNA. *Annu. Rev. Biochem.* **39:** 183–226.

Brimacombe R., G. Stoffler & H.G. Wittmann, 1978. Ribosome structure. *Annu. Rev. Biochem.* **47:** 217–249.

Craig N.C. 1974. Ribosomal RNA synthesis in eukaryotes and its regulation. In *Biochemistry of Nucleic Acids,* K. Burton (ed.), vol. 6 in MTP International Review of Science, pp. 255–288. Butterworths, London.

Davies J. & M. Nomura, 1972. The genetics of bacterial ribosomes. *Annu. Rev. Genet.* **6:** 203–234.

Kurland C.G. 1972. Structure and function of bacterial ribosomes. *Annu. Rev. Biochem.* **41:** 377–408.

Kurland C.G. 1977. Structure and function of bacterial ribosomes. *Annu. Rev. Biochem.* **46:** 173–200.

Kurland C.G. 1977. Aspects of ribosome structure and function. In *Molecular Mechanisms of Protein Biosynthesis.* H. Weissbach and S. Pestka (eds.), pp. 81–116. Academic Press, New York.

Maden B.E.H. 1976. Ribosomal precursor RNA and ribosome formation in eukaryotes. *Trends in Biochem. Science* **1:** 196–199.

Maden B.E.H., M. Salim & D.F. Summers. 1972. Maturation pathway for ribosomal RNA in HeLa cell nucleolus. *Nature New Biol.* **237:** 5–9.

Nomura M. 1969. Ribosomes. *Sc. Amer.* **221:** 28–35.

Nomura M. 1970. Bacterial ribosome. *Bacteriol. Rev.* **34:** 228–277.

Nomura M. 1973. Assembly of bacterial ribosomes. *Science,* **179:** 864–873.

Nomura M., A. Tissieres & P. Lengyel (eds.), 1974. *Ribosomes.* Cold Spring Harbor Laboratory, New York.

Russell P.J., J.R. Hammett & E.U. Selker, 1976. *Neurospora crassa* cytoplasmic ribosomes: ribosomal ribonucleic acid synthesis in the wild type. *J. Bacteriol.* **127:** 785–793.

Udem S.A. & J.R. Warner. 1972. Ribosomal RNA synthesis in *Saccharomyces cerevisiae. J. Mol. Biol.* **65:** 227–242.

Wellauer P.K., I.G. Dawid, D.D. Brown & R.H. Reeder, 1976. The molecular basis for length heterogeneity in ribosomal DNA from *Xenopus laevis. J. Mol. Biol.* **105:** 461–486.

Topic 8
Protein Biosynthesis (Translation)

OUTLINE
Protein components
Peptide bond
Protein structure
Protein synthesis
Polypeptide chains are made in N-terminal to C-terminal direction
The steps of protein synthesis in prokaryotes
  initiation
  elongation
  termination
  polysomes
  relationship of transcription and translation
Protein synthesis in eukaryotes
  initiation
  elongation
  termination
  protein synthesis and cellular compartmentation.

In the previous Topic, we discussed the transcription of the genes in DNA into RNA molecules. All three of the RNA classes are involved in the protein synthesis process. The mRNAs are transcripts of the so-called structural genes which code for the production of proteins which are composed of amino acids. The mRNA attaches to the ribosome upon which it is *translated*; that is the genetic content of the mRNA contained in the sequence of ribonucleotides is converted into a linear sequence of amino acids. The resulting protein has enzymatic, structural or regulatory function within the cell. The conversion process from the four nucleotide language of nucleic acids to the 20 amino acid language of proteins is mediated by the genetic code. Thus the sequence of three nucleotides (a codon or triplet) in the mRNA codes for the insertion of one amino acid into a growing polypeptide chain. This ribosome-localised event involves the previously discussed tRNA molecules to which the amino acids are covalently attached.

PROTEIN COMPONENTS

The basic building blocks of proteins (polypeptides) are amino acids. There are 20 naturally occurring amino acids and their structures are shown in Fig. 8.1.

With the exception of proline, all of the amino acids have a common structure consisting of a central carbon atom (the alpha-carbon) to which is bonded an alpha-amino group ($-NH_3^+$), an alpha-carboxyl ($-COO^-$), and a hydrogen atom (proton). The other part of the amino acid is called the R group which varies from one amino acid to another. It is the R group which gives the amino acid its chemical properties and the sequence of amino acids in a protein gives the protein its overall properties. Thus, the general structure of an amino acid is as depicted in Fig. 8.2.

PEPTIDE BOND

Proteins consist of long chains connected together to form polypeptide chains. The amino acids are attached by peptide bonds which are formed by the interaction of an alpha-carboxyl group of one amino acid with an alpha-amino group of another with the elimination of one molecule of water (Fig. 8.3).

Each polypeptide, then, has an amino end, with a free amino group, and a carboxyl end, with a free carboxyl group. These are often called the N- and C-terminals of the molecule, respectively. All of the other amino groups and carboxyl groups of the amino acids have been subsumed into the peptide bonds (Fig. 8.4).

As was the case with the nucleic acids, the backbone of the polypeptide chain carries no information; rather it is the order of the R groups and their chemical properties which give the protein its characteristics.

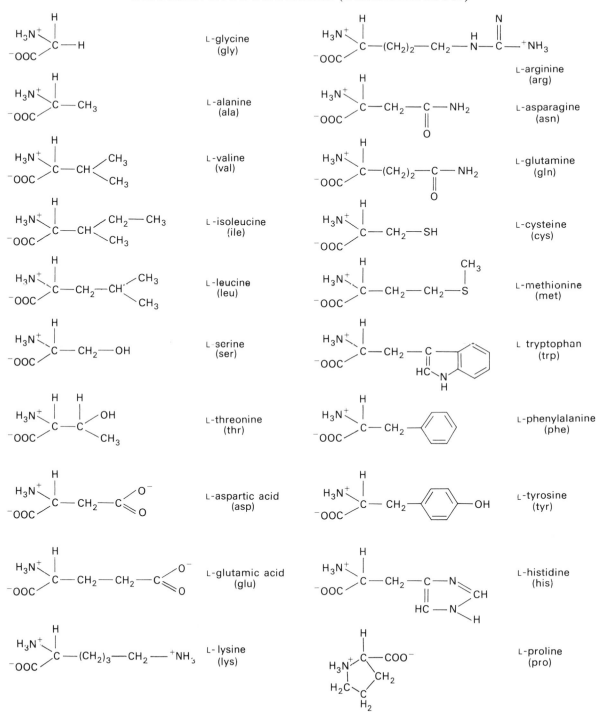

Fig. 8.1 Structures of the 20 naturally-occurring amino acids.

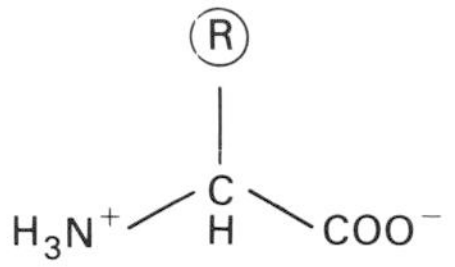

Fig. 8.2 General formula for an amino acid.

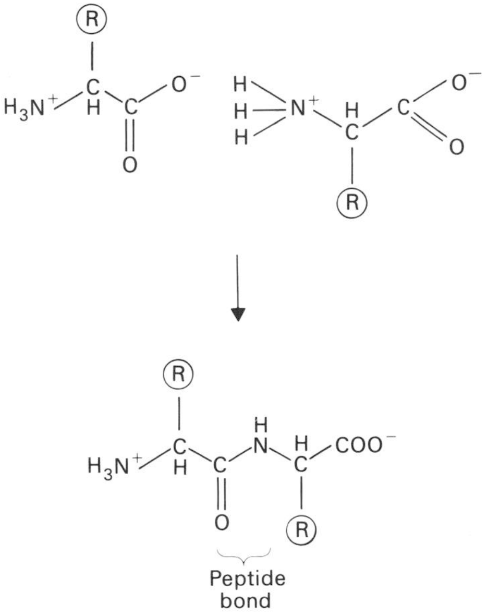

Peptide
bond

Fig. 8.3 The formation of the peptide bond.

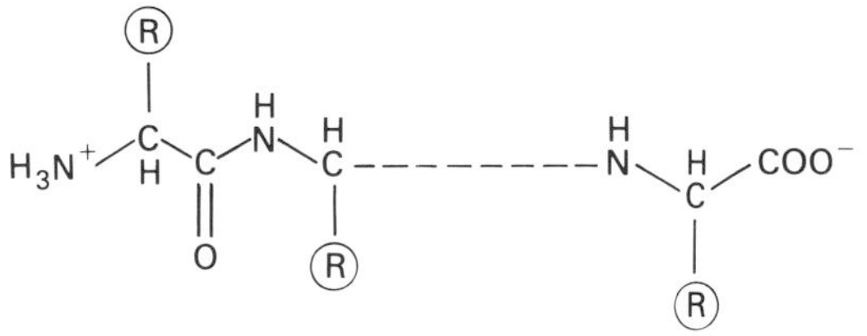

N-terminus   C-terminus

Fig. 8.4 General structure of a polypeptide chain.

PROTEIN STRUCTURE

There are four levels of protein structure:

Primary structure. This is the amino acid sequence of the polypeptide chain. This sequence determines the secondary and tertiary structure of the polypeptide chain.

Secondary structure. This is the folding of the chain into simple patterns as a result of the formation of electrostatic bonds (e.g. between carboxyl and amino groups) and hydrogen bonds between amino acids relatively close in the chain. The so-called alpha-helix found in many parts of the polypeptide chains is an example of secondary structure (Fig. 8.5).

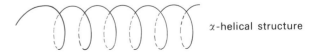

α-helical structure

Fig. 8.5 Diagrammatic representation of an α-helix segment in a polypeptide chain.

Tertiary structure. This is the way the helices and other parts of the polypeptide are folded to make a compact globular molecule. The tertiary structure of a polypeptide (protein) often places hydrophobic (literally 'water-hating') groups on the inside and hydrophilic ('water-loving') groups on the outside.

Quaternary structure. This is the way polypeptide chains are packed into the whole protein molecule if a number of chains is involved. For example, the haemoglobin molecule consists of four polypeptide chains, two alpha and two beta, associated in the quaternary structure required for the molecule to function. (From this we can come to the realisation that a protein is considered in a functional sense and may, in fact, contain more than one polypeptide chain.)

PROTEIN SYNTHESIS

A polypeptide chain is synthesised from the N-terminal to the C-terminal end. The mRNA coding for the polypeptide is moved through the ribosome (the site of protein synthesis) starting with the 5′ end. The nucleotide sequence is read in groups of three (called codons or triplets) such that one amino acid is inserted into the polypeptide for each codon on the mRNA. A given codon is complementary (in a base-pairing sense) to an anticodon of a tRNA molecule, and each tRNA molecule carries a specific amino acid such that the correct amino acid is put into the polypeptide chain when a particular codon is exposed at the ribosome. As we shall see when we discuss the genetic code, there are 64 possible codons, and 61 of these code for amino acids. Thus there is usually more than one codon for a given amino acid, and there must be at least 61 tRNA molecules with the appropriate anti-codons. As we mentioned before, all tRNA molecules have a similar tertiary structure which is suited for their function in protein synthesis. Nonetheless it is necessary that the correct amino acid become attached to a given tRNA so that 'reading' of the codons is of high fidelity. There are twenty enzymes, called amino acyl synthetases, involved in the so-called charging reaction of tRNA. Each enzyme is specific for the attachment of one amino acid to a tRNA molecule. Thus all of the tRNA molecules for a particular amino acid (even though different anticodons may be involved) must have a common nucleotide sequence and/or three-dimensional structure that can be recognised by the requisite, specific, amino acyl synthetase.

The link between an amino acid and a tRNA molecule is a covalent bond between the alpha-carboxyl group of the amino acid and the terminal ribose of the 3′ adenine nucleotide of the tRNA (Fig. 8.6).

The linkage between the two involves a high energy bond and hence it is common to talk about 'activated' or 'charged' tRNA. The energy in this bond is used in the formation of a peptide bond during polypeptide chain growth. The source of energy for the bond is ATP and the sequence of steps in forming the amino acyl-tRNA is summarised in Fig. 8.7.

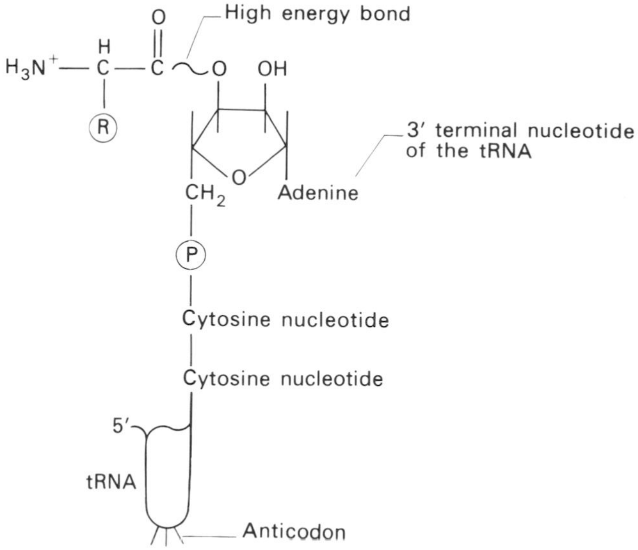

Fig. 8.6 An amino acyl-tRNA.

POLYPEPTIDE CHAINS ARE MADE IN N-TERMINAL TO C-TERMINAL DIRECTION

A polypeptide chain is synthesised commencing at the amino-terminal end. This was shown by H. Dintzis in 1960 in his work with reticulocytes of rabbits. Reticulocytes are young red blood cells that make haemoglobin as their sole protein product. In radioactive-labelling experiments, Dintzis showed that an in-vitro culture of reticulocytes made the beta polypeptide of haemoglobin

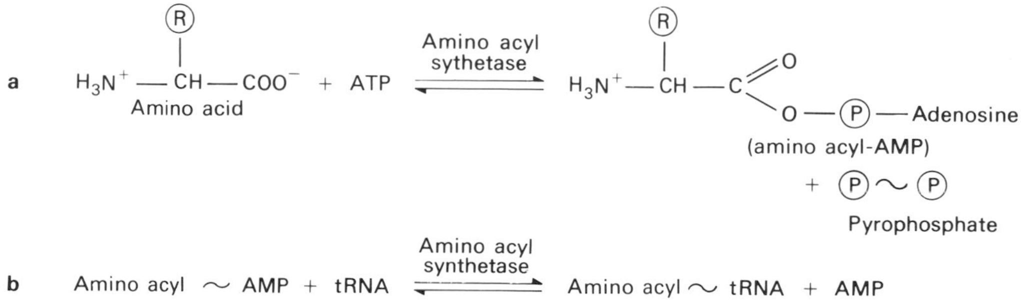

Fig. 8.7 Reactions catalysed by amino acyl synthetase in the synthesis of amino acyl-tRNA.

in about one minute. Then in a separate experiment he added radioactive leucine to the cell suspension for about 30 seconds, after which he stopped protein synthesis by quickly cooling the cells. He isolated the complete haemoglobin molecules ($\alpha_2\beta_2$) and purified them, thus removing incomplete chains from consideration. Then he separated the two types of polypeptides and broke the chains into fragments (peptides) with the aid of the enzyme, trypsin. The fragments were separated electrophoretically and the distribution of radioactivity among the fragments was examined. The reasoning here was that, since the label was present for only about half the time it took to make a complete chain, and since only whole chains were examined, then the radioactivity would only be found in peptides corresponding to the end of the chain synthesised last. Thus, if synthesis is from N- to C-terminus, the results depicted in Fig. 8.8 would be expected.

Here the prediction was that the label would be found towards the C-terminal end, and *no* label would be found at the N-terminal end. Indeed, data were obtained of this kind, thus supporting the N- to C-terminal polarity of polypeptide synthesis.

More recent work, using in-vitro protein synthesising systems, has confirmed Dintzis' conclusions.

THE STEPS OF PROTEIN SYNTHESIS IN PROKARYOTES

There are three major steps in protein synthesis: initiation, elongation and termination. These will be discussed in turn.

Initiation

Initiator tRNA
The first amino acid in the synthesis of all bacterial polypeptides is N-formyl-methionine (fmet), which is a modified methionine amino acid in which the alpha-amino group is 'blocked' and therefore cannot participate in peptide bond formation. The formyl group is added to the methionine after the amino acid has become attached to a specific tRNA, called tRNA·fmet or tRNA·f. This reaction is catalysed by the enzyme transformylase (Fig. 8.9). In many cases the fmet that starts a polypeptide chain is subsequently removed by enzymatic action.

Biochemical analysis has shown that at least two species of tRNA that can be charged with methionine are present in all prokaryotic organisms: one is the tRNA involved with initiation and the other species is responsible for the insertion of methionine elsewhere in the polypeptide

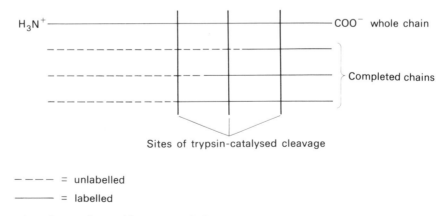

Fig. 8.8 Demonstration that polypeptides are made in the N-terminal to C-terminal direction.

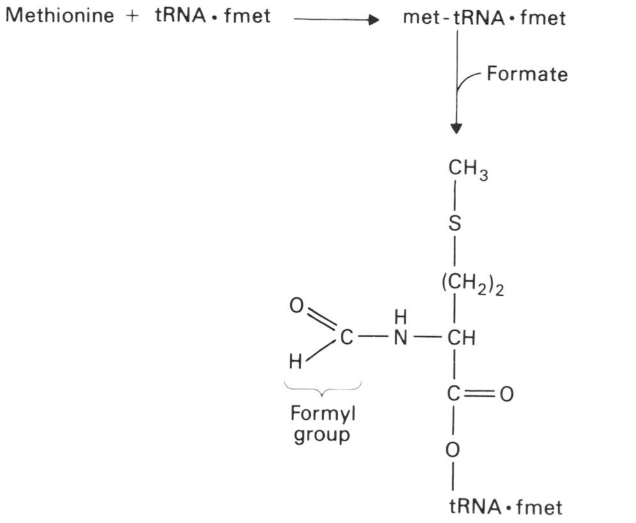

Methionine + tRNA·fmet ⟶ met-tRNA·fmet

| Formate

CH_3

|

S

|

$(CH_2)_2$

O=C—N—CH

H   H |

C=O

|

O

|

tRNA·fmet

Formyl group

N-formyl-methionyl-tRNA·fmet

Fig. 8.9 Synthesis of N-formyl-methionyl-tRNA.

chain. The latter is designated tRNA·mmet or tRNA·m. Both of these tRNAs in bacteria are amino acylated by the same enzyme, but only tRNA·fmet is a substrate for the transformylase-catalysed reaction. Both tRNAs read AUG (the only methionine codon), but in addition tRNA·fmet can recognise GUG and UUG codons. Sequencing studies have shown that both molecules have an anticodon that is complementary to AUG. The two tRNA molecules do differ in some other properties. For example, the binding of fmet-tRNA·f to ribosomes is catalysed by an initiator factor (discussed later) whereas the binding of met-tRNA·m is catalysed by elongation factors. The two tRNAs apparently bind to the ribosome at different sites. Thus, it is clear that fmet·tRNA·f must have a structure that is specific for its role in initiation.

Ribosome binding sites
In bacteria, the first step in initiation is the formation of a complex between the 30S ribosomal subunit, fmet-tRNA and an mRNA molecule. The 50S subunit adds on later to form the active 70S ribosome (monosome). The mRNA may contain information for one to several distinct polypeptide chains. For each of the segments coding for a poly-

peptide, there is a specific nucleotide sequence for orienting the mRNA correctly and in the right reading frame on the ribosome. These sequences are called the ribosome binding sites. When the 30S subunit is bound to the mRNA at the binding site, the RNA is resistant to digestion by ribonuclease action. This has allowed the protected regions to be sequenced and we now know the ribosome binding sequences for a variety of mRNA molecules, for example that coding for the enzyme beta-galactosidase (Fig. 8.10). The initiation codon AUG is found at the 3′ end of the sequence. The rest of the sequence varies from mRNA to mRNA with the exception that an AGGA sequence (boxed), or a very similar sequence, is found in nearly all molecules.

5′ — ACA |AGGA| AACAGCU AUG ———

β-galactosidase mRNA ribosome binding site

Fig. 8.10 Nucleotide sequence of the ribosome binding site of *E. coli* β-galactosidase mRNA showing the AGGA sequence (boxed) that is common to most mRNA molecules, and the initiation codon (underlined).

Initiation factors and initiation
In addition to mRNA, fmet-tRNA and ribosomal subunits, three protein initiation factors (IF-1, IF-2, and IF-3) and GTP are required for the initiation process to occur. We will discuss the properties of the initiation factors and then present the scheme proposed for the initiation process in protein synthesis.

1. IF-3. The IF-3 factor weighs 23 000 daltons and functions in binding mRNA to the 30S subunit. It also acts as a dissociation factor for separating the 30S and 50S subunits after polypeptide synthesis is complete. Like all of the IFs, IF-3 is found bound to free 30S subunits and can be released by washing the subunits in 0.5M ammonium chloride.

Experiments with radioactive IF-3 have shown that it is capable of binding to both 30S subunits and to mRNA molecules. In an in-vitro protein synthesising system, IF-3 enhances the binding of fmet-tRNA to mRNA · 30S subunit complexes. It is attractive to suppose that IF-3 recognises

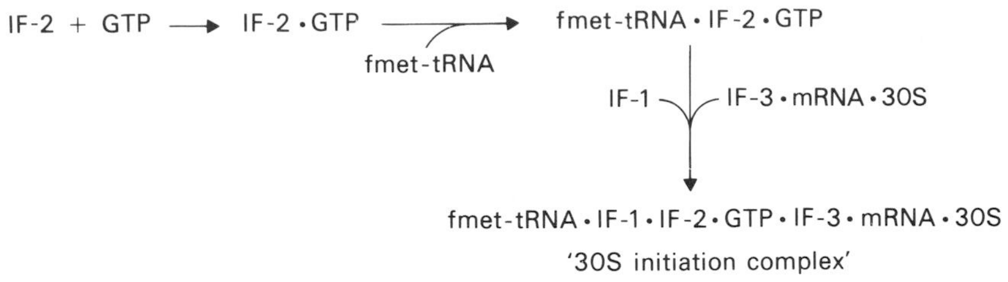

IF-2 + GTP $\longrightarrow$ IF-2·GTP $\xrightarrow{\qquad\qquad}$ fmet-tRNA·IF-2·GTP

fmet-tRNA

IF-1 $\diagdown$ $\diagup$ IF-3·mRNA·30S

fmet-tRNA·IF-1·IF-2·GTP·IF-3·mRNA·30S

'30S initiation complex'

Fig. 8.11 Initiation of protein synthesis: steps in the formation of the 30S initiation complex.

mRNAs by the AUG or GUG initiation codons, but there is no solid evidence on this point.

Parenthetically, the studies with isolated IF-3 illustrate an interesting technical point. Early work suggested the factor was very unstable, whereas it is now known to be very stable. The early results may simply be explained by the fact that IF-3 sticks to glass.

In summary, then, the initiation reaction in which IF-3 is involved is:

IF-3 + mRNA + 30S subunit →
          (IF-3·mRNA·30S) complex.

2. IF-2. The 80 000 dalton IF-2 protein is involved with the binding of the initiator tRNA to the IF-3·mRNA·30S complex. The 'high energy' molecule GTP is used in this reaction. In-vitro experi-

ments have shown that IF-2 and GTP will bind to form a complex which is stabilised when it, in turn, forms a complex with fmet-tRNA. This latter complex then binds with the IF-3 · mRNA · 30S complex and the IF-1 protein factor (9000 daltons) to form the 30S initiation complex (Fig. 8.11).

3. Dissociation of initiation factors from the initiation complex. The initiation factors function to bring fmet-tRNA, mRNA and 30S subunits into a stable association. The next step is the addition of a 50S subunit to form a 70S initiation complex. This leads to the hydrolysis of GTP to GDP + Ⓟ and the release of three initiation factors (Fig. 8.12). The factors can then be used for further initiation reactions on the same or different mRNA.

A summary of the initiation steps is given in Fig. 8.13.

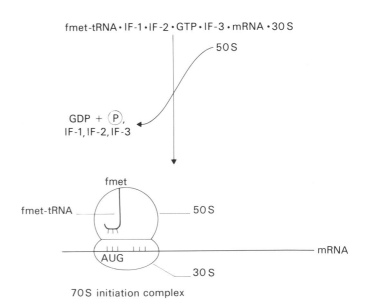

Fig. 8.12 Initiation of protein synthesis: addition of 50S ribosomal subunit to 30S initiation complex leads to formation of 70S ribosome in frame on the mRNA.

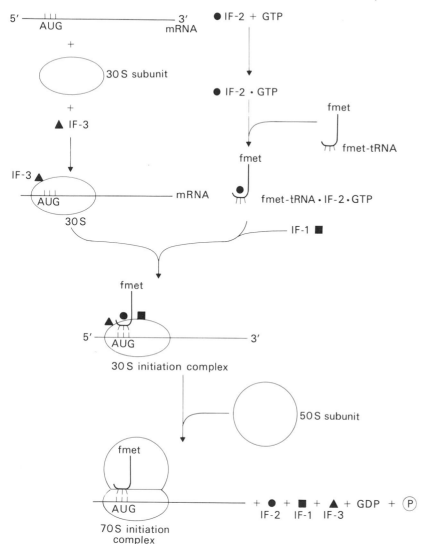

Fig. 8.13 Summary of the steps in the initiation of protein synthesis in prokaryotes. (After J. D. Watson, 1977, *The Molecular Biology of the Gene*. W. A. Benjamin, Menlo Park.)

Elongation

The 70S ribosome has two sites for binding amino acyl-tRNA. In protein synthesis, charged tRNA binds first to the 'A' (amino acyl) site and then the amino acid it carries becomes joined to the growing polypeptide chain carried by the tRNA at the 'P' (peptidyl) site by the formation of a peptide bond.

It is not known whether the fmet-tRNA enters the 'A' site and then moves to the 'P' site, or whether it enters the 'P' site directly. Before further protein synthesis can occur, however, the fmet-tRNA must become located in the 'P' site hydrogen-bonded to the start codon on the mRNA. Once this has occurred a cyclic sequence of events commences in which one amino acid at a time is added to the growing polypeptide chain. This is called elongation, and is summarised in Fig. 8.14. These steps will now be discussed in more detail.

Binding of amino acyl-tRNA

The charged tRNA with the complementary anti-codon to the codon in the reading frame of the 'A'

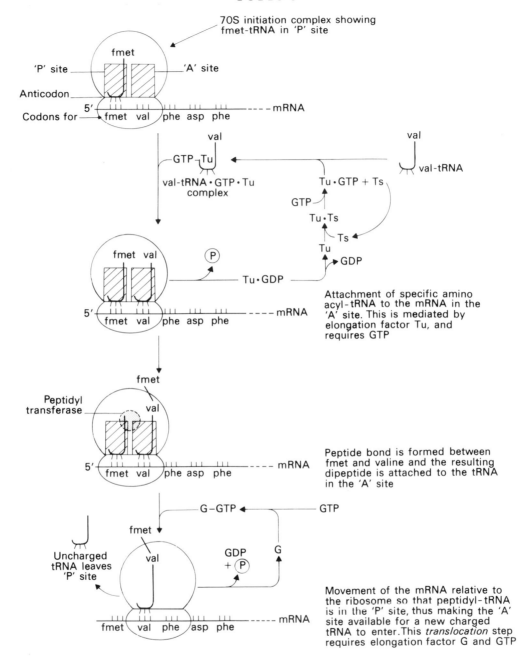

Fig. 8.14 Summary of the elongation (peptide bond formation) and translocation steps in protein synthesis. (After J. D. Watson, 1977, *The Molecular Biology of the Gene.* W. A. Benjamin, Menlo Park.)

site becomes bound to that site of the ribosome in a reaction requiring elongation factor T (EF-T) and GTP. This factor can be isolated from the soluble proteins of *E. coli* and by column chromatography it can be separated into two polypeptides: Ts, which is stable and weighs about 30 000 daltons, and Tu, which is unstable and weighs 42 000 daltons.

EF-T has been shown to bind with GTP and this is postulated to bring about the dissociation of the factor into the two polypeptides, resulting in the formation of an EF-Tu·GTP complex and releasing free EF-Ts. The next step in the elongation process is the binding of aminoacyl-tRNA to the complex to produce an amino acyl-tRNA·Tu·GTP complex. There is evidence that this complex is an intermediate in aminoacyl-tRNA binding to ribosomes. Once the charged tRNA is bound in the 'A' site, GTP is hydrolysed as a result of the enzymatic action of one or more 50S ribosomal proteins. This hydrolysis causes the release of EF-Tu in a complex with GDP. The latter is released and the elongation factor can reassociate with EF-Ts. The process can then be repeated with another aminoacyl-tRNA. In summary, these events are depicted in Fig. 8.15.

Experiments with an analogue of GTP which cannot be hydrolysed have shown that GTP hydrolysis is required for release of EF-Tu from the ribosome but it is not needed for amino acyl-tRNA binding to the ribosome. Other experiments have shown that binding of fmet-tRNA to the ribosome does not require EF-Tu.

Peptide bond formation

At the beginning of this stage, a tRNA carrying the growing polypeptide chain is located in the 'P' site, and an amino acyl-tRNA is located in the 'A' site. These tRNAs are maintained in a juxtaposition conducive for peptide bond formation by the hydrogen bonds between the respective codons and anticodons, and by the tertiary structure of the ribosome. The peptide bond is formed with the aid of the enzyme, peptidyl transferase, which is a ribosomal protein of the 50S subunit. The end result of the reaction is that the polypeptide chain is one amino acid longer, and the entire polypeptide chain has been transferred from the tRNA in the 'P' site to the tRNA in the 'A' site. The reactions are presented schematically in Fig. 8.16.

Translocation

Once the peptide bond has been formed and the polypeptide chain is now on the tRNA in the 'A' site, the next step is advancement of the ribosome precisely one codon (three nucleotides) down the mRNA. During this translocation, the peptidyl-tRNA remains attached to the mRNA by codon-anticodon pairing properties and thus becomes located in the 'P' site. The 'A' site is then vacant and the amino acyl-tRNA specified by the new codon there becomes bound by the process already described. The uncharged tRNA left in the 'P' site after peptide bond formation is also released from the ribosome during translocation. Elongation factor G (EF-G), a 72 000–84 000-dalton protein and

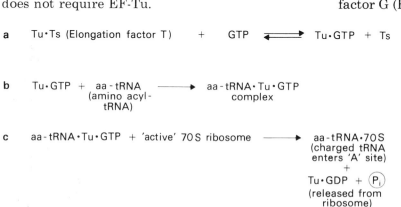

a   Tu·Ts (Elongation factor T)   +   GTP ⇌ Tu·GTP + Ts

b   Tu·GTP + aa-tRNA ⟶ aa-tRNA·Tu·GTP
        (amino acyl-   complex
        tRNA)

c   aa-tRNA·Tu·GTP + 'active' 70S ribosome ⟶ aa-tRNA·70S
                                                (charged tRNA
                                                enters 'A' site)
                                                +
                                                Tu·GDP + Pᵢ
                                                (released from
                                                ribosome)

d   Tu·GDP + Ts ⟶ Tu·Ts

Fig. 8.15 The reactions in protein synthesis involving the prokaryotic elongation factors EF-Tu and EF-Ts.

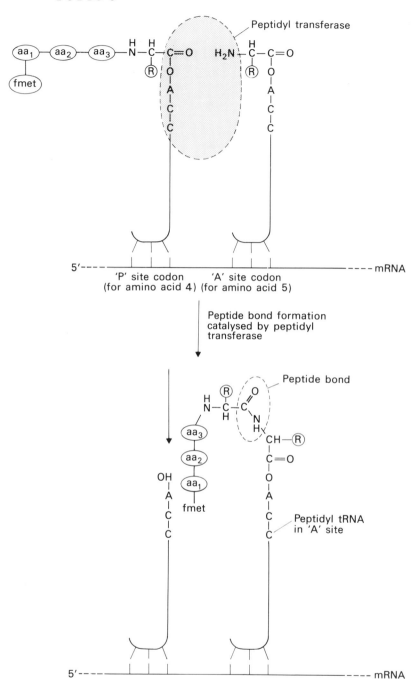

Fig. 8.16 Diagrammatic representation of peptide bond formation on ribosomes as catalysed by peptidyl transferase.

GTP hydrolysis are needed for translocation to occur and, at the present time, it is not known how the translocation mechanism works. One GTP molecule is hydrolysed for each translocation event. It appears that EF-G leaves the ribosome after translocation, since EF-Tu and EF-G cannot interact with the ribosome at the same time.

Termination

The end of the polypeptide chain is indicated on a mRNA molecule by a specific chain terminating (stop) codon. Three such codons are known: UAA, UAG, and UGA. No naturally occurring tRNA has an anticodon for any of these stop codons, and therefore no amino acid can be put into the polypeptide. Chain growth does not cease merely by running out of amino acids to insert. In fact, three specific termination factors have been shown to be involved in reading the stop codon. They differ in their codon specificity and GTP requirement (Table 8.1).

Table 8.1 Properties of prokaryotic termination factors.

Termination factor	Molecular weight (daltons)	Stop codons recognised	GTP requirement
RF1	44 000	UAA & UAG	No
RF2	47 000	UAA & UGA	No
RF3 (which was formerly S)	46 000	None	Yes

As can be seen, RF1 and RF2 have overlapping specificity for the stop codons. They have been shown to interact with the termination codons by interaction at the 'A' site. The RF3 factor apparently plays a stimulatory role in RF1 and RF2 activity. There is some evidence for a GTP requirement in the RF3 factor's activity. In any event, chain termination, as mediated by these factors, involves the cleavage of the carboxyl group of the C-terminal end of the polypeptide chain from the tRNA in the 'P' site. This results in the release of the polypeptide and the now uncharged tRNA. The ribosome will then move along the mRNA until a new initiation sequence is encountered,

or it will dissociate from the mRNA. IF-3 serves to keep the two subunits apart. Thus, when a new 70S initiation complex is formed, the two subunits are drawn randomly from the free pools of 30S and 50S subunits.

As the polypeptide chain is being synthesised, the primary sequence of amino acids directs the three-dimensional shape. In other words, the elongating chain begins to assume its final shape as it is being made. Indeed, some enzyme activity can be detected on ribosomes which have not yet completed the synthesis of the polypeptide.

Polysomes

Efficient translation of an mRNA molecule cannot be achieved by a single ribosome moving along it. In fact the amount of space a ribosome takes up on a mRNA is relatively small and thus several ribosomes can work on the mRNA at once. The association of a number of ribosomes on a single mRNA chain is called a polyribosome or polysome, and this allows several polypeptide chains to be made from each template. The length of the polypeptide chain on a given ribosome will be directly proportional to how far the ribosome has moved along the mRNA from the 5' end of the molecule. The existence of polysomes explains why a cell needs so little mRNA, while at the same time it contains so much more protein (Fig. 8-17).

Relationship of transcription and translation

There is good evidence in bacteria that the mRNA becomes associated with ribosomes while its synthesis is continuing. This is possible owing to lack of a nuclear membrane so that, as the 5' end of the growing mRNA molecule is displaced from the DNA as the double helix reforms, the ribosome binding site becomes available. Ribosomes then load onto the mRNA in rapid sequence, the first being close behind the RNA polymerase (Fig. 8.18).

To give some idea of the rates of these processes, the mRNA for the tryptophan biosynthetic operon (see Topic 19) is transcribed at a rate of about 1000 nucleotides per minute and the translation process proceeds at about the same rate. Thus approximately 350 amino acids can be polymerised into polypeptide chains each minute.

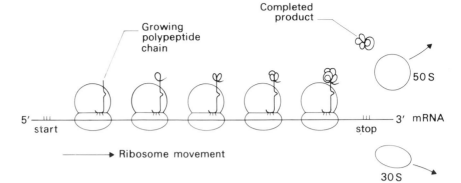

Fig. 8.17 Diagrammatic representation of a polysome engaged in protein synthesis.

→ Ribosome movement

Another important aspect of translation is the lability of the mRNA. As mentioned in an earlier Topic, prokaryotic mRNAs are considered to be short lived in that degradation of the molecule by 5'-exonuclease action competes with the ribosome-mediated initiation of protein synthesis. Continued mRNA synthesis, then, is necessary for continued polypeptide synthesis.

PROTEIN SYNTHESIS IN EUKARYOTES

The steps and mechanisms of protein synthesis are very similar in both eukaryotes and prokaryotes. As we have discussed, the ribosomes are different and this is also the case with the soluble protein factors.

Initiation

Initiation of protein synthesis involves the binding of mRNA to the ribosomes. In eukaryotes it has now been firmly established that the 5'-cap structure is necessary in order to get efficient binding but the 3' poly(A) sequence apparently is not needed. There is good evidence, though, that the poly(A) sequence stabilises the mRNA during the translation process. The precise mechanism whereby the eukaryotic message binds to the ribosome is not known, although it is most likely that RNA–RNA and RNA–protein interactions are involved. At the present, the nucleotide sequence has been determined to the 5' side of the AUG start codon for a number of eukaryotic mRNAs and there are very few common features. Thus it appears that, unlike the case in prokaryotes, there is not a fixed nucleotide sequence that plays a role in ribosome binding to the message.

As in prokaryotes, the initiation codon is AUG, and there is a special initiator methionyl-tRNA that recognises that signal in the message. Unlike its prokaryotic counterpart the methionine carried by the tRNA does not become formylated since the appropriate enzyme system does not exist in eukaryotic cells. The initiator tRNA can be distinguished from the met-tRNA that reads AUG codons elsewhere in the message, however, by the fact that it can be formylated *in vitro* in the presence of an *E. coli* extract. Hence in eukaryotes it is also appropriate to define the two methionine-accepting tRNAs as tRNA·f and tRNA·m.

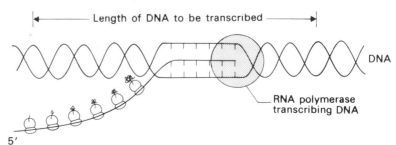

Fig. 8.18 Schematic of the possible translation of a mRNA while it is still being transcribed in prokaryotes.

Table 8.2 Properties of some of the many mammalian initiation factors.

Initiation factor	Molecular weight (daltons)	No. of subunits	Activity in initiation
eIF-1	15 000	1	40S complex formation
eIF-2	150 000	3 (unequal)	met-tRNA·f binding; GTP binding
eIF-3	300 000–500 000	many	mRNA binding; dissociation of subunits
eIF-4A	50 000	1	mRNA binding
eIF-4B	80 000 (?)	1	recognition of 5′cap on mRNA; tRNA binding; joining of ribosomal subunits
eIF-4C	17 000	1	stabilisation of initiation complex
eIF-5	125 000	1	80S ribosome formation; GTPase

In eukaryotes, or at least in mammals, there are many more initiation factors than in prokaryotes. In many cases the proteins have not been purified to homogeneity and thus their absolute roles in protein synthesis are a little uncertain. Table 8.2 summarises what is known about the number, molecular weight, and functions for initiation factors that have been found in mammalian systems. Between them all, they carry out the initiation events performed by the three prokaryotic IFs. It remains to be seen to what extent the situation in mammals can be generalised throughout eukaryotes.

Elongation

As in prokaryotes, there are two elongation factors, eEF-1 (equivalent to prokaryotic EF-T) and eEF-2 (equivalent to prokaryotic EF-G). The former has been studied in a number of systems and in general it exists in multiple forms. Purified eEF-1 from rabbit reticulocytes, for example, has a molecular weight of 186 000 and consists of three subunits weighing 62 000 daltons each. In some systems the subunits aggregate to produce molecules of greater than 1 million daltons. Regarding the function of eEF-1, much less is known than in the prokaryotes. The eEF-1 from rabbit reticulocytes, for example,

has been shown to bind to amino acyl-tRNA and to GTP, and thus to facilitate binding of the amino acyl-tRNA to the 'A' site in ribosomes. During this step, GTP is hydrolysed to GDP as a result of GTPase activity of the elongation factor and an eEF-1·GDP complex is released from the ribosome.

The eukaryotic eEF-2 is very similar to prokaryotic EF-G although the two are not interchangeable in in-vitro systems. The factor from rabbit reticulocytes has a molecular weight of 96 500–110 000 daltons and, after binding with GTP, binds to the ribosome. This event results in hydrolysis of GTP to GDP, translocation of the ribosome one codon down the message and release of an eEF-2·GDP complex. All of these steps are very similar to the events that take place in prokaryotes, one exception being that the eukaryotic factor forms a stable complex with GTP whereas the prokaryotic factor does not.

Termination

The same chain termination codons are operant in eukaryotes as in prokaryotes. One release factor has been identified in rabbit reticulocytes and this has a molecular weight of 115 000 daltons and may be a dimer. This factor recognises all three chain termination codons and it requires GTP in order to carry out its function. No stimulatory factor analogous to the prokaryotic RF-3 has been found in eukaryotes.

Protein synthesis and cellular compartmentation in eukaryotes

In prokaryotes there is no nuclear membrane to separate the transcription process from the translation process. The presence of a nuclear membrane and the various modification processes peculiar to mRNAs in eukaryotes, present many levels at which the regulation of gene expression can be effected. These include transcription itself, the processing of the primary transcript to produce mature mRNA, RNA–RNA splicing, and the movement of mRNA from the nucleus to the cytoplasm. In this section we will discuss the fate of the mRNA after it has entered the cytoplasm.

In eukaryotic cells, the cytoplasm contains a network of interconnecting channels bounded by membranes. This is called the *endoplasmic reticulum* (ER) and the membranes involved are continuous and may in fact connect to the nuclear membrane and to the cell membrane. Close examination of the ER reveals that it is differentiated into two types, *smooth* (SER) and *rough* (RER), which are distinguished by the fact that the latter has ribosomes bound to it (hence the 'rough' appearance) whereas the former does not. Thus, when one considers the ribosomes present in the cytoplasm, they fall into two classes: those that are free and those that are membrane bound. These two classes of ribosomes have different roles in the cell in that the former synthesise proteins that remain free in the cytoplasm whereas the latter synthesise proteins that are ultimately secreted by the cell.

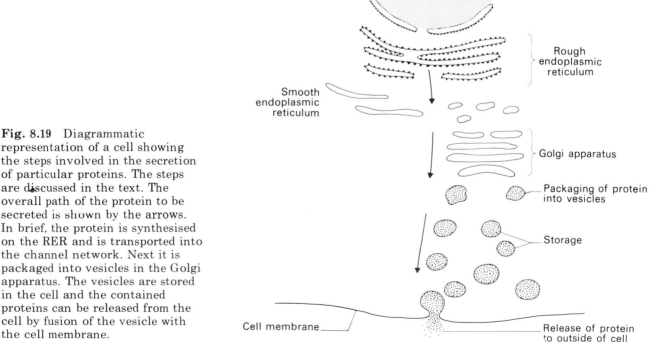

Fig. 8.19 Diagrammatic representation of a cell showing the steps involved in the secretion of particular proteins. The steps are discussed in the text. The overall path of the protein to be secreted is shown by the arrows. In brief, the protein is synthesised on the RER and is transported into the channel network. Next it is packaged into vesicles in the Golgi apparatus. The vesicles are stored in the cell and the contained proteins can be released from the cell by fusion of the vesicle with the cell membrane.

Thus, for example, pancreatic cells that secrete enzymes into the intestine are extremely rich in RER. We shall now describe some of the events associated with the secretion of proteins from cells.

Fig. 8.19 presents a diagrammatic view of the secretion system of a eukaryotic cell. The proteins to be secreted are made on the ribosomes of the RER and then transferred across the membrane into the channel system. Clearly, then, the mRNAs for the secreted proteins must somehow become associated specifically with the ribosomes of the RER. G. Blobel and B. Dobberstein have proposed a 'signal' hypothesis to explain this. They suggest that there is a unique sequence of codons located to the 3' side of the initiator AUG codon which is present only in mRNAs for proteins that must be transferred across membranes. Translation of these codons results in a specific amino acid sequence at the N-terminal end of the protein. Then it is postulated that the special end of the protein facilitates the attachment of the ribosome to the membrane so that the protein can be transferred across it. Once the protein has been completed, it is proposed that the ribosome dissociates from the ER.

Once in the channels of the RER the proteins to be secreted by the cell move towards the perimeter of the cell, becoming concentrated in the Golgi apparatus which are stacks of flat, membranous sacs. In the Golgi, the aggregated proteins are wrapped by a single membrane and the resulting vesicles migrate towards the cell surface where, by fusing with the cell membrane, the contents (i.e. the protein to be secreted) are released to the outside of the cell. Clearly this is much more complicated than the situation in prokaryotic cells.

REFERENCES

General

Blobel G. 1972. Protein tightly bound to globin mRNA. *Biochem. Biophys. Res. Commun.* **47**: 88–93.

Blobel G. & B. Dobberstein. 1975. Transfer of proteins across membranes. I. Presence of proteolytically processed and unprocessed nascent immunoglobulin light chains on membrane-bound ribosomes of murine myeloma. *J. Cell Biol.* **67**: 835–851.

Cold Spring Harbor Symposium for Quantitative Biology, 1970, vol. 35. *Transcription of Genetic Material*. Cold Spring Harbor Laboratory, New York.

Felicetti L. & F. Lipmann. 1968. Comparison of amino acid polymerization factors isolated from rat liver and rabbit reticulocytes. *Arch. Biochem. Biophys.* **125**: 548–557.

Ganoza M.C. & C.A. Williams. 1969. In vitro synthesis of different categories of specific protein by membrane-bound and free ribosomes. *Proc. Natl. Acad. Sci. USA*, **63**: 1370–1376.

Hardesty B., R. Arlinghaus, J. Shaeffer & R. Schweet. 1963. Hemoglobin and polyphenylalanine synthesis with reticulocyte ribosomes. *Cold Spring Harbor Symp. Quant. Biol.* **28**: 215–222.

Hardesty B., W. Culp & W. McKeehan. 1969. The sequence of reactions leading to the synthesis of a peptide bond on reticulocyte ribosomes. *Cold Spring Harbor Symp. Quant. Biol.* **34**: 331–344.

Haselkorn R. & L.B. Rothman-Denes. 1973. Protein synthesis. *Annu. Rev. Biochem.* **43**: 397–438.

Huez G., G. Marbaix, E. Hubert, M. Leclercq, U. Nudel, R. Salomon, B. Lebleu, M. Revel & U.A. Littauer. 1974. Role of the polyadenylate segment in the translation of globin messenger RNA in *Xenopus* oocytes. *Proc. Natl. Acad. Sci. USA*, **71**: 3143–3146.

Humphries S., M. Doel & R. Williamson. 1973. The translation of mouse globin mRNA from which the polyadenylic acid sequence has been removed in a reinitiating protein synthesis system. *Biochem. Biophys. Res. Commun.* **58**: 927–931.

Pestka S. 1976. Insights into protein biosynthesis and ribosome function through inhibitors. *Progr. Nucl. Acid Res. Mol. Biol.* **17**: 217–245.

Revel M. 1977. Initiation of messenger RNA translation into proteins and some aspects of its regulation. In *Molecular Mechanisms of Protein Biosynthesis*, H. Weissbach and S. Pestka (eds.), pp. 246–321. Academic Press, New York.

Revel M. & Y. Groner. 1978. Post-transcriptional and translational controls of gene expression in eukaryotes. *Annu. Rev. Biochem.* **47**: 1079–1126.

Schreier M.H. & H. Noll. 1971. Conformational changes in ribosomes during protein synthesis. *Proc. Natl. Acad. Sci. USA*, **68**: 805–809.

Shafritz D.A. 1974. Protein synthesis with messenger ribonucleic acid fractions from membrane-bound and free liver polysomes. Translation characteristics of liver polysomal ribonucleic acids and evidence for

albumin production in a messenger-dependent reticulocyte cell-free system. *J. Biol. Chem.* **249**: 81–93.

Shatkin A.J., A.K. Banerjee, G.W. Both, Y. Furuichi &
S. Muthukrishnan. 1976. Dependence of translation
on 5′-terminal methylation of mRNA. *Fed. Proc.* **35**:
2214–2217.

Smith A.E. 1976. *Protein Biosynthesis.* Chapman and
Hall, London.

Spirin A.S. 1969. A model of the functioning ribosome:
Locking and unlocking of the ribosome subparticles.
Cold Spring Harbor Symp. Quant. Biol. **34**: 197–207.

Sussman M. 1970. Model for quantitative and qualitative
control of mRNA translation in eukaryotes. *Nature,*
225: 1245–1248.

Travers A. 1973. Control of ribosomal RNA synthesis
in vitro. Nature, **244**: 15–18.

Weissbach H. & S. Ochoa. 1976. Soluble factors required
for eukaryotic protein synthesis. *Annu. Rev. Biochem.* **45**: 191–216.

Weissbach H. & S. Pestka (eds.). 1977. *Molecular
Mechanisms of Protein Biosynthesis.* Academic Press,
New York.

Yanofsky C., G.R. Drapeau, J.R. Guest & B.C. Carlton.
1967. The complete amino acid sequence of the tryptophan synthetase A protein (or subunit) and its
colinear relationship with the genetic map of the
A gene. *Proc. Natl. Acad. Sci. USA,* **57**: 296–298.

Initiation

Caskey C.T., B. Redfield & H. Weissbach. 1967. Formy-
lation of guinea-pig liver methionyl-sRNA. *Arch.
Biochem. Biophys.* **120**: 119–123.

Gupta N.K., B. Chatterjee, Y.C. Chen & A. Majumdar.
1975. Protein synthesis in rabbit reticulocytes. A
study of met-tRNA·fmet binding factor(s) and met-
tRNA·fmet. *J. Biol. Chem.* **250**: 853–862.

Miller, D.L. & H. Weissbach. 1977. Factors involved in
the transfer of aminoacyl-tRNA to the ribosome.
pp. 323–373 In *Molecular Mechanisms of Protein Biosynthesis,* H. Weissbach and S. Pestka (eds.),
pp. 323–373. Academic Press, New York.

Petrissant G. 1973. Evidence for the absence of the
G-T-ψ-C sequence from two mammalian initiator
transfer RNAs. *Proc. Natl. Acad. Sci. USA,* **70**:
1046–1049.

Revel M., M. Herzberg & H. Greenspan. 1969. Initiator
protein dependent binding of messenger RNA to the
ribosome. *Cold Spring Harbor Symp. Quant. Biol.*
34: 261–275.

Steitz J.A. & K. Jakes. 1975. How ribosomes select
initiator regions in mRNA: Base pair formation between the 3′ terminus of 16S rRNA and the mRNA
during initiation of protein synthesis in *E. Coli. Proc.
Natl. Acad. Sci. USA,* **72**: 4734–4738.

Elongation

Arlinghaus R., G. Favelukes & R. Schweet. 1963. A
ribosome-bound intermediate in polypeptide synthesis. *Biochem. Biophys. Res. Commun.* **11**: 92–96.

Brot N. 1977. Translocation. In *Molecular Mechanisms
of Protein Biosynthesis,* H. Weissbach and S. Pestka
(eds.), pp. 375–411. Academic Press, New York.

Brot N., E. Yamasaki, B. Redfield & H. Weissbach. 1972.
The properties of an *E. coli* ribosomal protein re-
quired for the function of factor G. *Arch Biochem.
Biophys.* **148**: 148–155.

Caskey C.T. 1977. Peptide chain formation. In *Molecular
Mechanisms of Protein Biosynthesis,* H. Weissbach
and S. Pestka (eds.), pp. 443–465. Academic Press,
New York.

Jaskunas S.R., L. Lindahl, M. Nomura & R.R. Burgess.
1975. Identification of two copies of the gene for the
elongation factor EF-Tu in *E. coli. Nature,* **257**:
458–462.

Rohrbach M.S., M.E. Dempsey & J.W. Bodley. 1974.
Preparation of homogeneous elongation factor G
and examination of the mechanism of guanosine
triphosphate hydrolysis. *J. Biol. Chem.* **249**: 5094–5101.

Tocchini-Valentini G.P. & F. Mattocia. 1968. A mutant
of *E. coli* with an altered supernatant factor. *Proc.
Natl. Acad. Sci. USA,* **61**: 146–151.

Weissbach H., B. Redfield & J. Hackman. 1970. Studies
on the role of factor Ts in aminoacyl-tRNA binding
to ribosomes. *Arch. Biochem. Biophys.* **141**: 384–386.

Weissbach H., B. Redfield & H. Moon. 1973. Further
studies on the interaction of elongation factor 1 from
animal tissues. *Arch. Biochem. Biophys.* **156**: 267–275.

Termination

Brot N., W.P. Tate, C.T. Caskey & H. Weissbach. 1974.
The requirement for ribosomal proteins L7 and L12 in
peptide-chain termination. *Proc. Natl. Acad. Sci.
USA,* **71**: 89–92.

Goldstein J., G. Milman, E. Scolnick & C.T. Caskey.
1970. Peptide chain termination VI. Purification and
site of action of S. *Proc. Natl. Acad. Sci. USA,*
65: 430–437.

Milman G., J. Goldstein, E. Scolnick & C.T. Caskey.
1969. Peptide chain termination, III. Stimulation of
in vitro termination. *Proc. Natl. Acad. Sci. USA,*
63: 183–190.

Shine J. & L. Dalgarno. 1974. The 3′-terminal sequence
of *Escherichia coli* 16S ribosomal RNA: complementarity to nonsense triplet and ribosome-binding
sites. *Proc. Nat. Acad. Sci. USA,* **71**: 1342–1346.

Tompkins R.K., E.M. Scolnick & C.T. Caskey. 1970.
Peptide chain termination, VII. The ribosomal and
release factor requirements for peptide release. *Proc.
Natl. Acad. Sci. USA,* **65**: 702–708.

Topic 9
The Genetic Code

OUTLINE
Evidence for three-letter code
Elucidation of the genetic code
Characteristics of the genetic code
'Wobble'
Mutations and the genetic code.

EVIDENCE FOR THREE-LETTER CODE

As we have mentioned in previous Topics, the information for the amino acid sequence in polypeptides is coded in the nucleotide sequence of mRNA. This genetic code was 'cracked' in 1961 by F. Crick and co-workers. Here the evidence for the nature of the code will be presented.

It was reasoned *a priori* that the code must be at least a triplet code. Specifically, a one-letter code in which one nucleotide codes for one amino acid could not deal with the 20 amino acids known to occur in proteins. A two-letter code would have $4 \times 4 = 16$ 'words' which is still insufficient coding capacity. A three-letter code, however, generates $4 \times 4 \times 4 = 64$ code words which is more than enough to code for the 20 amino acids. There is irrefutible evidence that the code is indeed a triplet code. This came from studies with phage T4 which, as we have discussed before, infects and lyses *E. coli*. The experiments involved two types of strains of T4, the wild type and *rII* strains, which can be distinguished by their plaque morphology and their host range phenotype. That is to say, when phages are added to a lawn of bacteria grown on solid medium in a petri dish, the successive rounds of infection and lysis produce cleared areas in the lawn called plaques. Wild-type T4 ($r+$) produces small turbid plaques whereas *rII* mutants produce large clear ones (Fig. 9.1).

Fig. 9.1 Diagrammatic representations of the plaque morphologies of $r+$ and *rII* strains of phage T4 grown on *E. coli*.

Regarding host-range properties, wild type can grow on both the *B* and *K12* (*lambda*) strains of *E. coli* whereas *rII* strains only grow on *B*. These two properties are summarised in Table 9.1.

Table 9.1 Host range properties of $r+$ and *rII* strains of phage T4.

Phage genotype	Plaque morphology on *E. coli*	
	Strain B	Strain K12(λ)
$r+$	$r+$ type (turbid)	$r+$ type
rII	r type (clear)	no growth

Crick and co-workers started with an *rII* mutant which had been induced by proflavin treatment which has the effect of causing either the addition or deletion of a base-pair in the DNA. They treated the *rII* mutant with proflavin and isolated a number of $r+$ revertants which could be detected by their ability to grow on *K12* (*lambda*). Genetic analysis showed that several of the revertants actually resulted from a second mutation within the *rII* gene so that the combination of the two mutations gave an almost wild-type (pseudo-wild) phenotype. The second mutation alone also resulted in a *rII* phenotype, and indeed by treating a strain carrying only that mutation with proflavin, a new series of revertants was obtained, many of which carried two mutations. The explanation for these revertants is very straightforward. Let us suppose that the wild-type DNA sequence is transcribed to produce a mRNA with the nucleotide sequence shown in Fig. 9.2. If read in groups of three, we will

mRNA: CAG CAG CAG CAG CAG CAG CAG
Amino acid: 1   1   1   1   1   1   1

Fig. 9.2 An hypothetical mRNA with a repeating triplet coding for a polypeptide chain made up of one amino acid type.

get a seven-amino acid polypeptide with each amino acid being the same. If the original proflavin-induced *rII* mutation involved a deletion (−) of a nucleotide pair in the DNA the mRNA might be as shown in Fig. 9.3. Here only the first amino acid

⁻C

mRNA: CAG AGC AGC AGC AGC AGC AGC
Amino acid: 1   2   2   2   2   2   2

Fig. 9.3 Consequences of the deletion of a nucleotide pair on the codon and amino acid sequence of Fig. 9.2.

will be the same as the wild type and the rest will be changed. All pseudo-wild revertants of this strain presumably involved an addition (+) mutation close to the first (−) mutation. As can be seen, this restores the reading frame with only a few amino acids in between being erroneous ones. The resulting protein is likely to be almost as functional as the wild-type protein (Fig. 9.4).

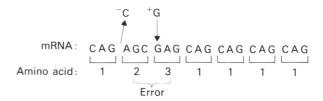

⁻C   ⁺G

mRNA: CAG AGC GAG CAG CAG CAG CAG
Amino acid: 1   2   3   1   1   1   1
                         Error

Fig. 9.4 Restoration of reading frame in the message by the addition of a nucleotide pair near the original deletion.

Similarly, all pseudo-wild revertants of the secondary (+) mutation must be deletion (−) mutations close by. In this way, then, a collection of (+) and (−) strains can be isolated. (Note that the (+) and (−) designations are arbitrary as from the experiments of Crick and his colleagues it was not possible to ascertain whether a particular

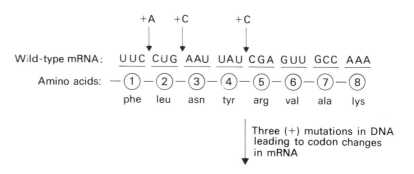

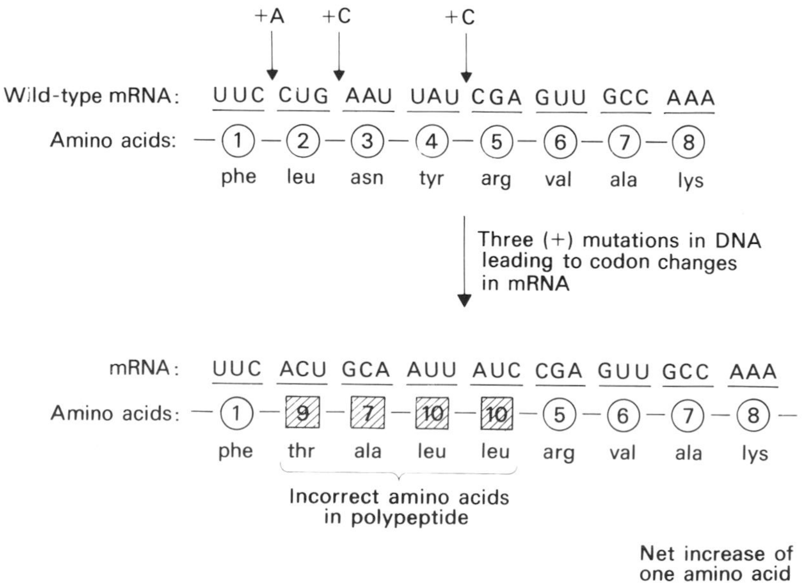

+A   +C   +C

Wild-type mRNA: UUC CUG AAU UAU CGA GUU GCC AAA
Amino acids: —①—②—③—④—⑤—⑥—⑦—⑧
               phe leu asn tyr arg val ala lys

Three (+) mutations in DNA leading to codon changes in mRNA

mRNA: UUC ACU GCA AUU AUC CGA GUU GCC AAA
Amino acids: —①—⑨—⑦—⑩—⑩—⑤—⑥—⑦—⑧—
               phe thr ala leu leu arg val ala lys

Incorrect amino acids in polypeptide

Net increase of one amino acid

Fig. 9.5 Illustrations of how the reading frame in a mRNA can be restored by three separate, closely-linked nucleotide pair additions in the DNA. The result is a net increase of one amino acid in the polypeptide.

mutation involved an addition or a deletion.) Then it was possible to combine a number of (+) mutations into one strain by genetic crosses. If the (+) mutations were close enough in the DNA (testable by genetic analysis), recombinants containing three (+) mutations sometimes gave functional products as evidenced by the strains' abilities to grow in *K12 (lambda)*. A similar result was obtained with certain sets of three (−) mutations, whereas combinations of two (−) or two (+) did not produce pseudo-wilds. The conclusion was drawn, therefore, that the code is a triplet code, since the triple mutation strains restored the reading frame (Fig. 9.5).

ELUCIDATION OF THE GENETIC CODE

Elucidation of the actual genetic code in terms of which triplets code for which amino acids, was largely based on the use of a cell-free protein synthesising system. M. Nirenberg in 1961 found that an extract of *E. coli* containing ribosomes, tRNAs, amino acyl synthetases, mRNA, amino acids and other ingredients was able to incorporate amino acid into protein. This reaction only goes on for a few minutes but protein synthesis does restart if fresh mRNA is added (Fig. 9.6).

This was an exceptionally important discovery since, once the natural mRNA was used up, an artificial, enzymatically-synthesised mRNA could be introduced into the system and it was found to function, poorly. In these experiments, the ionic conditions were altered from the normal physiological state so that translation began randomly without the need for a specific start codon. In the first experiments, Nirenberg and Matthaei set up a series of reaction mixtures each containing the 20 amino acids, but with a different amino acid radioactively labelled in each. Each mixture also contained all other ingredients for in-vitro protein synthesis, except natural mRNA. The assay mixtures were then primed with a synthetic mRNA, for example, a polynucleotide containing only one base, such as poly(U). These synthetic messengers were made using the activity of the enzyme poly-

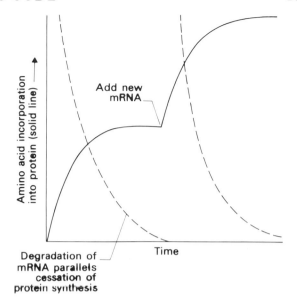

Fig. 9.6 Diagrammatic representation of results showing the dependence of an in-vitro protein synthesising system on the presence of mRNA. Protein synthesis decreases as mRNA degrades and resumes when fresh mRNA is added. (After the work of M. Nirenberg and J. Matthaei.)

nucleotide phosphorylase which catalyses the reaction shown in Fig. 9.7. Normally the reaction goes from left to right but in the presence of high concentrations of the diphosphate, the reaction is forced towards the left and an RNA is formed.

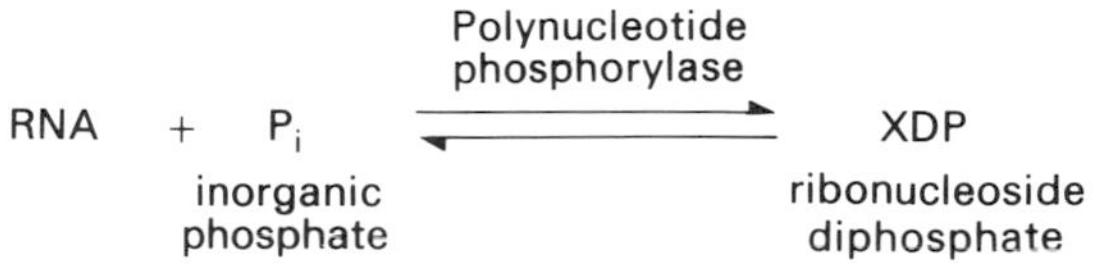

$$\text{RNA} + \underset{\substack{\text{inorganic}\\\text{phosphate}}}{P_i} \; \underset{\xleftarrow{\hspace{1.5cm}}}{\overset{\text{Polynucleotide phosphorylase}}{\xrightarrow{\hspace{1.5cm}}}} \; \underset{\substack{\text{ribonucleoside}\\\text{diphosphate}}}{\text{XDP}}$$

Fig. 9.7 Reaction catalysed by polynucleotide phosphorylase.

After incubation of the reaction mixtures with the RNA, the proteins in each tube were precipitated by treatment with tricholoroacetic acid (TCA) and the precipitates were then assayed for the presence of radioactivity which would indicate the incorporation of a radioactive amino acid into the protein. Using this procedure they found that poly(U) caused the incorporation of labelled phenylalanine into the TCA-insoluble protein. Thus UUU must be a codon in the mRNA for the amino acid phenylalanine. Similar experiments showed that AAA is a codon for lysine and CCC is a codon for proline.

The rest of the codon assignments were determined by a number of in-vitro experimental approaches including the use of mRNAs made randomly from mixtures of two nucleotides, and the use of mRNAs with alternating bases, such as UCUCUCUCU———, or a repeating series of three bases such as AAGAAGAAGAAG———. In the latter, for example, because of the randomness with which translation will begin, three polypeptides will be produced by reading the three different possible repeating codons: polylysine (AAG), polyarginine (AGA) and polyglutamic acid (GAA).

In an elegant approach, M. Nirenberg and P. Leder developed the tRNA binding technique in which the addition of synthetic trinucleotides of known sequence to a reaction mixture of ribosomes, tRNA, amino acyl synthetases, amino acids, GTP, etc. caused the formation of a complex between an amino acyl-tRNA, ribosome and tri-

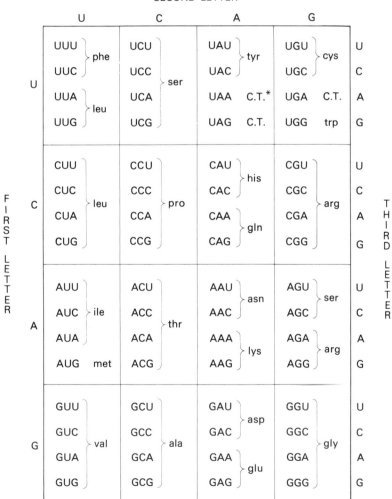

Fig. 9.8 The genetic code.

*C.T. = chain termination codon

nucleotide. More specifically the trinucleotide binds to the 30S subunit as does a mRNA, thus facilitating the binding of the amino acyl-tRNA with the complementary anticodon. Thus, for example, if the trinucleotide is 5'-GAG-3', then glutamic acid-tRNA will become bound to the ribosome. This complex can be separated from the uncomplexed amino acyl-tRNAs by using a filter with pores large enough to let them through but small enough to retain the complex. Thus if 20 assay mixtures are set up, with a different radioactive amino acid in each, in only one tube will radioactivity be retained on the filter and that will indicate the amino acid coded by the trinucleotide being tested. For example, radioactive glutamic acid will be retained on the filter when GAG is the triplet.

None of the procedures alone produced completely unambiguous results. However, by a combination of these techniques 61 of the 64 possible codons were assigned to specific amino acids. The three other codons we now know to be chain terminating triplets which do not code for any amino acid. The codon assignments for the amino acids and for chain termination in protein synthesis is called the genetic code and this is presented in Fig. 9.8.

CHARACTERISTICS OF THE GENETIC CODE

1. The code is comma-free, that is the message is read three nucleotides at a time in a non-overlapping way.
2. The code is universal, that is all organisms use the same language. This means that a message from an animal cell will produce the same protein whether it is translated with animal cell or *E. coli* protein synthesis machinery.
3. The code is degenerate, that is, with two exceptions (AUG and UGG) there is more than one codon for each amino acid. Indeed certain patterns become apparent in this regard, that is when the first two nucleotides are identical, the third nucleotide can be either C or U and the codon may still code for the same amino acid. A similar situation prevails for A and G in the third position.

4. The code uses specific start and stop codons. AUG is the start signal and its absence from the 5' end of the synthetic messengers explains their inefficient translation. The chain termination codons are UAG, UAA and UGA and, in many natural mRNAs, there are adjacent stop codons presumably to ensure a stop in polypeptide synthesis.

'WOBBLE'

It was first thought that 61 different tRNA molecules would be needed with specific anticodons for each of the amino acid-coding triplets. There is now evidence that purified tRNA species can recognise several different codons. Before we get to that, from sequence analysis of tRNA molecules, it was found that several of them have inosine as one of the anticodon bases. The structure of inosine is shown in Fig. 9.9. This has a number of base-pairing possibilities as will become apparent.

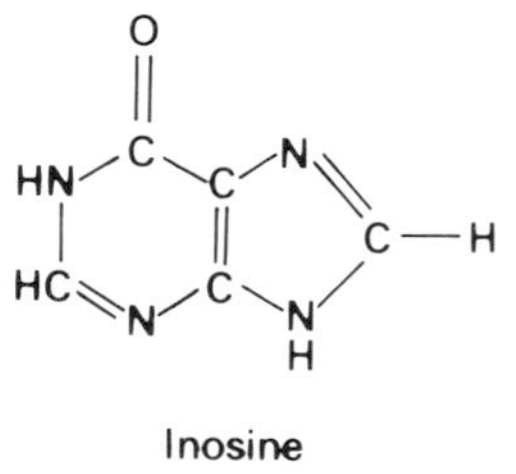

Inosine

Fig. 9.9 Structure of inosine, a base found in the anticodon of some tRNAs.

Sequence analysis also showed that the base at the 5' end of the anticodon (complementary to the third letter of the codon) is not as sterically confined as the other two thus allowing it potentially to pair with more than one base at the 3' end of the codon. This is called base-pairing 'wobble' and only certain pairings are possible in this regard (Table 9.2). As can be seen from the table, 'wobble' does not allow any single tRNA molecule to recognise four different codons. As an example, Fig. 9.10 shows how a single tRNA can pair with different codons. Here the anticodon 5'-UAA-3' (conventionally written in the 5' to 3' direction) can recognise both leucine codons UUA and UUG.

Fig. 9.10 An example of base-pairing 'wobble.' Here a single leu-tRNA recognises two different leucine codons.

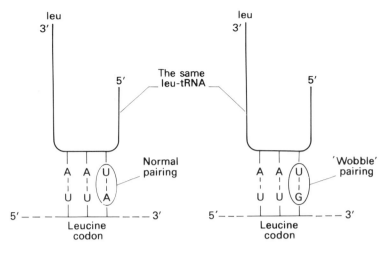

Table 9.2 A summary of base pairing 'wobble.'

Base at 5′ end of anticodon	pairs with	Base at 3′ end of codon
G		U or C
C		G
A		U
U		A or G
I (inosine)		A, U, or C

MUTATIONS AND THE GENETIC CODE

In a previous Topic, we discussed the mechanism of action of a variety of chemical mutagens. In brief they bring about base-pair transitions, deletions or additions, and the effect of these changes depends on whether or not an amino acid change is produced in the polypeptide that diminishes or abolishes its function. Three basic types of mutations can be defined with this in mind. All three produce a non-functional protein which results in a detectable mutant phenotype:

Missense mutation. Here a single base-pair change in the DNA leads to a different codon in the mRNA which may code for a different amino acid in the polypeptide. If an altered amino acid results the effect on the function of the polypeptide will depend on the location of the amino acid in the chain.

Nonsense mutation. It is possible that a single base-pair change in the DNA will lead to the generation of a premature chain termination codon in the mRNA. The result will be a shorter-than-normal polypeptide chain which is likely to be non-functional.

Frameshift mutation. A base-pair deletion or addition in the DNA will produce a mRNA in which the reading frame is shifted one space. Past this point in the mRNA, the codons will code for a completely different set of amino acids thus resulting, most likely, in a non-functional protein (unless the change is very close to the C-terminal end of the polypeptide). This was discussed in more detail earlier in this Topic.

REFERENCES

Cold Spring Harbor Symposia for Quantitative Biology, vol. 31, 1966, *The Genetic Code.* Cold Spring Harbor Laboratory, New York.

Crick F.H.C. 1966. Codon-anticodon pairing: the wobble hypothesis. *J. Mol. Biol.* **19**: 548–555.

Crick F.H.C., L. Barnett, S. Brenner & R.J. Watts-Tobin. 1961. General nature of the genetic code for proteins. *Nature,* **192**: 1227–1232.

Garen A. 1968. Sense and nonsense in the genetic code. *Science* **160**: 149–159.

Khorana H.G. 1966–67. Polynucleotide synthesis and the genetic code. *Harvey Lectures* **62**: 79–105.

Khorana H.G., H. Buchi, H. Ghosh, N. Gupta, T.M.

Jacob, H. Kossel, R. Morgan, S.A. Narang, E. Ohtsuka & R.D. Wells. 1966. Polynucleotide synthesis and the genetic code. *Cold Spring Harbor Symp. Quant. Biol.* **31**: 39–49.

Morgan A.R., R.D. Wells & H.G. Khorana. 1966. Studies on polynucleotides, LIX. Further codon assignments from amino acid incorporation directed by ribopolynucleotides containing repeating trinucleotide sequences. *Proc. Natl. Acad. Sci. USA,* **56**: 1899–1906.

Nichols J.L. 1970. Nucleotide sequence from the polypeptide chain termination region of the coat protein cistron in bacteriophage R17 RNA. *Nature* **225**: 147–151.

Nirenberg M. & P. Leder. 1964. RNA code words and protein synthesis. *Science* **145**: 1399–1407.

Nirenberg M. & J.H. Matthaei. 1961. The dependence of cell-free protein synthesis in *E. coli* upon naturally occurring or synthetic polyribonucleotides. *Proc. Natl. Acad. Sci. USA,* **47**: 1588–1602.

Nirenberg M., T. Caskey, R. Marshall, R. Brimacombe, D. Kellog, B. Doctor, D. Hartfield, J. Levin, F. Rottman, S. Pestka, M. Wilcox & F. Anderson. 1966. The RNA code and protein synthesis. *Cold Spring Harbor Sump. Quant. Biol.* **31**: 11–24.

Streisinger G., Y. Okada, J. Emrich, J. Newton, A. Tsugita, E. Terzaghi & M. Inouye. 1966. Frameshift mutations and the genetic code. *Cold Spring Harbor Symp. Quant. Biol.* **31**: 77–84.

Topic 10
Phage Genetics

OUTLINE
Phage T4
   life cycle
   genetic recombination
   genetic fine structure
   unit of function
Phage $\phi \times 174$
   life cycle
   genetic organisation.

We now move from the area of molecular genetics to more classical areas of genetics, centering on transmission genetics. With the background already presented in molecular genetics, it should be possible to consider all of the material of this and future Topics in molecular terms.

PHAGE T4

In a previous Topic we talked about phages, their chromosomes, and how they infect and lyse bacterial cells. In this Topic we will concentrate our attention on phages and in particular the phage T4 which is one of the virulent phages, meaning that it will always enter the lytic cycle when it infects its host, *E. coli*.

T4 life cycle
In outline, a phage particle makes contact with the host bacterium and, after adsorbing to the wall or membrane by means of the tail spikes and tail fibres, the DNA is injected into the host. The DNA is replicated and transcribed and the resulting phage-specific proteins and DNA assemble into progeny phage. At the same time a bacterial lysing enzyme (lysozyme) is made and eventually this causes the cells to break open, releasing the progeny phage. About 200 progeny phage are produced per infected bacterium.

The life cycle of phage T4 was studied by M. Delbruck in 1940. He reasoned that kinetic analysis of the steps in the life cycle required large numbers of cells in which phage infection occurred synchronously. In Delbruck's experiment, the phages were allowed to infect the bacteria for a short time and then the unadsorbed phages were removed by treatment with antibodies made against the phage particles. The infected bacterial population was incubated at 37°C and, at various times, samples were taken and added to a lawn of bacteria on a petri dish. The plaques that were produced were counted and that number was graphed as a function of time, resulting in the one-step growth curve shown in Fig. 10.1.

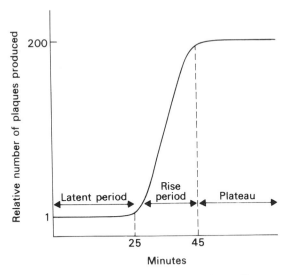

Fig. 10.1 One-step growth curve for phage T4 infection of *E. coli*.

The curve is divisible into three distinct regions:

1. Latent period. During this phase the bacteria have been infected and the biosynthesis of phage components is proceeding. No progeny phage are

106

released (by definition) during this time. Thus, when samples from this period are plated on the lawn of indicator bacteria, it is actually infected bacteria that are dispersed over the plate. Each infected bacterium will eventually lyse, releasing progeny phage which will infect surrounding bacteria on the plate. After a number of rounds of this, a plaque will be apparent on the indicator lawn (Fig. 10.2).

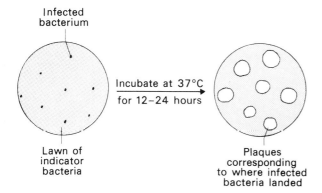

Fig. 10.2 Diagrammatic representation of the production of phage plaques on a bacterial lawn when phage-infected bacteria are spread on it.

2. Rise period. At the beginning of the rise period, some infected cells have lysed, releasing progeny phage. For the next 20 minutes at 37°C the remainder of the infected cells lyse synchronously until all have lysed. Platings during this period include a mixture of infected bacteria and progeny phase.
3. Plateau period. All of the infected bacteria have lysed and thus plating of the culture involves entirely progeny phage.

From this one-step growth experiment, it was concluded that 25–45 min at 37°C are required for one life cycle of phage T4. Further experiments by A. Doermann in which cells were broken open at various times during the latent period, showed that about 12 min are required before any mature phage particles are found. That period involves the synthesis of the phage-specific molecules. From about 12 to 25 min, mature phages are accumulating in the cells. Cell lysis then occurs when the lytic enzyme reaches a high enough concentration.

Genetic recombination in phage T4
Phage T4 is haploid and hence one does not find lethal mutations in this organism. The useful mutations fall into two main classes: firstly, those that result in visible phenotypes, usually altered plaque morphology (the disadvantage here is that relatively few of the 80 or so genes of T4 mutate to give plaque-type mutants); secondly, those that are conditionally-lethal, that is, heat-sensitive or cold-sensitive mutants in which no mature progeny phage particles are produced at the respective non-permissive temperatures. These two groups of mutants have led to the establishment of a fairly complete genetic map of T4 by genetic recombination techniques.

Genetic recombination in T4 was discovered by M. Delbruck and A. Hershey in 1946. They did genetic crosses by infecting *E. coli* cells simultaneously with two types of visible mutants; one, a host-range (*h*) mutant, and the other, a rapid lysis (*r*) mutant. After a round of replication, four types of progeny phage particles were released, namely, the two parental types (*h* and *r*), and two recombinant types (the double mutant *hr* and a wild type). In these genetic studies, it became apparent that recombination occurs between a large pool of phage DNA molecules. This, and the fact that recombination occurs throughout the DNA replication phase of the life cycle, makes the genetic analysis more in the realm of statistics than genetics. Nevertheless, we can at least consider this process at a simple level here (Fig. 10.3).

In the example, genetic recombination will occur at random along the length of the T4 genome. Any recombination event that occurs between the two gene sites will generate two recombinant phage types, *ab* and + +, in this case. If recombination does not occur between the genes, two parental progeny phage types, *a*+ and +*b*, will be produced. Since the recombination event is random, the probability of it occurring between two genes under investigation is a function of how far apart the two genes are on the DNA. Thus if 2% of the phages are recombinant types, we would say that the two genes are relatively close to one another (i.e. two map units). This type of analysis, of course, requires procedures to distinguish all four progeny types.

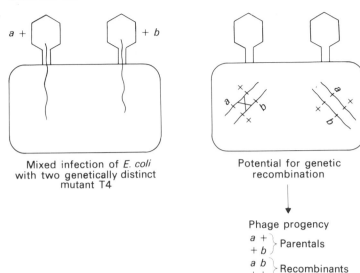

Fig. 10.3 Scheme for mapping genes in bacteriophages.

By doing a number of two-factor crosses such as that described, or by doing three-factor crosses (e.g. *abc* × +++), the genetic map was constructed. This map is circular, although the chromosome is known to be linear. This is a result of circular permutation of the genome as was discussed in an earlier Topic.

The genetic analysis procedures described for phage T4 are also applicable to other bacteriophages such as lambda, T2, T7, etc. All that is needed are genetic mutants with phenotypes that are distinguishable alone and in combinations; then two- and three-factor crosses may be carried out. Particularly useful are temperature-sensitive host-range, morphological and nonsense mutants. Such genetic analyses have led to the identification of most of the genes of the respective phage genomes and the construction of detailed genetic maps.

Genetic fine structure

The classical view of a gene was that it is the unit of genetic material that can mutate to alternative forms, recombine with other genes, and function in the organism. Indeed, many people considered genes to be 'indivisible beads on a string' with, for example, recombination occurring between the beads. This classical view was drastically modified in the 1950s when S. Benzer did a series of elegant experiments designed to elucidate the fine structure of the gene. More specifically, he set out to define the unit of mutation, the unit of recombination and the unit of function at the molecular level. For his studies, Benzer chose the phage T4 because of the large number of progeny produced in a short time. He worked with the *rII* mutants that we discussed before, since they produced plaques with a morphology easily distinguishable from that of the wild type, and because of the host-range properties; namely, the inability of *rII* mutants to grow in *E. coli K12 (lambda)* (Table 10.1).

Table 10.1 Host-range and plaque morphology phenotypes of r^+ and *rII* strains of T4.

	Plaque morphology on	
Strain	*B*	*K12(λ)**
r+ (+)**	+ type (turbid)	+
rII	r-type (clear)	no plaques

* This will be abbreviated to *K* throughout the discussion.

** The symbol + is an abbreviation for wild type.

Benzer realised that the growth defect of *rII* on *K* was a powerful selective tool for detecting the presence of a very small proportion of r^+ genotypes within a large population of *rII* mutants. In particular it is possible to score selectively for the very rare r^+ recombinants that arise in genetic crosses between two very closely linked *rII* mutants.

Initially 60 *rII* mutants were isolated and these were crossed in all pairwise combinations. The progeny produced from each cross were analysed for the frequency of recombination in order to construct a genetic map. The experimental procedure was as shown in Fig. 10.4.

The progeny phage, then, are plated on both *B* and *K*. Appropriate dilutions of the phage suspension are made in order to get a reasonable number of plaques on the plates so that mathematical accuracy can be ensured. The dilution factor is taken into consideration in the calculations, of course. All of the progeny phage will produce plaques on *B* and thus the number of plaques gives information about the total number of progeny, usually expressed as plaque-forming units per millilitre (pfu/ml) of suspension. On the other hand, the plaques growing on *K* can only be the wild-type recombinants. In diagrammatic form, a hypothetical cross is presented in Fig. 10.5.

Four types of progeny phages will be produced:

$\left.\begin{array}{l} r7 \\ r12 \end{array}\right\}$ Parentals—clear plaques on *B*

$\left.\begin{array}{l} r7, r12 \\ r^+ \end{array}\right|$ Recombinants — *r7, r12* gives clear plaque on *B*; r^+ grows on *K* and on *B*

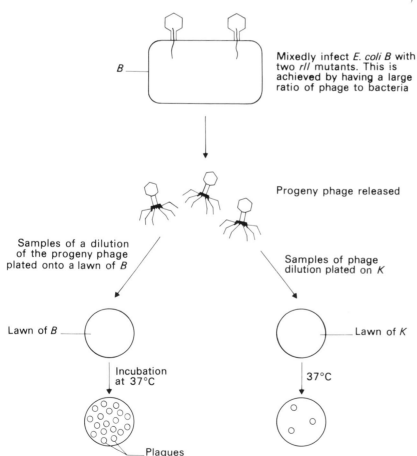

Mixedly infect *E. coli B* with two *rII* mutants. This is achieved by having a large ratio of phage to bacteria

Progeny phage released

Samples of a dilution of the progeny phage plated onto a lawn of *B*

Samples of phage dilution plated on *K*

Lawn of *B*

Lawn of *K*

Incubation at 37°C

37°C

Plaques

Fig. 10.4 Protocol for determining the number of wild-type recombinants ($= \frac{1}{2}$ the total number of recombinants) produced in a cross between two *rII* mutants of T4.

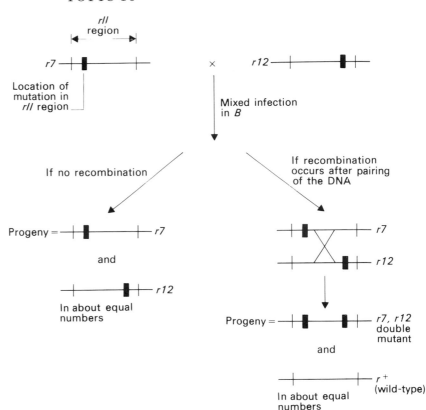

Fig. 10.5 Chromosomal diagram of the production of recombinants in a cross of two *rII* mutants, *r7* and *r12*, of phage T4.

Thus, the total number of progeny per millilitre of suspension is calculated from the number of plaques produced on *B*, and the number of wild-type recombinants is determined from the number of plaques on *K*. Each cross-over event that produces an r^+ recombinant will also generate an *r7, r12* double mutant which produces clear plaques on *B* but does not grow on *K*. Map distance between two mutations is given by the percentage of recombinant progeny among all progeny from crosses of two mutants. In the case of these *rII* crosses, the number of recombinants is twice the number of r^+ plaques on *K*. In our example, then, the map distance between *r7* and *r12* is given by:

$$\frac{2 \times \text{number of } r^+ \text{ plaques on } K}{\text{Total plaques on } B} \times 100\%$$

the number obtained being given in map units.

From the recombination data accumulated from the crosses of the *rII* mutants, Benzer constructed a preliminary fine structure genetic map in which the mutations were arranged in a linear order. Among the first set of 60 mutants, some pairs of crosses were encountered that produced no r^+ recombinants. This was interpreted to mean that the two mutations were extremely close or perhaps altered the same nucleotide pair. (The latter conclusion is a retrospective one since the work was done to define the lower limit for recombination). Such mutations are considered to be *homoalleles*. Those pairs that did produce r^+ recombinants were considered to carry *heteroallelic* mutations. From the map, it was ascertained that the lowest frequency with which r^+ recombinants were found among progeny of crosses of two *rII* mutants was 0.01%, which was much higher than the 0.0001% limit of resolution in the experiment. Given this figure it is possible to make a rough calculation of the physical distance between the two mutations on the DNA. The percentage of recombination equals $2 \times 0.01\%$, corresponding to 0.02 map units between the two mutations. The total T4 genome consists of 700 map units and 2×10^5 nucleotide pairs. The 0.02 map units, then, is equivalent to:

$$\frac{0.02}{700} \times 2 \times 10^5 = 6 \text{ nucleotide pairs.}$$

At that time, this was the minimum recombination distance found. We now know that genetic recombination can occur between adjacent nucleotide pairs on the DNA and thus a nucleotide pair is the unit of recombination.

Unit of mutation

The smallest mutational unit is the nucleotide pair. A change of a single nucleotide pair is called a point mutation and each should occupy only a single site on the genetic map. Each point mutation in the *rII* region should be able to yield *r+* recombinants in crosses with other point mutations, unless both are changed in the same nucleotide pair. Also, point mutations should be revertible to the wild-type state.

Some *rII* mutants do not behave as point mutants since they do not yield *r+* recombinants in crosses with two or more *rII* mutants previously identified as nonallelic point mutants. These *rII* mutants also do not revert. M. Nomura and S. Benzer discovered that these new types of *rII* mutants involved deletions of several nucleotide pairs of the DNA. Genetic experiments were used to provide proof that deletions were involved and an example follows. The mutant *r1695* was proposed to be a deletion since it produced no wild-type recombinants in crosses with known point mutations and because it did not revert. In the experiment two point mutations (*r168* and *r924*) were used that mapped on opposite sides of *r1695* but were not covered by it (i.e. no *r+* recombinants were produced in crosses of either with *r1695*). The frequencies of *r+* recombinants were as shown in the map presented in Fig. 10.6.

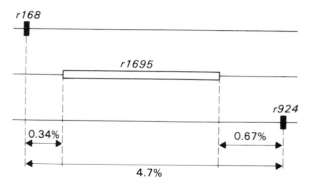

Fig. 10.6 Genetic structure of three *rII* mutants. *r168* and *r924* are two point mutants 4.7 map units apart, and *r1695* has a deletion of most of the DNA between the two.

Then, by genetic crosses, the double mutants *r168,r1695* and *r1695,r924* were constructed. These two were crossed in *E. coli* B and the progeny were isolated. It was reasoned that if *r1695* is a deletion (as we have assumed), then the distance between *r168* and *r924* should be shorter than the 4.7 map units determined from *r168* × *r924* crosses. The results showed that the distance between the two was now 1.0 map units, thus showing that *r1695* is a deletion. Indeed 800 nucleotides have been lost in the deletion strain (Fig. 10.7).

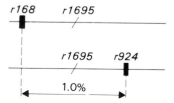

Fig. 10.7 Map distance between *r168* and *r924* is 1.0 map units in a cross of *r168,r1695* × *r1695,r924* providing evidence that *r1695* is a deletion of 3.7 map units of DNA (see Fig. 10.6 for comparison).

Note that in this experiment it is not possible to generate *r+* recombinants since every progeny type has in common the *r1695* deletion, and thus they all have the *rII* phenotype. Therefore all progeny from the cross had to be cloned and backcrossed to the three grandfathers *r168*, *r1695*, and *r924* to determine the genotypes so that parental and recombinant types could be sorted out and map distance computed (Table 10.2).

Table 10.2 Results of crossing progeny of *r168 r1695* ×
r1695 r924 with grandparents. – = no *r+* recombinants;
+ = *r+* recombinants formed.

	Parentals		Recombinants	
	r168, r1695	*r1695, r924*	*r1695 r168, r1695, r924*	
r168	–	+	+	–
r1695	–	–	–	–
r924	+	–	+	–

The patterns of *r+* recombinants formed or not in
the crosses indicated the genotype of the progeny
phage involved.

Thus, mutation in the phage genome can involve
a change in a single nucleotide pair or the deletion
of up to many hundreds of nucleotide pairs.

Mapping using deletion mutants
Deletion mutants made it possible to map mutations within the *rII* region (or in any other gene)
in a way that made it unnecessary to cross every
mutant with every other mutant. Benzer used a
series of overlapping deletions whose ends were
mapped by crosses with point mutants of known
locations. In the early experiments, the *rII* region
was divided into a number of segments, each of
which was defined by the length of the map covered
by one particular deletion but not by another
(segments A1 to A6 and B in Fig. 10.8).

Given such a set of overlapping deletions, it is
relatively simple to make a general placement of
each new *rII* point mutant into its appropriate
segment by crossing the mutant to each deletion
mutant and establishing the deletions with which
the mutant does and does not produce wild-type
(*r+*) recombinants. For example, a point mutation
in region A4 would give *r+* recombinants with
deletions V, VI and VII, but not with the rest.
Obviously in 'real life', the argument is made in
the reverse direction to localise the mutation. Once
mutations were localised to the seven regions,
Benzer used a number of other deletions which
ended at various points within the seven segments
discussed. In two further rounds of crosses with
deletions involving increasing levels of subdiv-
ision, then, a point mutation could be localised
to one of 47 distinct segments of the *rII* region.
The final step in constructing a fine structure map
was then pairwise crosses of all mutants within
each of these segments to establish mutation order
and map distance.

In summary the recombination test for *rII*
mutants is:

1. Cross two *rII* point mutants in *B*.
2. Test sample of progeny on *B* to determine total
number of progeny.
3. Test sample of progeny on *K12* (*lambda*) to
determine number of *r+* recombinants.
4. Map distance = 2 × frequency of *r+* recombinants.

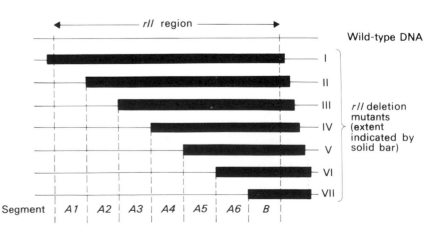

Fig. 10.8 Division of *rII* region
into segments by overlapping
deletion mutants.

Unit of function

The three basic aspects of genes are mutation, recombination and function. Here we shall discuss the role of a gene as the unit of function. As we know, many genes code for polypeptides which may have a structural, regulatory, or functional role in the cell. The properties of a particular polypeptide are derived from the three-dimensional shape and the amino acid sequence of the molecule. A mutation at a particular genetic site can therefore result in a mutant phenotype because it often results in an amino acid change at a corresponding site in the polypeptide, thereby potentially altering the latter's function. If the function of a polypeptide is altered, this may involve either a complete loss of activity or only a partial loss. This is certainly the case with point mutants. Deletion mutants usually result in a non-functional polypeptide.

Benzer designed experiments to elucidate the unit of function of the *rII* region. The fact that he had a number of *r* mutants that produced the same phenotype and mapped close to one another did not necessarily mean that the mutations they carried were all in the same functional unit (gene). To test the number of functional units in the *rII* region, Benzer adapted to his system the 'cis-trans' or *complementation test* previously developed by E. Lewis in his work with *Drosophila*. For the purposes of the discussion, we will start by stating that the *rII* region consists of two units of function, A and B, and a mutation of either will produce the *rII* phenotypes. Obviously this conclusion followed the experiments that will now be described.

The initial observation of importance for this test was that if a *K12 (lambda)* cell is infected with an *rII* mutant and an *r+* phage at the same time, both genomes will replicate and both phage types will be found among the progeny (Fig. 10.9).

The explanation for this is that the normal *rII* gene (A in the example) of the wild type is able to supply the function necessary for growth in *K* of both phages. This result is obtained for *rII* mutations either in A or in B.

The next step is to infect *K* with pairs of *rII* mutants to see if any progeny phage can be produced. If that occurs, the two mutants are said to have complemented, and the two mutations must be in different functional units. If no progeny phage are produced, the mutants do not complement, and the mutations are in the same functional unit (Fig. 10.10).

In example (a), complementation occurs since the strain with a mutation in A makes a functional B product and the B mutant makes a functional A product. Between the two mutants, then, both products necessary for growth in *K* are produced and thus progeny phage can be assembled and released. Each mutant makes up for the other's defect. In example (b), no complementation occurs since both strains have in common the lack of the A function, thus preventing growth in *K*.

On the basis of this sort of test, Benzer found that *rII* mutants fall into the two functional groups, A and B. That is, all *rIIA* mutants complement all *rIIB* mutants but *rIIA* mutants do not complement *rIIA* mutants and *rIIB* mutants do not complement other *rIIB* mutants. These two groups of mutations can be assigned definite positions on the genetic map of the *rII* region. No A mutants are found in the B area and vice versa (Fig. 10.11).

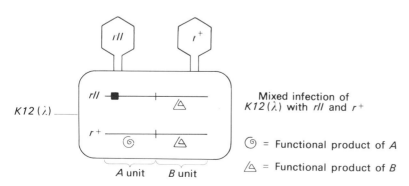

Mixed infection of
K12(λ) with *rII* and *r+*

◎ = Functional product of *A*

△ = Functional product of *B*

Fig. 10.9 Control experiment for testing for complementation of *rII* mutants. Here the non-permissive host, *K12*(λ), is co-infected with *r+* and a *rII* mutant. Phages of both types are produced showing that *r+* products permit the replication of both genomes.

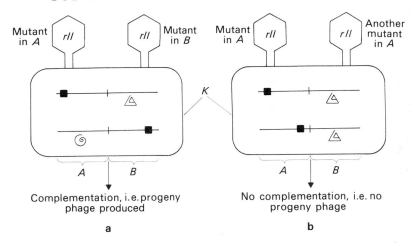

Fig. 10.10 Complementation tests for determining the units of function of *rII* mutants. If the non-permissive host *K12(λ)* is co-infected with an *rII* phage carrying a mutation in the *A* cistron and with an *rII* phage with a mutation in the *B* cistron, the two strains complement each other and progeny are produced. Co-infection with two different *rII A* or with two different *rII B* phages results in no progeny, i.e. complementation does not occur.

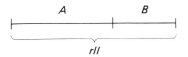

Fig. 10.11 Organisation of the *rII* region into two units of function (i.e. *A* and *B* cistrons).

A and B are called complementation groups. Since the cis-trans test was used in the studies, Benzer called each a *cistron*. A cistron is a pseudonym of a gene, in some senses, and is considered to code for a polypeptide. The *rIIA* cistron is 6 map units and 1700 nucleotide pairs long, and the *rIIB* cistron is 4 map units and 1100 nucleotide pairs long.

It is remarkable that, throughout this work, Benzer did not know the nature of the polypeptides coded by the A and B cistrons. It now seems likely that the cistrons code for proteins which bind to the cell membrane of the host bacteria and stimulate them to break open more easily.

PHAGE φ × 174

We shall conclude this Topic by briefly discussing the genetic organisation of bacteriophage φ × 174 since it has some rather unusual features.

φ × 174 is an icosahedral bacteriophage and thus roughly resembles the head of phage T4. The genetic material of φ × 174 is circular, single-stranded DNA and there is no redundancy of genes as there is in phages T2 and T4 for example.

Life Cycle of φ × 174

After infecting *E. coli*, the DNA of the phage attaches to the host cell membrane and a double-stranded circular DNA is produced by the production of the complementary DNA strand in polymerisation reactions catalysed by host cell enzymes. The infecting DNA strand is called the + strand since it is of the same 'sense' as the phage-coded mRNA molecules. Thus the complementary strand that is made is the − strand and the membrane-attached double-stranded DNA is called *replicative form 1* (RF1).

Once the RF1 has been produced, mRNAs are transcribed from it and these have the + sense like the infecting DNA strand. The mRNAs code for the proteins required for replication of phage DNA and the components needed for the assembly of progeny bacteriophage particles. In addition the parental RF1 repeatedly replicates by the rolling-circle method to produce many copies of non-membrane bound progeny RF molecules called RF2s. Then, again by the rolling-circle method of replication, the RF2s generate linear viral DNA which are + as was the original infecting strand. These linear progeny viral strands are circularised by ligase and then become associated with proteins to form the mature phage particles which are released from the cell when the host lyses.

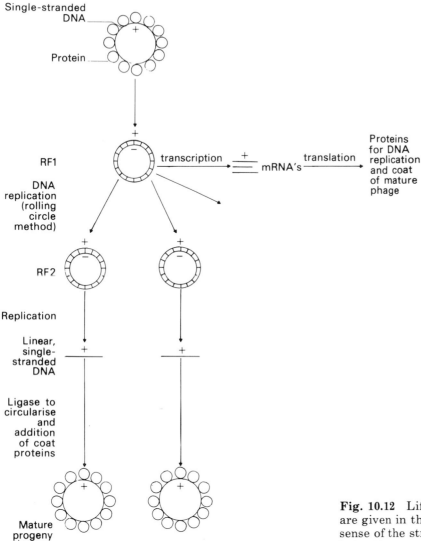

Single-stranded DNA

Protein

RF1

DNA replication (rolling circle method)

RF2

Replication

Linear, single-stranded DNA

Ligase to circularise and addition of coat proteins

Mature progeny $\phi \times 174$

transcription

mRNA's

translation

Proteins for DNA replication and coat of mature phage

Fig. 10.12 Life cycle of phage $\phi \times 174$. The details are given in the text. The + and − designate the sense of the strand, the original infecting strand being the same sense as the mRNAs produced (i.e. +).

Genetic organisation of $\phi \times 174$

Phage $\phi \times 174$ is a relatively small bacteriophage. Its single-stranded DNA genome is approximately 5375 nucleotides long and contains 9 known genes (as defined by genetic mutations) which code for the nine phage-specific proteins that have been identified in *E. coli* following infection by $\phi \times 174$. Protein A functions in double-stranded DNA replication, proteins B, C, and D function in the production of single-stranded progeny, proteins F,

G, H, and J are components of the phage particle (called the capsid), and protein E is responsible for the lysis of the host cell. The order of the genes on the genetic map is A B C D E J F G H, and the origin for DNA replication is located within gene A.

When the number of nucleotides needed to code for the known amino acid sequences of the nine proteins was determined, it was realised that the entire $\phi \times 174$ genome had approximately 700 too few nucleotides. In order to resolve this apparent paradox, F. Sanger and his colleagues applied the relatively new, rapid DNA sequencing techniques and consequently determined the entire nucleotide sequence of $\phi \times 174$. Having done that, they correlated the nucleotide sequence information and formed the following conclusions:

1. The protein coding sequence of gene B is totally contained within gene A, with the two reading frames of the mRNAs they code for staggered by one nucleotide. The A gene itself extends 85 nucleotides (28 amino acids) beyond the end of the B gene. The genes A and B, then, are examples of overlapping genes.
2. Gene C is located between genes B and D and this gene illustrates a second type of overlap. Specifically, the sequence of gene C for the initiation codon of the mRNA overlaps by one nucleotide the sequence coding for the termination codon of gene A as shown in Fig. 10.13a. A similar one nucleotide overlap is shown for the comparative sequences for genes D and J (Fig. 10.13b).

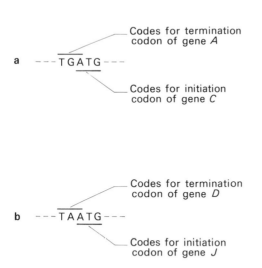

Fig. 10.13 Examples of overlapping genes in phage $\phi \times 174$. **a** The DNA sequence shown contains the sequence for the termination codon of gene A overlapping by one nucleotide the sequence for the initiation codon of gene C. **b** Overlap by one nucleotide of the sequences for the termination codon of gene D and for the initiation codon of gene J.

3. Genes D and E provide a second example of complete gene overlap with the sequence of the latter completely contained within the sequence of the former. Again the genes differ in reading frame. The products of the two genes actually have quite different functions within the cell, i.e. D codes for a protein needed for the production of single-stranded viral DNA and E codes for a lysing protein. The two genes behave completely independently by genetic tests. (Indeed, this was part of the evidence that two distinct reading frames are involved.) Thus nonsense mutants in E usually result only in a missense change in D or, because of code degeneracy, perhaps no change at all. The same situation has been shown also for nonsense mutations in gene E.
4. After the termination sequence of gene J there is a 39 nucleotide gap before the initiation of gene F. A 111 nucleotide gap is present between F and G, with a 66 nucleotide gap between H and A. The functions, if any, of these gaps are unknown. Genes F, G and H do not share nucleotides with any other genes.

In summary then, $\phi \times 174$ provides very good examples of overlapping genes, both where one gene is wholly within another, and where only one nucleotide is shared. This situation is not unique among organisms. For example, T. Platt and C. Yanofsky have found an example of the latter type of overlap in the mRNA of the tryptophan operon of *E. coli*. In addition, the DNA sequence of simian virus 40 (SV40), a double-stranded DNA virus that infects mammalian cells, has been determined by V.B. Reddy and his colleagues. Their sequence data show that this virus also has examples of limited overlap of genes. Further, the mRNAs coded for by SV40 are produced by RNA–RNA splicing of parts of long transcripts as has already been described for Ádenovirus-2 in Topic 7.

REFERENCES

Barrell B.G., G.M. Air & C.A. Hutchison. 1976. Overlapping genes in bacteriophage $\phi \times 174$. *Nature* **264**: 34–41.

Benbow, R.M., C.A. Hutchison, J.D. Fabricant & R.L. Sinsheimer. 1971. Genetic map of bacteriophage $\phi \times 174$. *J. Virol.* **7**: 549–558.

Benzer S. 1959. On the topology of the genetic fine structure. *Proc. Natl. Acad. Sci. USA*, **45**: 1607–1620.

Benzer S. 1961. On the topography of the genetic fine structure. *Proc. Natl. Acad. Sci. USA*, **47**: 403–415.

Delbruck M. 1940. The growth of bacteriophage and lysis of the host. *J. Gen. Physiol.* **23**: 643–660.

Doermann A.H. 1952. The intracellular growth of bacteriophages I. Liberation of intracellular bacteriophage T4 by premature lysis with another phage or with cyanide. *J. Gen. Physiol.* **35**: 645–656.

Eisenberg S., J.F. Scott & A. Kornberg. 1976. Enzymatic replication of viral and complementary strands of duplex DNA of phage $\phi \times 174$ proceeds by separate mechanisms. *Proc. Natl. Acad. Sci. USA*, **73**: 3151–3155.

Ellis E.L. & M. Delbruck. 1939. The growth of bacteriophage. *J. Gen. Physiol.* **22**: 365–384.

Hayes W. 1968. *The Genetics of Bacteria and Their Viruses*, 2nd edn. Wiley, New York.

Hershey A.D. and R. Rotman. 1949. Genetic recombination between host-range and plaque-type mutants of bacteriophage in single bacterial cells. *Genetics* **34**: 44–71.

Reddy V.B., B. Thimmappaya, R. Dhar, K.N. Subramanian, B.S. Zain, J. Pan, P.K. Ghosh, M.L. Celma and S.M. Weissmann. 1978. The genome of simian virus 40. *Science* **200**: 494–502.

Sanger F., G.M. Air, B.G. Barrell, N.L. Brown, A.R. Coulson, J.C. Fiddes, C.A. Hutchison, P.M. Slocombe & M. Smith. 1977. Nucleotide sequence of bacteriophage $\phi \times 174$. *Nature* **265**: 687–695.

Smith M., N.L. Brown, G.M. Air, B.G. Barrell, A.R. Coulson, C.A. Hutchison & F. Sanger. 1977. DNA sequence at the C terminal of the overlapping genes A and B in bacteriophase $\phi \times 174$. *Nature* **265**: 702–705.

Topic 11
Bacterial Genetics

OUTLINE
Conjugation in *E. coli*
   the sex factor
   Hfr strains
   transfer of DNA from donor to recipient
   conjugation mapping
Transduction
   generalised transduction
   specialised transduction
Transformation
   mechanism of transformation
   determination of gene linkage by transformation

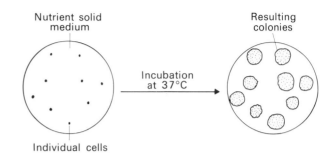

Fig. 11.1 Diagrammatic representation of the production of bacterial colonies on solid culture medium.

We now turn our attention from bacteriophages to bacteria which are much more complex in a genetic sense. In this Topic we will consider three ways that genetic material can exchange between different bacteria. These are *conjugation, transduction,* and *transformation.* Before we discuss these, it is appropriate to consider the ways of growing a bacterium such as *E. coli.*

Bacteria can be cultured either on a solidified **surface (usually containing agar)** or in a liquid medium. In both cases nutrients, necessary salts and minerals must be present. In liquid culture, the bacteria multiply exponentially until the nutrients are exhausted or until toxic products accumulate. The number of bacteria present at any time in a liquid culture can be determined quite simply. A small sample can be pipetted onto a petri plate containing a solid nutrient medium and spread over the surface using a spreader. Thus individual bacteria will be distributed evenly over the plate and, as each cell divides during the incubation period, a clone is produced that remains together in a clump called a *colony.* Each cell will give rise to a discrete colony (Fig. 11.1).

CONJUGATION IN *E. COLI*

Sexual conjugation in *E. coli* was discovered by J. Lederberg and E. Tatum and hinged on the appearance of prototrophs, i.e. $a^+ b^+ c^+ d^+ e^+ f^+$ (all wild-type alleles for genes a to f) from a mixed culture of two kinds of auxotrophs: $a^- b^- c^- d^+ e^+ f^+$ (requires a, b and c for growth) and $a^+ b^+ c^+ d^- e^- f^-$ (requires d, e and f for growth). Since three auxotrophic mutations were present in each strain, it was extremely unlikely that the prototrophs arose by reversion of one or other of the strains. Also, it was possible to use a U-tube apparatus to show that cell contact is required for prototroph formation (Fig. 11.2).

In this experiment nutrient medium is present in both sides of the tube and the two types of bacterial cells are separated by a filter through which medium, but not bacteria, can move. By alternately sucking and blowing on the side tube, the medium can be moved between compartments. As long as the filter is present, however, *no* prototrophs are formed. They will occur once the filter is removed thus indicating the necessity of cell contact for this phenomenon.

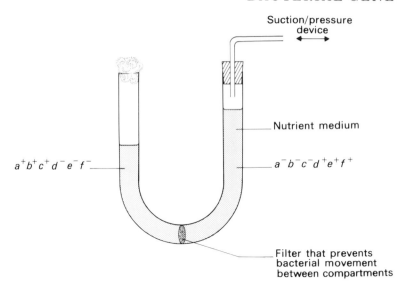

$a^+b^+c^+d^-e^-f^-$

$a^-b^-c^-d^+e^+f^+$

Fig. 11.2 Use of U-tube apparatus to demonstrate that cell contact between genetically-distinct *E. coli* strains is needed for prototroph formation.

The sex factor

W. Hayes proposed the hypothesis that the genetic transfer described was mediated by some kind of infectious vector that is found in so-called 'donor' (D) cells. The vector facilitates the transfer of genes from the donor cells to recipient (R) cells under certain conditions. The D and R cells differ by the presence (F^+) or absence (F^-), respectively, of an extrachromosomal sex or F factor. The Ds transmit the F factor in high frequency to R cells, which then become Ds. This transmission of F is independent of the transmission of chromosomal genes of the host (Fig. 11.3).

The F factor itself is circular, double-stranded DNA consisting of about 100 000 nucleotide pairs (about $\frac{1}{40}$ that of the host). The F^+ cell (or donor) contains one copy of the F factor per cell and this replicates in concert with replication of the host cell chromosome. For its replication, the F factor attaches to a unique site on the bacterial membrane. The F factor contains genes that result in the formation of hairlike cell-surface components called F-pili on cells containing F. The pilus is thin and flexible and may be very long. Other genes are responsible for forming the conjugal unions between D and R cells that are necessary for transfer of the F factor. Once that has occurred, replication

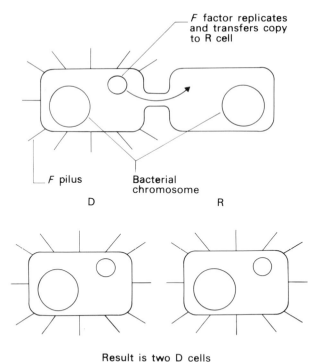

Fig. 11.3 Conjugation between F^+ (donor) and F^- (recipient) cells of *E. coli* results in transfer of a copy of the F factor and the conversion of the F^- to the F^+ state.

of the *F* factor independently of the host chromosome – accompanied by the movement of a copy of the *F* factor through the conjugation tube – results in the conversion of the *F⁻* recipient to an *F⁺* donor.

Hfr strains

L. Cavalli-Sforza and W. Hayes discovered that some *F⁺* populations gave rise to donor cells which could transmit chromosomal genes at high frequency. These were called high frequency recombinant, or *Hfr*, strains and physically they resemble *F⁺* cells in the presence of F-pili. In crosses of *Hfr* strains with *F⁻* strains, there is a gradation in frequencies in which various donor loci appear among recombinant cells. In addition, the recombinant progeny remain *F⁻* (Fig. 11.4).

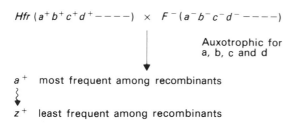

$$Hfr\,(a^+b^+c^+d^+----) \times F^-(a^-b^-c^-d^----)$$

Auxotrophic for a, b, c and d

a^+ most frequent among recombinants

z^+ least frequent among recombinants

Fig. 11.4 Hypothetical cross between an *Hfr* strain and a *F⁻* strain to show the sequence in which host genes are transferred from the donor to the recipient.

From this it was concluded that *Hfr* gene a^+ entered into the *F⁻* first and therefore had the greatest chance of recombination. The rest of the genes b^+ to z^+ entered sequentially in that order and their frequency of recombination was determined by the order of entry of the genes.

The difference between *F⁺* and *Hfr* cells is that, in the latter, the *F* factor has become integrated into the bacterial chromosome. Since the *F* genes are still in the cell, the *F* functions are all present, including the capacity to form unions with *F⁻* cells, and the ability to transfer the *F* factor to the *F⁻* cell. Now, since *F* is integrated into the chromosome, this process is accompanied by transfer of the host chromosome across the conjugation tube.

The integration of the *F* factor occurs by a genetic recombination event involving a single cross-over. A given *Hfr* can revert to the *F⁺* state with an extrachromosomal *F* factor by the reverse of this process (Fig. 11.5).

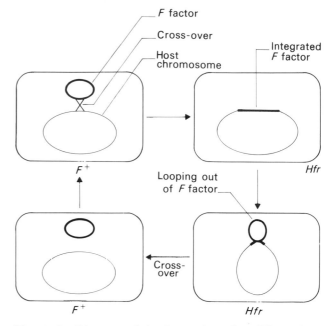

Fig. 11.5 Diagram of the formation of an *Hfr* strain by integration of the *F* factor, and 'reversion' of the *Hfr* to the *F⁺* state by the reverse process.

The *F* factor is able to integrate a number of sites over the host chromosome and thus the order of genes transferred to an *F⁻* cell varies from *Hfr* strain to *Hfr* strain. Indeed, since the *F* factor has polarity, different *Hfrs* transfer genes in different orientations (Fig. 11.6).

On the diagram, the arrowheads indicate a number of the known *Hfr* cells, and the direction of transfer of host genes. For example, *HfrH* (*H* stands for Hayes) transfers genes in the order $a^+\,b^+\,c^+\,d^+\,e^+\,----$, whereas *HfrJ4* which has an F integrated close to the *HfrH* site, transfers genes in the order $u^+\,t^+\,s^+\,r^+\,q^+\,----$, and so on.

For a given *Hfr* strain, not all of the host genes can be transferred to the *F⁻* since the conjugation tube is quite fragile and there is a good chance the

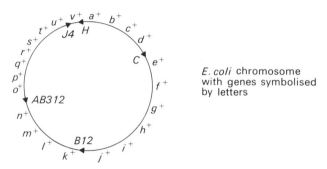

E. coli chromosome with genes symbolised by letters

Fig. 11.6 Stylised representation of the *E. coli* chromosome with letters designating genes and arrowheads signifying the points of integration and orientation of the *F* factor to produce various *Hfr* strains.

cells will break apart before the 90 min at 37°C that would be required for transfer of the whole chromosome. Thus to build the complete map of the *E. coli* chromosome, a number of *Hfr* strains with different start points and orientations were used. In the start of the process of gene transfer, the *F* factor breaks along its length so that part of the *F* factor is transferred first whereas the rest is at the tail end of the *E. coli* chromosome. It is for this reason that the recipients usually remain *F⁻* in phenotype since only after transfer of the entire chromosome would a complete *F* factor be found in the recipient cell.

Transfer of DNA from donor to recipient

F⁺ × F⁻ matings.
The two cells form a conjugal union and a copy of the *F* factor is transferred to the *F⁻* cell by replication of the *F* factor. During this event, a single

strand of the *F* factor passes through the conjugation bridge and the complementary strand is synthesised in the recipient, eventually resulting in a circularised *F* factor and a conversion of the *F⁻* to an *F⁺* (Fig. 11.7). Since the *F* factor is small (relative to the host chromosome) there is a likelihood of transfer of the complete factor copy before the cells pull apart.

This is an example of the rolling-circle model of DNA replication proposed by W. Gilbert and D. Dressler, which is summarised in Fig. 11.8.

Hfr × F⁻ matings.
Essentially the same sequence of events occurs as in *F⁺ × F⁻* matings. A conjugal union is formed and the *F* factor breaks into two, one end passing through the conjugation tube followed by chromosomal genes (Fig. 11.9). Only one strand is transferred, the complementary one being synthesised in the *F⁻* recipient cell. If the recipient has different alleles of the genes being transferred, recombinant cells can be detected and studied. The recombination event occurs by crossing-over. Since the rest of the *F* factor is not transferred until all the chromosome has moved across the bridge (which is very unlikely), the *F⁻* rarely becomes an *Hfr*.

In Fig. 11.9, the cross-overs shown would generate recombinants with an a⁺ phenotype and all descendants of the *F⁻* cell would have that phenotype. During the transfer, the donor chromosome can break anywhere so only partial transfer occurs. The chance of breakage is random and thus the genetic recombination between donor and recipient chromosomes is limited to the length of the chromosome transferred during the 'zygote' phase.

Even when a donor chromosome is present in the

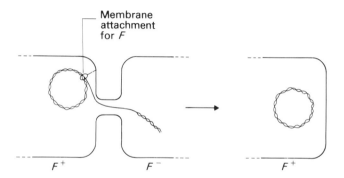

Membrane attachment for *F*

F^+ F^- F^+ $F^- \rightarrow F^+$

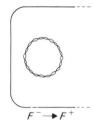

Fig. 11.7 Mechanism of replication of the *F* factor and its transfer to a *F⁻* strain by conjugation.

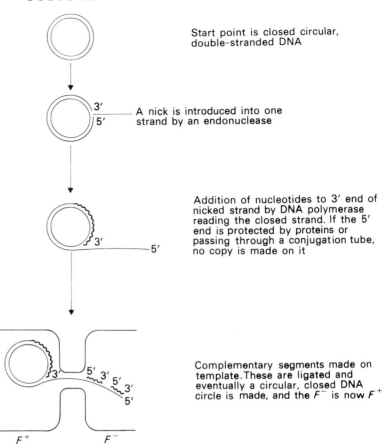

Start point is closed circular,
double-stranded DNA

A nick is introduced into one
strand by an endonuclease

Addition of nucleotides to 3' end of
nicked strand by DNA polymerase
reading the closed strand. If the 5'
end is protected by proteins or
passing through a conjugation tube,
no copy is made on it

Complementary segments made on
template. These are ligated and
eventually a circular, closed DNA
circle is made, and the F^- is now F^+

Fig. 11.8 The rolling-circle model
of DNA replication.

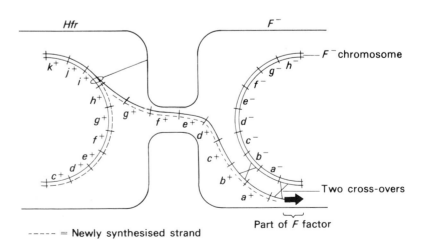

Fig. 11.9 Transfer of host genes
to the recipient in an $Hfr \times F^-$
mating.

----- = Newly synthesised strand

recipient, it does not necessarily mean that it is going to recombine with the recipient chromosome. Fortunately for genetic studies this *postzygotic coefficient of integration* is about the same for most chromosomal markers. When a bacterium contains its own chromosome plus a fragment transferred from a donor cell, it is called a merodiploid or merozygote.

Conjugation mapping

F. Jacob and E. Wollman realised that the random breakage of conjugal unions afforded a means of mapping gene sequences by timing the entry of different genes into a recipient. They used an interrupted mating technique in which *Hfr* and *F⁻* cells were mated and then at various intervals a sample of the culture was taken and blended in a Waring blender to separate the cells. Thus, the length of donor chromosome that entered the *F⁻* was controlled by timing the interval between the onset of conjugation and the blending treatments. The experiments were set up so that recombinant progeny for a number of genes could be scored at the various times. The relationships between genes and their positions on the chromosome could be mapped, therefore, in terms of time units. Note that one must take into account the fact that part of the *F* factor enters first and that this takes time. One must therefore only compare 'distance' in time units for chromosomal genes. Here is an example of an experiment done by Jacob and Wollman; involving the following cross:

$HfrH\ str^s \times F^-\ thr^-\ leu^-\ azi^-\ tonA\ lac^-\ gal^-\ str^R$

The *F⁻* cell carries a number of mutant genes: it is auxotrophic for threonine (*thr⁻*) and leucine (*leu⁻*), it is sensitive to inhibition by azide (*azi−*), it is sensitive to phage T1 infection (*tonA*), it is unable to ferment (and hence utilise as a sole carbon source) lactose (*lac⁻*) and galactose (*gal⁻*) and it is resistant to streptomycin, an antibiotic that will kill sensitive cells. The *Hfr* cell, on the other hand, has the wild-type alleles of all these genes and it is sensitive to streptomycin.

The two cell types are mixed together in nutrient medium to encourage mating. After a few minutes the culture is diluted so that no new mating pairs can form, thus ensuring some synchrony in the chromosome transfer for those mating pairs already established. Then, at various times, samples of the culture are taken, blended, and plated on a selective medium that is designed to allow particular recombinant types to grow while counterselecting against (killing) the *Hfr* and *F⁻* parental cells. In our example, the *thr⁺* and *leu⁺* genes are transferred first in the cross and thus the selective medium omitted both threonine and leucine, but other essential nutrients were included. On such a medium *thr⁺ leu⁺* recombinants could grow but the *F⁻* cell could not. The medium also would contain streptomycin in this case so that the *Hfr* parental would be killed. The *F⁻* recombinants, of course, would be unaffected by the antibiotic. In this way a selection of colonies each representing a clone of a *thr⁺ leu⁺* recombinant, can be isolated. Each must have arisen by recombination between a transferred D chromosome and the R chromosome. If one selects for a gene transferred early, there is a good chance of obtaining a large number of recombinants.

One can now test the *thr⁺ leu⁺* recombinants for whether or not they are also recombinant for the other marker genes of the *F⁻*, by using other selective platings or tests. The results of doing this for the cross we are discussing are shown in Fig. 11.10.

After 8 min, *azi⁺* recombinants have appeared and thus that gene on the donor chromosome must have entered the R cell. Similarly, the *ton⁺* gene entered at 10 min, the *lac⁺* gene at 17 min and the *gal⁺* gene at 23 min. It can be concluded, therefore, that the order of genes on the transferred donor chromosome is origin (part of *F* factor)—*thr–leu–azi–ton–lac–gal*. The distance between the genes can be computed from the times of entry, i.e. the times at which recombinant progeny for each gene is observed. Thus for example, the *ton* and *lac* genes are 7 min apart on the map.

As we mentioned before, a number of *Hfr* strains are necessary to map the entire circular chromosome of *E. coli* since the mating pairs generally break apart before all the chromosome has been transferred. There is very good agreement for the distance in time units between two particular genes

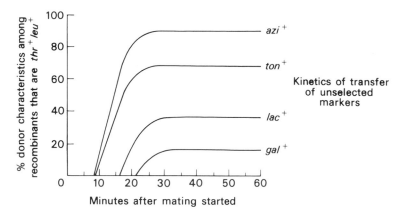

Fig. 11.10 The kinetics of transfer of unselected donor markers among selected thr^+ leu^+ recombinants as a function of time of conjugation. (After F. Jacob and E. L. Wollman, 1961, *Sexuality and the Genetics of Bacteria*. Copyright © Academic Press, New York.)

when different *Hfr* strains are used, thus lending credibility to this method of mapping. The entire map is 90 min long in *E. coli* and, conventionally, the threonine gene is placed at 0 min on this map.

TRANSDUCTION

The most widely used approach to mapping bacterial genes is transduction in which pieces of DNA are carried between bacteria by a phage vector. We shall consider two types of transduction — generalised and specialised — which differ significantly in the mechanisms involved. Before discussing these, we need to distinguish between two types of phage. The 'virulent' phages such as T2 or T4 always result in a lytic response, that is the production of progeny particles, when they infect their host. Other phages are 'temperate' — meaning that they can either elicit a lytic response or they can infect and remain dormant in a bacterial cell. This latter phenomenon is called the lysogenic response and the dormant phage in that state is called a prophage. At any time there can be a transition from the prophage to the vegetative state resulting in the lytic cycle and release of progeny phage particles. Lysogeny is an advantage to the bacterium because it also prevents super-infection with a second particle of the same phage type.

Generalised transduction

N. Zinder and J. Lederberg in 1952 showed that temperate phages may act as carriers for genes between one bacterial cell and another. Their motivation was to see if genetic exchange previously shown in *E. coli* (conjugation) also existed in the mouse typhus bacterium, *Salmonella typhimurium*. They mixed together two double amino acid auxotrophic cells, phe^- trp^- and met^- his^-, on minimal medium and looked for, and found, wild types. This did not occur unless the strains were mixed and was apparently the result of genetic exchange. The exchange did not occur with the same frequency for each strain examined, and the combination of strains LA22 and LA2 resulted in the frequency of wild types for many gene combinations.

Zinder and Lederberg used the U-tube apparatus we discussed before to test the mode of genetic exchange (Fig. 11.11). Wild-type bacteria appeared on the right side but not on the left side; suggesting that a 'filterable agent' was produced by LA2 that could produce wild-type recombinants in LA22. This filterable agent only arose when the two strains shared the same growth medium. The agent was shown not to be naked DNA or RNA since the transduction was not abolished either by DNase or RNase treatment. The formal explanation is that the filterable agent is the temperate phage P22 which can lysogenise the LA22 strain. In the experiment, progeny P22 particles were produced from some LA22 bacteria in which the prophage was converted to the vegetative state. These phages moved through the filter as the medium was moved between compartments by alternate suction and pressure. The phages then infected the non-lysogenic LA2 strain, and new phages were pro-

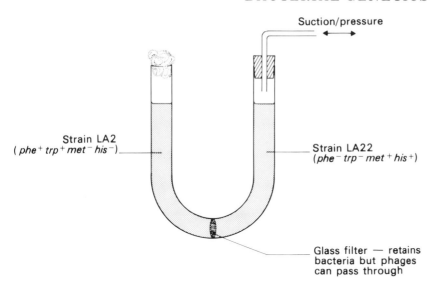

Fig. 11.11 Use of a U-tube apparatus to demonstrate transfer of genetic material between strains of *Salmonella typhimurium* does not require cell contact.

duced by the lytic cycle. During this process, occasionally a piece of host DNA can become wrapped up in a phage coat. The phage particles move back to the right side and the phages lysogenise the LA22 cells. This time some phage particles are associated with genetic material from LA2, some of which is wild type for the mutant genes of LA22. Thus wild-type recombinants are found among the LA22 cells as a result of recombination between the LA2 material carried by P22 and the LA22 chromosome. In other words the LA22 strain was *transduced*. No wild-type transductants were found in the LA2 side since that strain is nonlysogenic.

The frequency of transduction is very low. The relative efficiency of transduction of any P22 phage lysate is the ratio of the number of transductions produced to the number of P22 phage particles with which the recipient bacteria had been infected. This efficiency is around 10^{-5} 10^{-7}.

Transduction is not confined to *Salmonella typhimurium*. Genes can be transduced in *E. coli* with phage P1 and in *Bacillus subtilis* with phage SP10 for example.

How can generalised transduction be used in genetic analysis? The amount of DNA in P22 is about $\frac{1}{100}$ that of *Salmonella typhimurium* and thus the transducing particles can only carry a very small part of the host chromosome. Thus trans-

duction can provide information about whether or not two mutations are closely linked and it can also help ascertain gene order if three genes are being examined. An example, using the P1 transducing phage of *E. coli*, can illustrate this. Two strains of *E. coli* were used: the donor was *leu*$^+$ *thr*$^+$ *azi*^{r} and the recipient was *leu*$^-$ *thr*$^-$ *azi*^{s}. Phage P1 was grown on the donor cells and the progeny were used to transduce the recipient. In such an experiment, one can select for any of the donor markers in the recipient and then, as was done in the conjugation experiment, one can look for the presence of the other unselected markers among the transductants (Table 11.1).

In the experiment selecting for *leu*$^+$, the results show that the *leu* and *azi* genes are close together, and both are distant from the *thr* gene. The results of the experiment selecting for the *thr*$^+$ marker

Table 11.1 Co-transduction frequencies for markers in an experiment involving P1 phage, a *leu*$^+$ *thr*$^+$ *azi*^{r} donor and a *leu*$^-$ *thr*$^-$ *azi*^{s} recipient.

Selected marker	Unselected markers
leu$^+$	50% = *azi*^{r}
	2% = *thr*$^+$
thr$^+$	3% = *leu*$^+$
	0% = *azi*^{r}

show that the *leu* gene is closer to the *thr* gene than is the *azi* gene. Thus the order of genes is:

$$\underset{\text{thr} \qquad \text{leu} \quad \text{azi}}{\rule{0pt}{0pt}} $$

thr   leu   azi

In an experiment such as this, the recombinants result by recombination between the linear DNA fragment brought in from the donor by the phage particle and the circular DNA of the recipient. A double cross-over is necessary for each genetic exchange (Fig. 11.12).

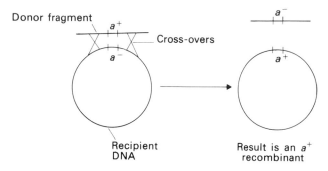

Fig. 11.12   Exchange of gene on transducing DNA with homologous gene on the host genome by a double cross-over event.

Specialised transduction

E. Lederberg discovered that *E. coli K12* is a lysogenic strain in that it can carry a temperate phage λ (*lambda*). The lysogeny of *K12* (thus *K12*(λ), c.f. Benzer's *rII* work) was found after nonlysogenic derivatives were accidentally isolated. Phage λ is a DNA-containing phage, whose genetic material consists of about 50 000 nucleotide pairs, which is about one-quarter that of T-even phage DNA. The DNA of phage λ is for the most part double-stranded, although the ends of the chromosome are single-stranded and complementary as a result of terminal redundancy. When *E. coli* is infected with λ, the DNA of the latter circularises and can either replicate and go through the lytic cycle, or it can integrate into the host chromosome to produce the prophage state. The integration step is similar to *F*-factor integration and involves a specific λ attachment site (*att-λ* locus) on the host DNA that is homologous with a site on the phage DNA (*b2*). The integration then results from a recombination event using both phage-specific and host enzymes (Fig. 11.13) and this places the λ genome between the *gal* (galactose) and *bio* (biotin) genes of the *E. coli* chromosome.

The lysogeny of the λ phage is brought about by a repressor that is coded by one of the genes of the phage itself. The repressor is a protein molecule consisting of four identical subunits each with a molecular weight of 38 000. The repressor acts by blocking the transcription of phage genes.

In 1956, J. Lederberg tested whether or not λ can transfer *E. coli* genes from donor to recipient cells. He took a lysogenic wild-type *K12*(λ) strain and induced the λ prophage with ultraviolet light. This destroys the repressor and thus the phage goes through the lytic cycle generating a lysate of λ phages. He then infected a variety of genetically marked nonlysogenic cultures of *K12* with the phages and plated them on selective media to see if any of the wild-type genes of the *K12*(λ) donor cells had been transferred to the now lysogenised mutant recipients. The results were mostly negative, the exception being that about 1 in 10^6 of λ-infected *gal$^-$* bacteria (unable to ferment galactose) had acquired the *gal$^+$* phenotype of the donor. Thus λ is capable of transduction but it is restricted to the *gal* genes in the vicinity of the *attλ* region.

Most of the tranductants are genetically unstable in that each *gal$^+$* colony contains about 1–10% of *gal$^-$* cells. The *gal$^+$* transductants, then, are actually *gal$^+$/gal$^-$* partial heterozygotes, that is, the *gal$^+$* donor fragment brought in by the transducing phage has been added to the recipient genome rather than exchanged for the *gal$^-$* gene (in this case). Thus there is a potential for losing the *gal$^+$* gene.

All of the events involved in specialised transduction can be summarised diagrammatically. The first step involves erroneous looping out of the λ genome when the prophage is induced (Fig. 11.14). A cross-over generates a circular DNA containing most, but not all, of the λ genome as well as some host genome — here including the *gal$^+$* genes. Enzyme action converts this circular DNA to a linear molecule which is then assembled into a phage particle. Thus, the result of these events is the production of a defective phage particle called

Steps in λ integration

a Infection of *E. coli*

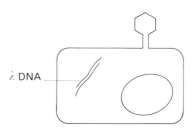

λ DNA

b Circularisation of λ genome

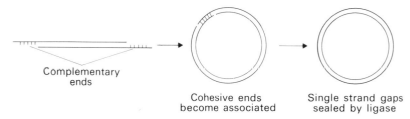

Complementary
ends

Cohesive ends
become associated

Single strand gaps
sealed by ligase

c Integration

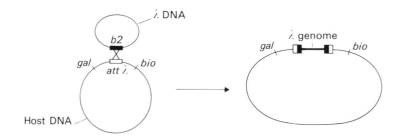

λ DNA

b2

gal   bio
att λ

Host DNA

λ genome

gal   bio

Fig. 11.13 Steps of integration of
phage λ genome into the *E. coli*
cell at the normal recognition site
att λ.

λ*dg* (λ-defective-galactose) which carries bacterial *gal*⁺ genes. This incorrect looping out occurs only very rarely among prophage excisions.

The λ*dg* phage is a transducing phage since it can transfer the *gal*⁺ genes to a nonlysogenic recipient cell. When it infects such a cell, the λ*dg* DNA can integrate into the recipient chromosome by crossing-over at the homologous *gal* regions (Fig. 11.15).

Here the cross-over generates a continuous DNA containing a defective prophage (λ*def*) between two bacterial *gal* genes. In the example, the dominance of the donor *gal*⁺ gene over the recipient *gal*⁻ gene results in a *gal*⁺ phenotype. Reversal

of the above event in the transductant will produce a *gal*⁻ segregant.

As we said, the number of λ*dg* phages present in a phage lysate is very small. Therefore if the bacterial infection is done with a relatively large number of phages compared with bacterial cells, then it is possible that a λ⁺ phage will coinfect with the λ*dg* phage. In this case the λ⁺ can integrate at the normal *att λ* site to produce a double lysogenic *K12(λ)(λdg)* (Fig. 11.16).

If the *K12(λ)(λdg) gal*⁺/*gal*⁻ transductant (which has a *gal*⁺ phenotype) is induced with ultraviolet light then, by the reversal of the processes described, about equal proportions of λ⁺ and λ*dg*

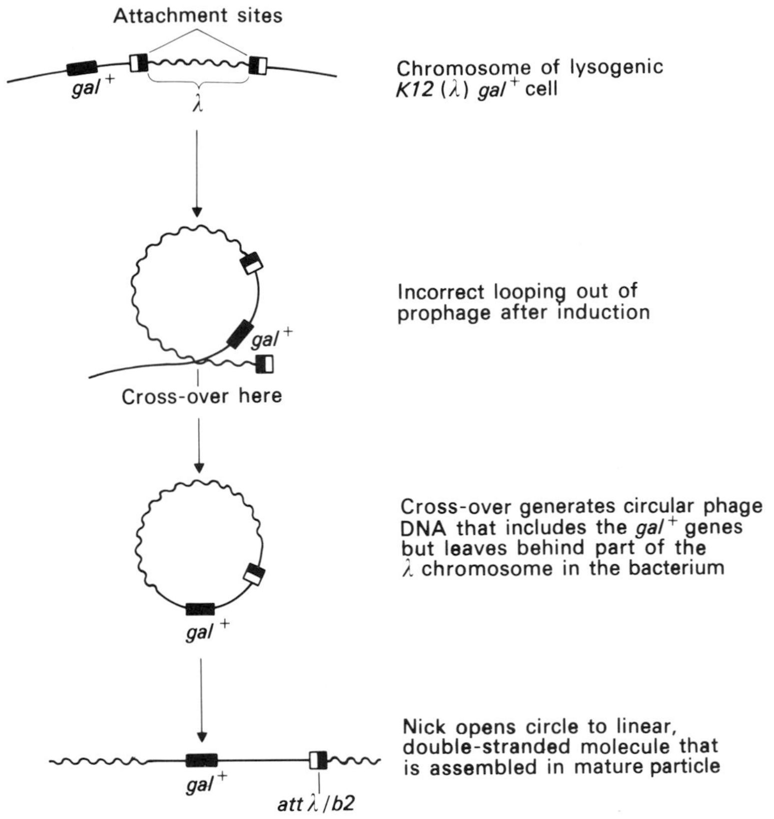

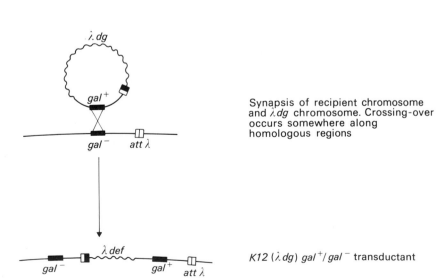

Attachment sites
gal+
λ
Chromosome of lysogenic K12 (λ) gal+ cell
Incorrect looping out of prophage after induction
gal+
Cross-over here
Cross-over generates circular phage DNA that includes the gal+ genes but leaves behind part of the λ chromosome in the bacterium
gal+
Nick opens circle to linear, double-stranded molecule that is assembled in mature particle
gal+
att λ/b2

Fig. 11.14 Generation of λ transducing phage (λdg) carrying host gal^+ genes by incorrect looping out of the prophage from the *E. coli* chromosome.

λ dg
gal+
gal−
att λ
Synapsis of recipient chromosome and λdg chromosome. Crossing-over occurs somewhere along homologous regions
λ def
gal−
gal+
att λ
K12 (λ dg) gal+/gal− transductant

Fig. 11.15 Integration of λdg transducing phage into the *E. coli K12* chromosome by crossing-over in the *gal* region. This results in a *K12* (λdg) gal^+/gal^- transductant.

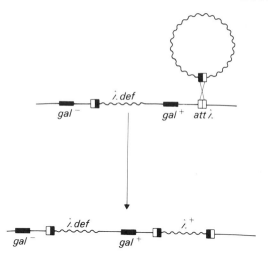

Pairing of λ^+ at the normal attachment site. (It is also possible to get pairing between homologous parts of the λ^+ and $\lambda\,def$ genomes to get integration)

Crossing-over generates a normal and defective λ prophage and two gal genes. This is a $K12\,(\lambda)(\lambda dg)\,gal^+/gal^-$ transductant

Fig. 11.16 Production of a $K12\,(\lambda)(\lambda dg)\,gal^+/gal^-$ transductant by simultaneous integration of λ and λdg DNA into the *E. coli* chromosome. The λdg integrates as in Fig. 11.15 and the λ DNA integrates at the $att\lambda$ site.

phages will be produced. The resulting lysate will have a very high capacity for transducing gal^- recipient cells and hence it is called an HFT (high frequency transducing) lysate to contrast it with the LFT (low frequency transducing) lysate described earlier where only 1 in 10^6 phages carried donor gal genes.

How can specialised transduction be used? One example is that complementation tests can be performed for mutations in the gal region to define the number of cistrons. (Indeed the gal region is an operon consisting of three cistrons.) Thus one can have lysogenic donor cells that carry one type of gal^- mutation and generate transducing phages that are used to infect nonlysogenic recipient cells carrying a different gal^- mutation. If the two mutations are in different cistrons (complementation groups), complementation will occur and the recipient will become gal^+. If the two mutations are in the same cistron, no gal^+ transductants will be found.

TRANSFORMATION

The previous sections in this Topic have demonstrated how gene maps can be constructed in bacteria that conjugate and/or that have transducing phages. There is a third method of DNA transfer between bacteria — transformation — that can be used for gene mapping in certain bacteria, including some that are not amenable to conjugation or transduction. Indeed we have described one example of transformation in Topic 1 when we discussed evidence that DNA is the genetic material. In this section we will discuss the general principles of transformation as it pertains to gene mapping.

The mechanism of transformation

The basic mechanism of transformation involves bacteria taking up fragments of DNA that may then exchange with the homologous DNA of the cells. Thus, as in our discussions of conjugation and transduction, it is appropriate to talk about donor and recipient cells and, for the purposes of mapping, the two strains are usually manipulated to have different genotypes. Experimentally, DNA is extracted from donor cells and the fragmented molecules are added to the recipient cell population. The recipient, then, takes up random pieces of DNA.

Not all bacterial species have the ability to take up DNA and even those that do may need to be in a particular phase of the growth cycle or in a particular growth medium in order for them to be competent to take in the DNA. In this regard the bacterial species *Diplococcus pneumoniae* (see Topic 1) and *Bacillus subtilis* are relatively easy to render competent whereas *E. coli* must be genetically deficient for two exonucleases and be incubated in the presence of a high concentration of calcium chloride in order to make the membrane permeable to DNA. The transformation of *E. coli*, then, has not been utilised for gene mapping; conjugation or transduction being preferable procedures to use. However, as will be discussed in the next Topic, transformation of *E. coli* is an integral part of recently developed recombinant DNA technology.

For efficient transformation of a bacterium such as *Bacillus subtilis,* the DNA must be double-stranded and of relatively high molecular weight ($1–8 \times 10^6$ daltons). As the DNA crosses the membrane of a competent bacterium, one of the strands of the DNA is broken down thus providing energy for the transfer of the DNA. The resulting single-stranded DNA then may exchange with the homologous region of the recipient's chromosome and this event can be detected given appropriate genetic differences between donor and recipient cells.

Determination of gene linkage by transformation

Let us suppose we have a donor bacterial strain that is $a^+ \; b^+$ and a recipient strain that is $a^- \; b^-$ and we wish to determine whether or not the two genes are linked. The first step is to use donor DNA to transform the recipient cell under selective conditions whereby the frequency of transformation (i.e. the proportion of recipients that are transformed to the wild-type state) can be calculated for the *a* and *b* genes separately. In such an experiment the frequency of transformation of a single gene will be quite rare, ranging from 1 in 10^6 to 1 in 10^3 depending on how much DNA was used. (Therefore, it is important to use the same amount of DNA when comparing frequencies of transformation.) In the next step, the experiment is repeated but this time the frequency of $a^+ \; b^+$ transformants (or, in other words, the co-transformation frequency) is determined. If the two genes a^+ and b^+ are not closely linked in the donor cell, then the co-transformants that are formed can do so only by taking up at least two separate pieces of DNA, with one carrying the a^+ gene and one carrying the b^+ gene. The probability

of such co-transformation in this case, then, is the product of the probabilities of transformation for the two genes separately. So, if the observed co-transformation frequency is significantly greater than this calculated frequency, the two genes must be closely linked on the chromosome. Indeed if the observed frequency is very close to the transformation frequency for a single gene, then the genes must be very closely linked on the chromosome such that there is a high likelihood that a piece of transforming DNA will carry both genes into the recipient.

In summary then, the linkage of genes can be established by determining the frequency of co-transformation as compared with the frequencies of transformation of the single markers alone. With the appropriate number of gene differences between the donor and recipients, a gene order and hence a gene map can be constructed by these sorts of experiments.

REFERENCES

Conjugation

Campbell A. 1969. *Episomes.* Harper and Row, New York.

Curtiss R. 1969. Bacterial conjugation. *Annu. Rev. Microbiol.* **23**: 69–136.

Gilbert W. & D. Dressler. 1969. DNA replication: the rolling circle model. *Cold Spring Harbor Symp. Quant. Biol.* **33**: 473–484.

Susman M. 1970. General bacterial genetics. *Annu. Rev. Genet.* **4**: 135–176.

Vielmetter W., F. Bonhoeffer & A. Schutte. 1968. Genetic evidence for transfer of a single DNA strand during bacterial conjugation. *J. Mol. Biol.* **37**: 81–86.

Wollman E.L., F. Jacob & W. Hayes. 1962. Conjugation and genetic recombination in *E. coli K-12. Cold Spring Harbor Symp. Quant. Biol.* **21**: 141–162.

Transduction

Campbell A.M. 1962. Episomes. *Adv. Genetics,* **11**: 101–145.

Jacob F. 1955. Transduction of lysogeny in *Escherichia coli. Virology,* 1: 207–220.

Jacob F. &. E.L. Wollman. 1961. *Sexuality and the Genetics of Bacteria.* Academic Press, New York.

Lennox E. 1955. Transduction of linked characters of the host by bacteriophage P1. *Virology* 1: 190–206.

Morse M.L., E.M. Lederberg & J. Lederberg. 1956. Transduction in *Escherichia coli K-12*. *Genetics*, **41**: 142–156.

Ozeki H. & H. Ikeda. 1968. Transduction mechanisms. *Annu. Rev. Genet.* **2**: 245–278.

Zinder N.D. & J.L. Lederberg. 1952. Genetic exchange in *Salmonella*. *J. Bacteriol.* **64**: 679–699.

Transformation

Archer L.J. 1973. *Bacterial Transformation*. Academic Press, New York.

Dubnau D., D. Goldthwaite, I. Smith & J. Marmur. 1967. Genetic mapping in *Bacillus subtilis*. *J. Mol. Biol.* **27**: 163–185.

Goodgal S.H. 1961. Studies on transformation of *Hemophilus influenzae*. IV. Linked and unlinked transformations. *J. Gen. Physiol.* **45**: 205–228.

Hotchkiss R.D. & M. Gabor. 1970. Bacterial transformation, with special reference to recombination processes. *Annu. Rev. Genet.* **4**: 193–224.

Hotchkiss R.D. & J. Marmur. 1954. Double marker transformations as evidence of linked factors in deoxyribonucleate transforming agents. *Proc. Natl. Acad. Sci. USA*, **40**: 55–60.

Lacks S., B. Greenberg & M. Neuberger. 1974. Role of a deoxyribonuclease in the genetic transformation of *Diplococcus pneumoniae*. *Proc. Natl. Acad. Sci. USA*, **71**: 2305–2309.

Ravin A.W. 1961. The genetics of transformation. *Adv. Genet.* **10**: 61–163.

Tomaz A. 1969. Some aspects of the competent state in genetic transformation. *Annu. Rev. Genet.* **3**: 217–232.

Topic 12
Recombinant DNA

OUTLINE
Restriction endonucleases
   the restriction-modification phenomenon
   properties of restriction enzymes
Cloning vehicles
Construction and cloning of recombinant DNA molecules
   insertion of DNA into the plasmid vehicle
   cloning of recombinant DNA
   selection of specific recombinant DNA clones
Applications of recombinant DNA technology.

In recent years, experimental procedures have been developed that have allowed researchers to construct recombinant DNA in the test tube, that is, a single DNA molecule combining genetic material from two different sources. This has opened the way for new and exciting research possibilities and affirms the plausibility of genetic engineering. In this Topic we will discuss the techniques by which recombinant DNA can be made and we will present some examples of how recombinant DNA technology is furthering our knowledge of the structure and function of prokaryotic and eukaryotic genomes.

In outline, recombinant DNA is made in the following way. A piece of DNA from the organism of interest is spliced into either a plasmid or a lambda phage (called the *vehicle* or *vector*) and the resulting chimaeric molecule is used to transform or inject, respectively, a host bacterial cell. The latter is often a special strain of *E. coli* that is unable to reproduce without special culture conditions. (This is done to minimise the risk of exposing the human population to new gene combinations that normally would not occur naturally and whose effects on an organism are unknown.) Reproduction of the *E. coli* results in the cloning of the recombinant DNA molecule (and the process is thus also called *molecular cloning*), thus producing many copies for analysis. At the outset of this type of experiment, it was recognised that there could be risks in carrying them out, especially where eukaryotic DNA was being cloned in the bacterial system. Action by concerned scientists has lead to the formulation in the USA of guidelines for carrying out this type of research. These guidelines apply to research scientists who are working under US Research Grants. Since the guidelines are constantly being reviewed as more information is found concerning the possible risks, they will not be considered here. Similar guidelines are also in effect in other countries.

RESTRICTION ENDONUCLEASES

The restriction-modification phenomenon

One of the reasons for the rapid development of recombinant DNA technology was the discovery of a variety of enzymes that catalyse the cleavage of DNA at a small number of reproducible sites. These enzymes are called restriction endonucleases or restriction enzymes. (Recall from previous discussions that an endonuclease results in a cut within a nucleic acid chain.)

In the 1950s, S. Luria and his colleagues found that the progeny phage produced after infection of a particular strain of *E. coli* — *B/4* — by phage T2 could no longer reproduce in the normal host strain of *E. coli*. They reasoned that the phage released from *B/4* had become modified in some way such that the normal host was no longer permissive for growth of T2. This phenomenon was called host-controlled modification and restriction.

In the 1960s, W. Arber and D. Dussoix shed some light on the restriction phenomenon. They grew phage *lambda* on *E. coli K* and used the progeny to infect *E. coli B*. with the result that most of the

DNA was broken into pieces. Apparently a few DNA strands remained intact since some progeny phage were produced from the injection of strain *B*. These could now infect *E. coli B* normally and about 2% of them could also infect *E. coli K*. To investigate this in more detail, Arber's group grew phage *lambda* in *E. coli K* which was growing in a culture medium containing the heavy isotopes ^{15}N and ^{2}H (deuterium) so that the DNA of the resulting progeny phage was more dense than normal phage DNA. The 'heavy' phages were then used to infect *E. coli B* growing in normal 'light' medium and the densities of the progeny phage's DNA were determined by caesium chloride density gradient centrifugation. Most of the progeny phages contained normal 'light' DNA which had been entirely synthesised from new material. These phages could not grow in *E. coli K*. A few of the progeny phage were 'heavy' and these retained the ability to grow in *E. coli K*. They reasoned, therefore, that phages grown in particular bacterial strains become modified in a way specific for that particular strain. Further, the modification occurred on the DNA itself and involved a covalent bond since it was not lost by passage through the bacterial host. It now appears that in many bacteria the modification involves the addition of methyl groups on certain bases, in particular DNA sequences in the genome.

Then in 1970 a major breakthrough occurred. This was the identification of restriction enzymes which play a role in the restriction phenomenon. These enzymes recognise a specific nucleotide sequence in DNA and this may then result in cleavage. The way this relates to the modification-restriction phenomenon is as follows. A bacterial cell has two enzymes (or sets of two enzymes depending on the number of modifications of which it is capable) both of which have a recognition site for a particular nucleotide pair sequence. One of them, the modification enzyme, catalyses the addition of a modifying group (often a methyl group) to one or more nucleotides in the sequence. The other enzyme, the restriction enzyme, can only recognise the unmodified nucleotide sequence and hence serves to digest invading DNA unless it, too, is modified like the host DNA. To date a number of restriction-modification systems have been ident-

ified in bacteria. However, it should be pointed out here that not all of the enzymes that are being used in recombinant DNA experiments have been shown to have a role in a restriction-modification system in the bacterium from which they have been isolated, even though they are called restriction enzymes.

Properties of restriction enzymes

Some 90 enzymes so far have been isolated that have the capacity to cleave DNA endonucleolytically. All of these enzymes have been found in prokaryotes; no similar enzymes have been identified

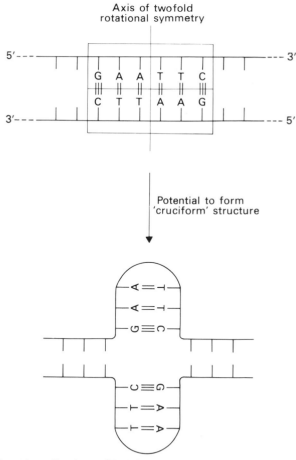

Fig. 12.1 Region of DNA showing two-fold rotational symmetry of nucleotide pair sequence. The sequence shown is actually the recognition site for the restriction endonuclease EcoRI.

in the few eukaryotic organisms that have been examined.

The restriction enzymes fall into two classes with regard to the way they cleave DNA. Class I enzymes recognise a specific nucleotide pair sequence and then cleave the DNA at a nonspecific site away from that recognition site. The enzymes involved in the *E. coli K* and *B* restriction system belong to this class. Class II enzymes, on the other hand, cleave the DNA at the specific recognition site. For that reason, and the fact that the recognition sites have two-fold rotational symmetry (Fig. 12.1), these enzymes are very valuable for constructing recombinant DNA molecules, as we shall see.

Class II restriction endonucleases have been isolated from a large number of microorganisms. All of them are sequence specific and thus the number of cuts they make in a DNA molecule is dependent upon the number of times the particular sequence is present in the DNA. The cleavage sites for a number of restriction enzymes are shown in Table 12.1 along with the number of cuts each enzyme makes in the commonly tested DNAs from phage *lambda*, Adenovirus-2 (Ad2) and SV40. This information can be used as a guideline for choosing an enzyme for a particular application. As can be seen, some enzymes cut both strands of DNA between the same nucleotide pair, whereas others make staggered cuts in the symmetrical nucleotide pair sequence.

Table 12.1 Characteristics of some restriction endonucleases. (After R.J. Roberts, 1976. *Crit. Rev. Biochem.* **4**: 123.)

Enzyme name	Organism from which enzyme isolated	Recognition sequence and position of cut	Number of cleavage sites in DNA from:		
			λ	Ad2	SV40
BamHI	*Bacillus amyloliquefaciens H*	5′ G↓GATCC 3′*	5	3	1
BglIII	*Bacillus globigii*	A↓GATCT	5	12	0
EcoRI	*E. coli RY13*	G↓AATTC	5	5	1
HaeIII	*Haemophilus aegyptius*	GG↓CC	50	50	18
HhaI	*Haemophilus haemolyticus*	GCG↓C	50	50	2
HindIII	*Haemophilus influenzae R_d*	A↓AGCTT	6	11	6
HpaI	*Haemophilus parainfluenzae*	GTT↓AAC	11	6	5
PstI	*Providencia stuartii*	CTGCA↓G	18	25	3
SmaI	*Serratia marcescens*	CCC↓GGG	3	12	0
SalI	*Streptomyces albus G*	G↓TCGAC	2	3	0

* In this column only a single strand of DNA is shown with the site of cleavage since there is an axis of two-fold rotational symmetry. Thus, for example, the recognition sequence for BamHI in more detail is 5′ G↓GA | TCC 3′ where
3′ C CT | AG↑ G 5′
the same nucleotide sequence is found for both strands in the 5′ to 3′ direction. The vertical line shows the axis of two-fold rotational symmetry, and the arrows indicate the points of cleavage. The resulting DNA molecules have complementary single-stranded ends: 5′ G   and 5′ GATCC 3′. For the
3′ CCTAG 5′   G 5′
other enzymes, a similar situation prevails with the exception that HaeIII, HpaI and SmaI cut the DNA in such a way that the ends are double-stranded ('blunt' ends).

CLONING VEHICLES

In order to clone a piece of DNA for study, it must first be attached to a cloning vehicle (vector). One of the types of cloning vehicles that is used for these experiments is plasmids. The DNA is spliced into the plasmid DNA and the chimaeric molecule is used to transform a host bacterium such as *E. coli*. The plasmid then replicates within the host as the host grows and divides, and the piece of DNA of interest becomes cloned.

Plasmids are extrachromosomal genetic elements that replicate autonomously within bacterial cells. Their DNA is circular and double-stranded and they carry the genes required for replication of the plasmid as well as for the other functions that the plasmids have. In general the plasmid vehicles have a different buoyant density than that of the host DNA and thus they can easily be purified. Some plasmids have the ability to integrate into the host's chromosome, and these are called episomes. The *F* factor that is involved with conjugation of *E. coli* is an example of an episome. Further, only some of the plasmids are able to promote conjugation.

Two of the plasmid vehicles that are used for molecular cloning are pSC101 (Fig. 12.2a) and pBR322 (Fig. 12.2b). Both of these plasmids are of the nonconjugal type and each *E. coli* cell transformed with the plasmids will have 6–8 copies of them per host chromosome.

The pSC101 plasmid has a molecular weight of 5.8×10^6 daltons (the *E. coli* chromosome is approximately 2500×10^6 daltons) and consists of 9200 base pairs. This plasmid has one cleavage site each for the restriction enzymes EcoRI, HindIII, BamHI and SalI, the last two falling in the region of the plasmid which confers tetracycline resistance (Tc^R) upon *E. coli* cells which harbour it. The pBR322 plasmid is, in fact, derived from pSC101. It weighs 2.6×10^6 daltons and also contains one cleavage site each for the four restriction enzymes mentioned above, as well as a site for PstI. This plasmid carries genes for tetracycline resistance (Tc^R) and for ampicillin resistance (Ap^R). The BamI and SalI sites are in the Tc^R region and the PstI site is within the Ap^R region. The EcoRI and

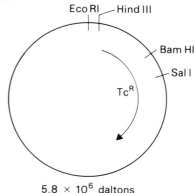

a pSC101

5.8×10^6 daltons

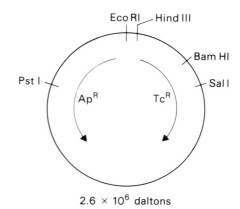

b pBR322

2.6×10^6 daltons

Fig. 12.2 Characteristics of non-conjugal plasmids commonly used in recombinant DNA research showing, in particular, drug-resistance genes and the recognition sites for restriction enzymes. (a) Plasmid pSC101, (b) Plasmid pBR322.

HindIII sites are between the two antibiotic resistance regions.

As will be described in more detail in the next section, the restriction enzyme cleavage sites in the plasmids are necessary for the insertion of 'foreign' DNA into the cloning vehicle. In order for this to work, only one site for each enzyme can exist in the plasmid. In many cases plasmids have been 'constructed' by researchers to have cleavage sites specific for restriction enzymes that cut out DNA segments of particular interest from genomes of the organisms they are studying. Clearly, to be useful at all, the enzyme cleavage and insertion of DNA at the restriction site must not inactivate the replicator genes that are required for the replication of the plasmid. Further, these events should not affect the genes that allow the experimenter to determine whether or not an *E. coli* cell has become transformed with the plasmid, such as Tc^R or Ap^R.

CONSTRUCTION AND CLONING OF RECOMBINANT DNA MOLECULES

We have already discussed plasmid vehicles and restriction enzymes. Here we describe how these two can be employed to prepare recombinant DNA

molecules, that is the insertion of foreign DNA into the plasmid vehicle.

Insertion of DNA into the plasmid vehicle

There are two commonly used general methods for splicing a piece of DNA into the plasmid vehicle.

1. Cohesive ends. In Table 12.1, it can be seen that a number of restriction endonucleases make staggered cuts at specific recognition sites. Thus, for example, foreign DNA can be cut into segments by a restriction enzyme that also makes one cut in the plasmid vehicle, converting the latter to a linear molecule. Fig. 12.3 illustrates this for EcoRI. By the very fact that EcoRI cleaves at a specific site in this way, the single-stranded ends of the linear plasmid vehicle are complementary to the single-stranded ends of the EcoRI-generated segments of

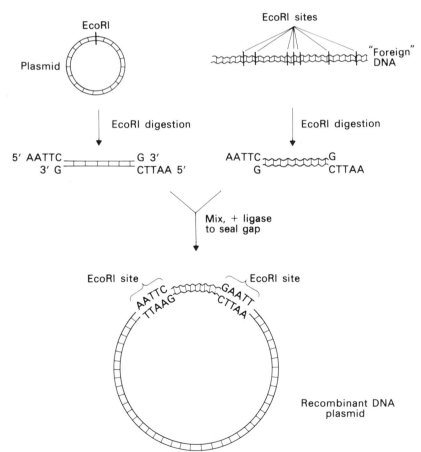

Fig. 12.3 Construction of a recombinant DNA plasmid through the use of the restriction enzyme EcoRI. Details of the procedure are given in the text.

foreign DNA. In solution the two DNAs can come together to produce a larger circular DNA held together by hydrogen bonding of the complementary ends. In the presence of the enzyme polynucleotide ligase, the single-stranded gaps in the sugar-phosphate backbones are sealed and the structure is stabilised. Since the ends of the DNA pieces produced by EcoRI digestion are all identical, the foreign DNA can insert into the plasmid vehicle in two orientations, and this will occur in a random way. This orientation may in fact have effects on transcription of the genes or gene fragments on the foreign DNA since the initiation of transcription is likely to depend on the location of promoter and other controlling sites on the vehicle.

2. Synthetic complementary terminals (dA:dT method). The number of single-stranded nucleotides on the DNA after digestion with a restriction enzyme that makes a staggered cut is small and thus the probability of the complementary sequences finding one another in solution is relatively small. Further, some restriction enzymes do not make staggered cuts and some methods for producing the DNA fragment to be cloned result in completely double-stranded (blunt-ended) DNA. In these cases it is possible to synthesise a single-stranded polynucleotide chain on the DNA molecules using the enzyme terminal deoxynucleotidyl transferase. Thus, for example, in the presence of dATP, this enzyme will catalyse the production of a poly(dA) tail on each 3′ end of the DNA. To apply this to the insertion of a DNA fragment into the plasmid vehicle, poly(dA) tails (approximately 100 nucleotides long) can be polymerised on the linearised plasmid, and poly(dT) tails of the same length can be polymerised onto the foreign DNA (Fig. 12.4). Then, by mixing the DNAs in solution and adding polynucleotide ligase, a recombinant DNA molecule can be produced. Note that in this procedure, the only circular DNA that can result is one in which the foreign DNA has been inserted into the plasmid.

Cloning of recombinant DNA

Once a recombinant DNA molecule has been formed, the next step is to transform an *E. coli* (or other bacterial) strain with it. As mentioned in the previous Topic, transformation of *E. coli* with relatively large pieces of DNA can be facilitated by treating the cells with $CaCl_2$ to make them permeable. With this procedure, the transforming DNA enters the cell intact and the host bacterium remains viable. Then, during growth of the *E. coli* cell, the plasmid replicates under the control of its genes as discussed previously.

Ideally one needs to have a way of determining whether or not the host cell has been transformed with the plasmid. Further, since it is the foreign DNA that one is interested in studying, it would be very useful if one could isolate just those transformed cells that carry recombinant plasmids. The presence of a plasmid in an *E. coli* cell can be shown by the fact that it will confer antibiotic resistance on the bacterium as a result of the plasmid specific genes it carries. If the recombinant plasmid had been constructed by the dA:dT method, this would also mean that a segment of DNA was present in the host. However, if the recombinant plasmids were made as a result of the complementarity of the restriction enzyme generated single-stranded sequences, then the population of DNA molecules used to transform the cells will contain a large number of non-recombinant plasmids, i.e. the plasmid vehicles alone that have spontaneously recircularised. In this case it may not be as easy to determine whether or not a clone of the transformed *E. coli* cell carries a piece of DNA that the experimenter wishes to study. One way to overcome this problem is to select a plasmid vehicle such as pBR322 which carries resistance genes for both ampicillin and tetracycline (see Fig. 12.2b). If a restriction enzyme is used that makes a cut in one of the antibiotic resistance genes, then the insertion of a foreign DNA fragment will split that gene apart and the recombinant plasmid is no longer able to confer resistance to that antibiotic on the bacterial cell that it transforms. Thus, for example, if PstI is used to generate DNA fragments and to open the pBR322 plasmid, then any recombinant plasmid that is produced will make the cell it transforms tetracycline resistant but *not* ampicillin resistant. Therefore, if the transformed cells are cloned by plating them and allowing them to form colonies on solid culture

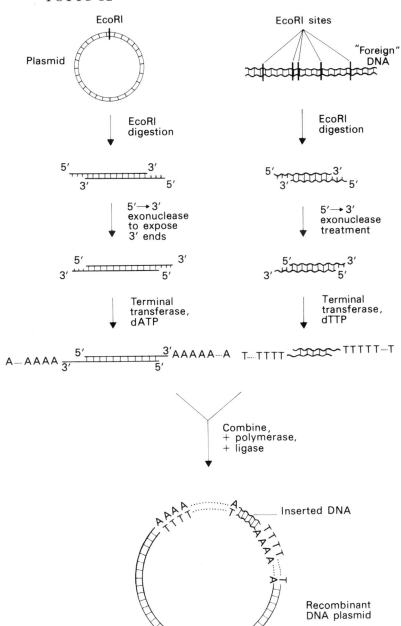

Fig. 12.4 Construction of a recombinant DNA plasmid using the enzyme terminal transferase to synthesise complementary ends on the linearised plasmid and a restriction-enzyme generated fragment of foreign DNA. Details of the procedure are given in the text.

medium, this can easily be tested by using the replica-plating technique described in an earlier Topic. The cells transformed with non-recombinant plasmids, then, can be detected since the bacteria will be resistant to both antibiotics and these clones can be removed from further consideration.

Selection of specific recombinant DNA clones

In many instances the genome of an organism is digested with a restriction enzyme and the fragments produced are then cloned by the procedures described. By using different restriction enzymes, the genome can be cut up in a number of ways, and this results in the formation of an extremely large number of recombinant DNA clones. However, in general an investigator wishes to study a clone carrying a particular DNA segment. In cases where a specific RNA probe is available, the clone or clones can be identified in a relatively easy way. As an example, this can be used for the selection of clones carrying fragments of ribosomal DNA (rDNA). In brief, the recombinant DNA clones are distributed onto a dish of culture medium and they are then replica-plated onto a nitro-cellulose filter. The cells on the filter are lysed *in situ* and the DNA that is released is denatured and allowed to stick to the filter. The filter is then placed in a solution containing a radioactive (usually ^{32}P-labelled) RNA probe and, in this case, this is rRNA purified from isolated ribosomes. If any of the clones contained rDNA that coded for the rRNA, then the RNA will hybridise with it. After removing the RNA that has not hybridised, a piece of photographic film is laid over the filter. The exposed areas on the resulting autoradiogram show those clones that carry DNA complementary to the RNA probe and these can be studied further.

Another approach to constructing a recombinant DNA plasmid carrying a specific segment of DNA involves the isolation of the mRNA for a protein. This approach, then, is limited to those mRNAs that are produced in large quantities by a cell, e.g. globin mRNA from rabbit reticulocytes. Once the mRNA has been isolated, a complementary DNA (cDNA) copy of it can be made in reactions involving, first, RNA-dependent DNA polymerase and, then, *E. coli* DNA polymerase I. The former

enzyme was first identified separately by Dr Baltimore and H. Temin in 1970 as a component of the RNA tumour viruses, Rous' sarcoma virus (RSV) and mouse leukaemia virus (MLV), respectively. The genetic material of these viruses is RNA and they are able to transform a cell that they infect to the tumour state. This process apparently involves the production of a DNA copy of the viral genome by the RNA-dependent DNA polymerase (also called reverse transcriptase, since the process it catalyses is the opposite of transcription).

The use of RNA-dependent DNA polymerase in molecular cloning is quite straightforward. The mRNA of interest (globin mRNA in our example) is incubated in a reaction mixture containing the enzyme and the four deoxyribonucleoside triphosphates (some or all of which can be radioactively labelled if necessary). The product is a DNA-mRNA double-stranded molecule from which a double-stranded cDNA molecule can be produced in the second reaction catalysed by *E. coli* DNA polymerase I. The cDNA will have blunt ends and by the addition of poly(dT) tails at the 3′ ends, the cDNA can be inserted into a linearised plasmid (as described previously) and it can be cloned. The cloned cDNA may then be purified and used as a probe in any one of several interesting applications, e.g. (a) to obtain sequence information about the mRNA (its coding and non-coding sequences); (b) to identify precursors to the mRNA (such as heterogeneous nuclear RNA), and (c) to use as a hybridisation probe to identify those recombinant DNA clones made from sheared total DNA that contains parts or all of the DNA region coding for globin mRNA. In view of the recent discovery of intervening sequences, this can allow regions of interest, such as controlling sites or the intervening sequences themselves to be analysed by DNA sequencing.

APPLICATIONS OF RECOMBINANT DNA TECHNOLOGY

In the few years since recombinant DNA technology has been developed, it has been used in a large number of studies of both prokaryotic and eukaryotic genomes, and it is impossible to deal with them all adequately in this brief discussion. To give but three examples, recombinant DNA cloning has made it possible to obtain large amounts of particular genes, to determine the functions of segments of nuclear and organellar DNA, and to map genomes (e.g. $\phi \times 174$ and SV40 were mapped this way). Particularly useful applications of the techniques are studies in which the controlling regions for prokaryotic and eukaryotic genes are cloned with an eye towards understanding the regulation of gene expression at a very basic level. Recombinant DNA cloning also has a lot of potential benefits for mankind, for example, it can be used to synthesise large quantities of antibiotics, enzymes and hormones (such as insulin) much less expensively then present methods. Also, in the near future it should be possible to introduce nitrogen-fixing genes into the genomes of crop plants — this would eliminate their requirement for nitrate fertilizers.

REFERENCES

Aaij C. & P. Borst. 1972. The gel electrophoresis of DNA. *Biochim. Biophys. Acta* **269**: 192–200.

Arber W. 1965. Host-controlled modification of bacteriophage. *Annu. Rev. Microbiol.* **19**: 365–378.

Arber W. 1974. DNA modification and restriction. *Prog. Nucl. Acid Res. Mol. Biol.* **14**: 1–37.

Arber W. & D. Dussoix. 1962. Host specificity of DNA produced by *Escherichia coli*. I. Host controlled modification of bacteriophage lambda. *J. Mol. Biol.* **5**: 18–36.

Arber W. & S. Linn. 1969. DNA modification and restriction. *Annu. Rev. Biochem.* **38**: 467–500.

Baltimore D. 1970. Viral RNA-dependent DNA polymerase. *Nature,* **226**: 1209–1211.

Blattner F.R., B.G. Williams, A.E. Blechl, K. Denniston-Thompson, D.O. Kiefer, D.D. Moore, J.W. Schumm, E.L. Sheldon & O. Smithies. 1977. Charon phages: safer derivatives of bacterio,hage lambda for DNA cloning. *Science,* **196**: 161–169.

Boyer H.W. 1971. DNA restriction and modification mechanisms in bacteria. *Annu. Rev. Microbiol.* **25**: 153–176.

Chan H.W., M.A. Israel, C.F. Garon, W.P. Rowe & M.A. Martin. 1979. Molecular cloning of polyoma virus DNA in *Escherichia coli:* plasmid vector system. *Science,* **203**: 883–892.

Chang L.M.S. & F.J. Bollum. 1971. Enzymatic synthesis of oligodeoxynucleotides. *Biochemistry* **10**: 536–542.

Curtiss R. III. 1976. Genetic manipulation of microorganisms: potential benefits and biohazards. *Annu. Rev. Microbiol.* **30**: 507–533.

Danna K. & D. Nathans. 1971. Specific cleavage of simian virus 40 DNA by restriction endonuclease of *Haemophilus influenzae. Proc. Nat. Acad. Sci. USA,* **68**: 2913–2917.

Danna K.J., G.H. Sack & D. Nathans. 1973. Studies of simian virus 40 DNA. VII. A cleavage map of the SV40 genome. *J. Mol. Biol.* **78**: 363–376.

Dussoix D. & W. Arber. 1962. Host specificity of DNA produced by *Escherichia coli*. II. Control over acceptance of DNA from infecting phage lambda. *J. Mol. Biol.* **5**: 37–49.

Freifelder D. 1978. *Recombinant DNA: Readings From Scientific American.* W.H. Freeman and Co., San Francisco.

Kelley T.J. & H.O. Smith. 1970. A restriction enzyme from *Haemophilus influenzae*. II. Base sequence of the recognition site. *J. Mol. Biol.* **51**: 393–409.

Lee A.S. & R.L. Sinsheimer. 1974. A cleavage map of bacteriophage $\times$174. *Proc. Natl. Acad. Sci. USA,* **71**: 2882–2886.

Lobban P.E. & A.D. Kaiser. 1973. Enzymatic end-to-end joining of DNA molecules. *J. Mol. Biol.* **78**: 453–471.

Luria S.E. 1953. Host induced modification of viruses. *Cold Spring Harbor Symp. Quant. Biol.* **18**: 237–244.

Maxam A.M. & W. Gilbert. 1977. A new method for sequencing DNA. *Proc. Natl. Acad. Sci. USA,* **74**: 560–564.

Maxam A.M., R. Tizard, K.G. Skryabin & W. Gilbert. 1977. Promoter region for yeast 5S ribosomal RNA. *Nature,* **267**: 643–645.

Meselson M., R. Yuan & J. Heywood. 1972. Restriction and modification of DNA. *Annu. Rev. Biochem.* **41**: 447–466.

Mulder C., J.R. Arrand, H. Delius, W. Keller, U. Pettersson, R.J. Roberts & P.A. Sharp. 1974. Cleavage maps of DNA from adenovirus types 2 and 5 by restriction endonucleases EcoRI and HpaI. *Cold Spring Harbor Symp. Quant. Biol.* **39**: 397–400.

Roberts R.J. 1976. Restriction endonucleases. *Crit. Rev. Biochem.* **4**: 123–164.

Sanger F. & A.R. Coulson. 1975. A rapid method for determining sequences in DNA by primed synthesis with DNA polymerase. *J. Mol. Biol.* **94**: 441–448.

Sharp P.A., B. Sugden & J. Sambrook. 1973. Detection of two restriction endonuclease activities in *Haemophilus parainfluenzae* using analytical agarose-ethidium bromide electrophoresis. *Biochemistry* **12**: 3055–3063.

Sinsheimer R.L. 1977. Recombinant DNA. *Annu. Rev. Biochem.* **46**: 415–438.

Smith H.O. & K.W. Wilcox. 1970. A restriction enzyme from *Haemophilus influenzae*. I. Purification and general properties. *J. Mol. Biol.* **51**: 379–391.

Southern E.M. 1975. Detection of specific sequences among DNA fragments separated by gel electrophoresis. *J. Mol. Biol.* **98**: 503–517.

Struhl K., J.R. Cameron & R.W. Davis, 1976. Functional genetic expression of eukaryotic DNA in *Escherichia coli*. *Proc. Natl. Acad. Sci. USA*, **73**: 1471–1475.

Temin H.M. & S. Mizutani. 1970. RNA-dependent DNA polymerase in virions of rous sarcoma virus. *Nature*, **226**: 1211–1213.

Wu R. 1978. DNA sequence analysis. *Annu. Rev. Biochem.* **47**: 607–634.

Topic 13
Eukaryotic Genetics:
Mendel and his Laws

OUTLINE
Mendel and the history of classical genetics
Mendel's first law: The Principle of Segregation
Incomplete dominance
Molecular model of genetic dominance
Mendel's second law: The Principle of Independent
  Assortment.

In an earlier Topic we discussed the behaviour of
chromosomes in meiosis. We can now relate that
behaviour to genetics by discussing the segregation
of genes as it was first studied by Gregor Mendel.
Mendel's experiments provided the first contributions to our knowledge of chromosomal genetics
and it is ironic that the significance of his work
was not realised until almost 30 years after his
death.

Before discussing Mendel's work, I shall present
a synopsis of Mendel's life and describe some of the
significant developments in classical genetics from
his time to the 1920s, by which time most of the
basic concepts had been established.

MENDEL AND THE HISTORY OF
CLASSICAL GENETICS

1822. Gregor Mendel was born in Austria.

1843. Mendel was admitted as a novice at a
monastery in Brünn.

1847. Mendel became a priest.

1850. Mendel failed an examination for a teaching
certificate in natural science.

1854–55. Mendel obtained 34 strains of peas and
checked them for consistency of characteristics.

1856–63. Mendel conducted his famous pea experiments concerning gene segregation.

1865. Mendel read a paper on his results to the
Brünn Society of Natural History.

1866. A paper on his work was published in the
Proceedings of the Brünn Society of Natural
History. His work was largely ignored.

1884. Gregor Mendel died after having held an
administrative position for many years.

Mendel's work and the significance of it were
rediscovered at about the turn of the century.
Between 1866 and then a number of important
results, of relevance to genetics (Mendelism), were
obtained.

1875. O. Hertwig showed that the nucleus of the cell
was required for fertilisation and cell division, and
hence presumably contained information necessary for those activities.

1882–1885. E. Strasburger and W. Flemming
showed that nuclei contained chromosomes, and
A. Weissmann proposed a theory of heredity and
development in which chromosomes contained the
hereditary material. Incorrectly, he thought each
chromosome had the information for the whole
organism, but at least it was an hypothesis worthy
of testing.

Then around 1900, three different workers produced results that confirmed Mendel's work in the
segregation of factors responsible for characters
in organisms.

1900. H. de Vries found 'Mendelian' segregation
ratios for crosses of a number of different plant
species and published a paper mentioning Mendel's
results. C. Correns also published data for maize
and peas confirming Mendel's conclusions. Finally
E. von Tschermark, working with peas, obtained
ratios paralleling those found by Mendel.

From that time on, a large number of basic genetic principles were discovered in a variety of systems.

1902. W. Bateson showed that Mendelian principles apply to animals. He used chickens for his experiments. In addition, Bateson coined the terms 'genetics', 'zygote' and 'allelomorph' (now shortened to 'allele'), and proposed F1 and F2 as symbols for the two generations of progeny commonly followed in genetic crosses. In 1909 W. Johannsen proposed the term 'gene' to replace Mendelian 'factor'.

1902–3. As a result of his work with grasshoppers, W.S. Sutton proposed that different pairs of chromosomes become oriented at random in meiosis and that this was responsible for the independent segregation of separate pairs of genes. This was a very important hypothesis since it related chromosomes to Mendelian factors.

1905. W. Bateson and R.C. Punnett showed that not all genes segregated independently; that is, two genes in sweet pea showed incomplete linkage. Also in this year, N.M. Stevens and E.B. Wilson separately showed that there were different chromosomes in the two sexes of many insects: XX in females and XY in males.

1909. In a piece of work that is very important in genetics, F.A. Janssens investigated exchanges between chromosome strands and proposed that they followed the formation of chiasmata (singular = chiasma).

1910. The fruit fly, *Drosophila melanogaster* made its appearance in genetics laboratories. In this year T.H. Morgan found the first sex-linked gene, *white*, and soon many others were discovered. This was the beginning of a long period of work with *Drosophila* by T.H. Morgan and many colleagues that established a number of important genetic principles. Only some of their contributions to genetics will be considered in the following.

1911. Linkage between two sex-linked genes, *yellow* and *white* was shown. Morgan proposed that linkage was the result of the genes involved being located in the same chromosome pair. 'Breaking up' of the linkage to produce 'recombinant' flies

was proposed to be the result of 'crossing-over' between the genes. Further, Morgan reasoned that close linkage implied infrequent crossing-over between the genes. This was a major breakthrough since it tied together data following the inheritance of genes and data concerning chromosomes and meiosis. In addition, the conclusions fitted Janssens' chiasma hypothesis of 1909.

1913. Once the relationship of crossing-over and linkage was established, A.H. Sturtevant took the next logical step in constructing the first linear chromosome map of five sex-linked genes. This was the beginning of genetic mapping as we know it today.

1919. T.H. Morgan and C.B. Bridges discovered autosomal linkage in *Drosophila melanogaster*.

1927. H.J. Müller used X rays to generate mutations in *Drosophila* as a basis for further genetic studies.

1931. C. Stern working with *Drosophila*, and H.B. Creighton and B. McClintock working with maize, showed that genetic recombination was accompanied by a physical exchange of homologous chromosomes.

After that, genetics ballooned. Many other organisms became the subject of investigation, biochemical pathways were dissected using genetic mutants, DNA was confirmed as the genetic material and molecular genetics studies of cell function were begun. Mendel's work and some of the classical genetics experiments will now be discussed.

MENDEL'S FIRST LAW: THE PRINCIPLE OF SEGREGATION

Mendel studied the pea, *Pisum sativum*. This plant is normally self-fertilising but can be emasculated to enable controlled crosses to be made. Mendel examined only pure-breeding strains (homozygous strains). He looked at only one character at a time in the early experiments, and he counted everything, probably because of his training in the physical sciences.

In his pea strains, Mendel noticed a number of alternate phenotypes, for example round v. wrinkled seeds, green v. yellow pods, long v. short stems, etc. He made crosses between pure-breeding strains and examined the progeny for the parental phenotypes. An example is shown in Fig. 13.1.

P　　round　×　wrinkled

F1　　　round

Fig. 13.1　Phenotype of progeny of a cross between two pure-breeding strains of peas.

Since the F1 resembled only one of the two parents rather than showing a blend of the two characteristics, we can describe the round character as being dominant over the wrinkled character. Conversely, wrinkled is recessive to round. The F1 was then selfed to produce the results shown in Fig. 13.2.

F1　×　F1　　round　×　round

F2　　　　5474 round : 1850 wrinkled

Fig. 13.2　F2 phenotypes and ratios resulting from selfing the F1 of the cross shown in Fig. 13.1.

The ratio of F2 phenotypes is 2.96 : 1, or roughly a 3 : 1 ratio of round to wrinkled. To obtain more information about F2 progeny, Mendel selfed 565 of them and examined the progeny for the two characters. He found that 193 of the F2s gave only round progeny and 372 gave both round and wrinkled progeny. This is approximately a 1 : 2 ratio.

Mendel drew some interesting conclusions from the data. He recognised that to get pure-breeding lines, as his parentals were, both egg and pollen must be of the same type. When the F1, which showed only one parental character, was selfed, the F2 progeny showed *both* parental characters. This showed that the F1 must be carrying one copy of each character and the F2s are produced in the

ratios found if it is assumed that the gamete types occur in equal frequency and the uniting of egg and pollen occurs by chance. We can relate this now to chromosome behaviour. That is, somatic cells are diploid and contain one chromosome from each parent. Gametes are haploid and are produced by meiosis. The round × wrinkled cross can then be illustrated using genetic symbols (Fig. 13.3). (Note

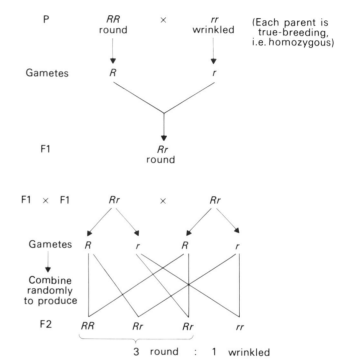

Fig. 13.3　Principle of segregation : Genotypic representation of the parental, F1 and F2 generations of the crosses shown in Figs. 13.1 and 13.2.

that these symbols are in common usage for plant systems and that alternative symbols are used in animal and other systems. These will be introduced as examples are presented.) The F1 is heterozygous and has a round phenotype owing to the dominance of the R allele over the r allele, and the F2 derived by selfing the F1 has a 3 : 1 phenotypic ratio of round to wrinkled.

A more mathematical way of representing the production of F2 by random association of gametes is to express the proportion of each gamete as a

P   *Rr*   ×   *Rr*

Gametes   ½ *R* ½ *r*   ½ *R* ½ *r*

Random combination of gametes results in

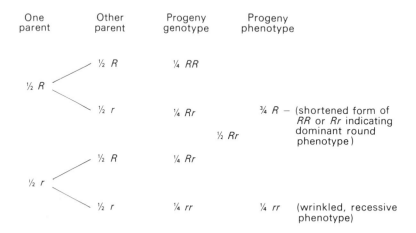

One parent	Other parent	Progeny genotype	Progeny phenotype

½ *R*

   ½ *R*   ¼ *RR*

   ½ *r*   ¼ *Rr*   ¾ *R* − (shortened form of *RR* or *Rr* indicating dominant round phenotype)

   ½ *Rr*

½ *r*

   ½ *R*   ¼ *Rr*

   ½ *r*   ¼ *rr*   ¼ *rr* (wrinkled, recessive phenotype)

Fig. 13.4 Calculation of the ratios of phenotypes in the F2 of the cross of Fig. 13.3 using a 'mathematical' approach.

fraction of the total number of gamete types. The F1 × F1 then becomes as shown in Fig. 13.4. The data also illustrate the fact that the genotypic ratio in the F2 is 1 *RR* : 2 *Rr* : 1 *rr*. As was indicated before, this was tested by selfing the F2 strains with the dominant phenotypes, thus showing a 1 : 2 ratio of *RR* : *Rr* (Fig. 13.5).

A more common way of testing for this, and one that overcomes the problem that most organisms do not 'self', is using the *testcross*, which is a cross of an unknown with a homozygous recessive parent. We can illustrate this for the same *RR*, *Rr* question (Fig. 13.6). Here 1/3 of the F2 round progeny will give all round progeny in a testcross and 2/3 of the F2 rounds will give 1/2 round and 1/2 wrinkled progeny.

From all of this, Mendel proposed his first 'law,' the *Principle of Segregation*: that (a) the gametes are pure in that they can carry only one allele of each gene (i.e. they are haploid), and (b) the gametes segregate from each other and new progeny are produced by the random combination of gametes from the two parents. Thus, in proposing the 'law' Mendel had distinguished between the factors that caused the traits and the traits themselves.

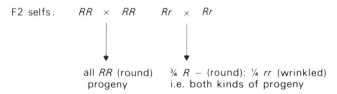

F2 selfs: *RR* × *RR*   *Rr* × *Rr*

all *RR* (round) progeny   ¾ *R* − (round): ¼ *rr* (wrinkled) i.e. both kinds of progeny

Fig. 13.5 Determination of the genotypes of the F2 round progeny of Fig. 13.4 by selfing.

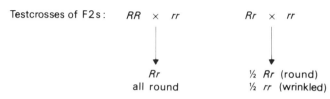

Testcrosses of F2s: *RR* × *rr*   *Rr* × *rr*

Rr all round   ½ *Rr* (round) ½ *rr* (wrinkled)

Fig. 13.6 Determination of the genotypes of the F2 round progeny of Fig. 13.4 by testcrossing with a homozygous recessive strain.

INCOMPLETE DOMINANCE

In the examples discussed up to now, there was complete dominance of one character over the alternative character such that heterozygotes are phenotypically indistinguishable from the homozygous dominant individuals. This is not always the case. That is, there are some genes that exhibit incomplete dominance and therefore the three genotypes *AA*, *Aa* and *aa* have three distinct phenotypes. One example of this is flower colour in snapdragons (*Antirrhinum*) where *AA* flowers are red, *aa* flowers are white and *Aa* flowers are pink. A cross of two pink-flowered plants will therefore, produce plants with red, pink and white flowers in a ratio of $1:2:1$, respectively. The simple explanation here is that *AA* individuals make red pigment, *aa* plants make white pigment, and *Aa* plants make both red and white pigments which blend to produce pink flowers.

Other examples of incomplete dominance are shown by Blue Andalusian chickens which are heterozygous for a pair of alleles one of which, in homozygous state, produces black feathers and the other of which, when homozygous, produces white feathers and frizzle fowl. In the latter, *FF* chickens have normal feathers and *ff* chickens have brittle and curly feathers that often break off. The *ff* individuals have abnormalities of a number of organs also. Here the *Ff* heterozygotes have a 'mild frizzle' phenotype that is intermediate between the phenotypes of the two homozygotes.

MOLECULAR MODEL OF GENETIC DOMINANCE

In this Topic we have discussed dominance and recessiveness with respect to genes and the characters they control. In most cases the mutant allele of a gene is completely recessive to the wild-type allele so that the heterozygote carrying the two alleles results in a wild-type phenotype. One explanation for this is the following. If the gene codes for an enzyme, then the mutant allele may produce a protein that lacks or has very little enzyme activity (e.g. a missense mutation) or it may produce no protein or a protein fragment (e.g. a nonsense mutation). Thus, barring regulation of transcription or translation rate, the heterozygote will produce approximately half the amount of active enzyme as the homozygous wild type. If that amount is all that is necessary for the cell or organism to carry out normal biochemical functions, then a wild-type phenotype will result and by definition the wild-type allele is completely dominant over the mutant allele. Presumably this situation has evolved through natural selection.

There are many examples of mutant alleles that are dominant over the wild-type allele. One explanation for this is that the locus codes for an enzyme that, in the mutant, has a greater affinity for the substrate of the reaction that it catalyses than does the wild-type enzyme. However, the mutant enzyme is unable to catalyse the reaction or it does so with very low efficiency. Thus, a mutant phenotype results either when the mutant allele is homozygous or when it is heterozygous with the wild-type allele. An example of this is the dominant *Stubble* (*Sb*) mutation of *Drosophila melanogaster* which results in short bristles. This is one of the genes used in an example of genetic mapping in the next Topic.

MENDEL'S SECOND LAW: THE PRINCIPLE OF INDEPENDENT ASSORTMENT

After having shown that each of the pairs of alternate characters behaved in the same way as the round/wrinkled pair, Mendel turned his attention to analysing data for the segregation of two gene pairs in the same cross. Fortunately he chose only those gene pairs that were unlinked genetically so genetic crossing-over was not a factor in the experiments. We can consider an example involving round/wrinkled and yellow/green characters, where round and yellow are the dominant alleles, respectively (Fig. 13.7).

As the figure shows, each F1 is a double heterozygote and produces gametes of 4 types: *RY*, *Ry*, *rY*,

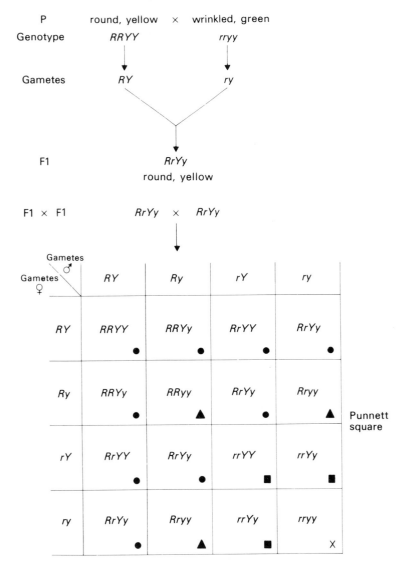

P   round, yellow × wrinkled, green
Genotype   *RRYY*   *rryy*

Gametes   *RY*   *ry*

F1   *RrYy*
                      round, yellow

F1 × F1   *RrYy* × *RrYy*

Punnett square

F2 phenotypes:
● Round, yellow 9
▲ Round, green 3
■ Wrinkled, yellow 3
X Wrinkled, green 1

Fig. 13.7 Principle of independent assortment: Demonstration of F2 9 : 3 : 3 : 1 phenotype ratio for two unlinked gene pairs by using Punnett square.

ry, each occurring with equal frequency. In the F1 × F1 self these four gamete types pair randomly in all possible combinations to give rise to the diploid progeny. Thus, in the above so-called 'Punnett square', 16 gamete combinations are produced. Owing to genetic dominance, only four distinct phenotypic classes are found, that is round, yellow; round, green; wrinkled, yellow; and wrinkled, green. These are predicted to occur with a relative proportion of 9 : 3 : 3 : 1, respectively, as a result of the random fusion of the gametes and the independent assortment of the two gene pairs into the gametes as a result of chromosome segregation during meiosis.

Generally geneticists concern themselves with the ratios of phenotypic classes. It is cumbersome to construct a Punnett square of gamete combinations and then count up the numbers of each phenotypic class. Actually it is not difficult when two gene pairs are being considered, but any more than that and it becomes complex. Therefore it is easier to deal directly with the expected ratios of phenotypic classes. If we consider the same example where the two gene pairs assort independently into the gametes, we can think about each pair in turn. We discovered earlier that an F1 self of a round/wrinkled (*Rr*) heterozygote gave 3/4 *R_* (round) and 1/4 *rr* (wrinkled) progeny. This is also the case with an F1 self of a *Yy* heterozygote. Since these two events occur independently, we can analyse the expected F2s more mathematically

(Fig. 13.8). The *independent assortment* of the gene pairs gives the 9/16 : 3/16 : 3/16 : 1/16 or 9 : 3 : 3 : 1 ratio of F2 phenotypes. This ratio will be obtained for double heterozygote selfs (*AaBb* × *AaBb*) whenever gene *A* is unlinked to gene *B*.

Finally, the testcross can be used to analyse the genotypes of the F1 or F2 progeny of a cross involving two gene pairs. The expected ratios of phenotypes among the progeny of such crosses are shown in Table 13.1. All of the patterns of gene segregation shown in the table are directly related to chromosome behaviour during meiosis.

Table 13.1 Proportions of phenotypic classes expected from testcrosses of strains with various genotypes for two gene pairs.

| Testcrosses | Proportion of phenotypic classes | | | |
	A_B_	A_bb	aaB_	aabb
$AABB \times aabb$	1	0	0	0
$AaBB \times aabb$	$\frac{1}{2}$	0	$\frac{1}{2}$	0
$AABb \times aabb$	$\frac{1}{2}$	$\frac{1}{2}$	0	0
$AaBb \times aabb$	$\frac{1}{4}$	$\frac{1}{4}$	$\frac{1}{4}$	$\frac{1}{4}$
$AAbb \times aabb$	0	1	0	0
$Aabb \times aabb$	0	$\frac{1}{2}$	0	$\frac{1}{2}$
$aaBB \times aabb$	0	0	1	0
$aaBb \times aabb$	0	0	$\frac{1}{2}$	$\frac{1}{2}$

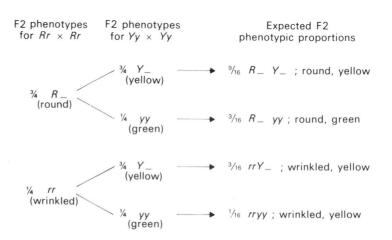

Fig. 13.8 As Fig. 13.7, but using the 'mathematical' approach.

REFERENCES

Bateson W. 1909. *Mendel's Principles of Heredity.* Cambridge University Press.

Mendel G., Experiments in Plant Hybridization (translation). In *Classic Papers in Genetics,* J.A. Peters (ed.). Prentice-Hall, Englewood Cliffs, New Jersey.

Sturtevant A.H. 1965. *A History of Genetics.* Harper and Row, New York.

Sutton W.S. 1903. The chromosomes in heredity. *Biol. Bull.* **4**: 213–251.

Tschermark-Seysenegg E. von. 1951. The rediscovery of Mendel's work. *J. Heredity,* **42**: 163–171.

Topic 14
Eukaryotic Genetics:
Meiotic Genetic Analysis in Diploids

OUTLINE

The testcross and genetic linkage
Sex chromosomes and sex linkage
Crossing over and genetic recombination
Gene mapping in diploids
Three-point testcross and gene mapping
Interference and coincidence
Summary statements.

THE TESTCROSS AND GENETIC LINKAGE

In the last Topic we discussed the principle of independent assortment which was discovered by Mendel. Simply stated, if an individual that is doubly heterozygous for two recessive, unlinked genes is crossed with another individual of the same genotype (i.e. $AaBb \times AaBb$), the *phenotypic* ratio in the resulting progeny will be 9 AB : 3 Ab : 3 aB : 1 ab. In a testcross $AaBb \times aabb$, the phenotypic ratio of the progeny in this case would be 1 AB : 1 Ab : 1 aB : 1 ab. The testcross, then, is a very useful cross to use to determine whether two genes are linked or unlinked since, if there is significant deviation from the 1 : 1 : 1 : 1 ratio such that too many parental genotypes and too few recombinant genotypes are found, then the two genes in question are considered linked (Fig. 14.1).

In the example, the recombinant types can only be produced by a physical exchange of homologous chromosome parts during meiosis in the double heterozygous F1 individual. As we have mentioned before, this event is called crossing-over, and if we assume for the moment that crossing-over is a random event along the chromosome, then the proportion of the total progeny that are recombinant types will be directly related to how far apart the two genes are on the chromosome. Genes very close together, for example, will be separated by crossing-

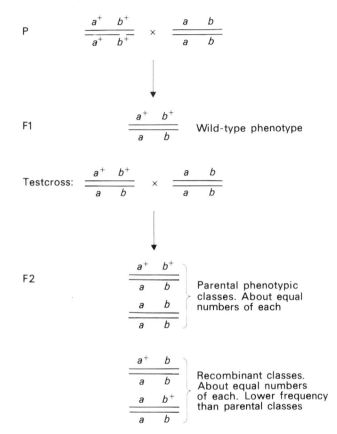

Fig. 14.1 Representation of a testcross when two genes are linked. Here a and b are linked genes on a chromosomes. The alleles a and b are recessive to the wild-type alleles a^+ and b^+. An excess of parental progeny over recombinant progeny indicates that the two genes are linked.

over quite rarely as compared with genes far apart. Indeed, as we shall develop here (and as we have already discussed for prokaryotic systems) we can use the frequency of recombinant types (the recombination frequency) to generate genetic maps. Note that the recombination frequency calculated for two genes from testcross data reflects physical exchanges of the chromosomes and is not related to the allelic forms of the genes. Therefore, the same percentage of recombinant types should be found either when both wild-type alleles are carried on the same chromosome in the F1 $\left(\dfrac{a^+\ b^+}{a\ b}\right)$ (a situation known as coupling) or when each chromosome carries one of the wild-type alleles $\left(\dfrac{a^+\ b}{a\ b^+}\right)$ (a situation known as repulsion).

Historically, T.H. Morgan working with the fruit fly *Drosophila melanogaster* found that many genes did not segregate independently of other genes during meiosis. Some genes, in fact, appeared to be associated in groups based on tests for genetic linkage such as those described above, and he identified four *linkage groups*. Significantly the number of linkage groups is the same as the haploid number of chromosomes of *D. melanogaster* and we can generalise this to all other organisms as well — the number of linkage groups is equivalent to the haploid number of chromosomes.

SEX CHROMOSOMES AND SEX LINKAGE

As was mentioned in an earlier Topic, the chromosome complement of animal cells consists of the autosomes and the sex chromosomes. In most cases the male of the species is XY and the female is XX with respect to the sex chromosomes. One of the distinguishing features about the sex chromosomes is that the X and Y are not homologous and in fact very few genes have been found on the Y chromosome that also occur on the X (the 'bobbed' gene of *Drosophila* is an example). For our purposes, then, we can consider the Y chromosome to be 'silent' in terms of discussing the dominance or recessiveness of genes on the X. Clearly the Y does contain genes and they play a very important role in, for example, development, but they are different genes from those on the X. In an XY individual the genes carried on the X are described as being hemizygous (literally 'half zygote') and

constitute the X-linkage group. Characters resulting from genes on the X chromosome are referred to as being sex-linked, although X-linked would be a better term for most animal systems that are studied experimentally. Sex-linked genes exhibit different segregation ratios from autosomal genes and this distinction will be made apparent in the following examples.

In *Drosophila* there is a recessive mutation called white (*w*) which results in individuals with white eyes instead of the red eyes characteristic of the wild type. When reciprocal crosses are done between wild type and the white strain the two F1s differ phenotypically (Fig. 14.2).

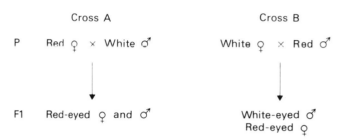

Fig. 14.2 Phenotypic results of reciprocal crosses of a strain carrying a sex-linked, recessive mutation (white eyes) with the wild type (red eye).

If the gene in question was located on an autosome, then both F1s would be expected to show the wild-type trait. The formal explanation here is that the white gene is located on the X chromosome and the two crosses can be rewritten using genotype symbols. The symbol ⌐ indicates the Y chromosome, and ‖ indicates the two homologous X chromosomes in this case (Fig. 14.3).

We showed earlier that a 9 : 3 : 3 : 1 ratio will result in crosses of strains carrying unlinked genes when the F1s are crossed together. When the two genes are autosomal, the 9 : 3 : 3 : 1 ratio will be found for both male and female F2 progeny. When a sex-linked gene and an autosomal gene are involved in a cross, the F2 ratios are different from the case of two autosomal genes. In this example, the gene symbolism will be abbreviated further and again the two mutant genes are recessive to the wild type: gene *x* is on the X chromosome and

Fig. 14.3 Genotypic diagrams of
the parental and F1 generations of
the reciprocal crosses of Fig. 14.2.

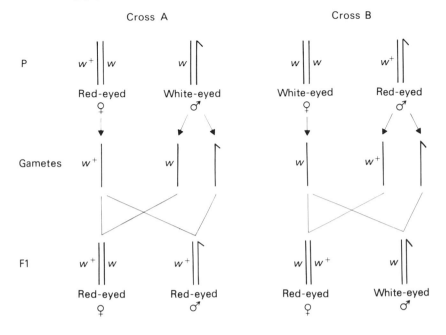

Cross A   Cross B

P

Gametes

F1

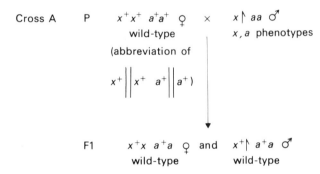

Cross A   P x^+x^+ a^+a^+ ♀   × x⚊ aa ♂
                  wild-type x, a phenotypes

                  (abbreviation of

                  x^+‖x^+ a^+‖a^+)

            F1 x^+x a^+a ♀   and x^+⚊ a^+a ♂
                  wild-type   wild-type

Fig. 14.4 Genotypes of the parental and F1
generations of a cross of a homozygous wild-type
female with a male carrying a sex-linked recessive and
an autosomal recessive mutation.

gene a is on an autosome and the example applies
to any organism with sex chromosomes. Again we
will follow the two reciprocal crosses, and calculate
the F2 ratios mathematically. Fig. 14.4 shows one
of the crosses.

Now, if we consider only the X-linked gene, the
F1 × F1 is: $x^+x \times x^+$⚊ and by random segregation
of the resulting gametes, this will result in F2 genotypes and phenotypes as shown in Fig. 14.5. And,
as was shown previously, the $a^+a \times a^+a$ cross will
generate 3/4 a^+ and 1/4 a type progeny. Since the
X chromosome and the autosome assort independently at meiosis, we can calculate the *phenotypic*
ratio of the F2 by multiplying the probabilities of
occurrence of the X and autosomal characters

Fig. 14.5 Generation of the F2
progeny with regard to the
sex-linked gene for the F1 × F1 of
Fig. 14.4.

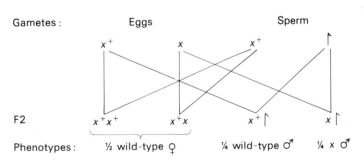

Gametes :   Eggs   Sperm

F2

Phenotypes :   ½ wild-type ♀   ¼ wild-type ♂   ¼ x ♂

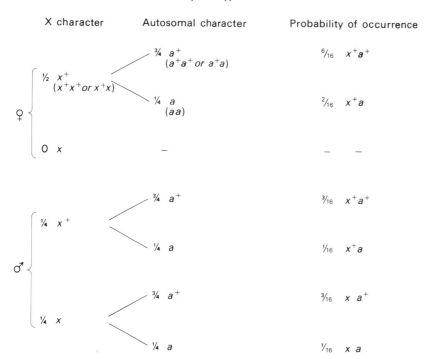

Fig. 14.6 Probability of occurrence of the F2 phenotypes for cross of Fig. 14.4.

together in all possible ways (Fig. 14.6). In this the females and males are considered separately, and x^+ and a^+ indicate the wild-type phenotypes and x and a indicate the mutant phenotypes for the X-chromosome and autosome characters, respectively. Thus for the four different phenotypes, the ratios for males and females separately, and for the two combined are shown in Fig. 14.7.

Phenotype ratios

	x^+a^+		x^+a		$x\ a^+$		$x\ a$
♀	6	:	2	:	0	:	0
♂	3	:	1	:	3	:	1
♀ and ♂	9	:	3	:	3	:	1

Fig. 14.7 Phenotypic ratios for F2 males and females of cross shown in Fig. 14.4.

As can be seen, a $9:3:3:1$ phenotypic ratio is the case when males and females are combined but quite different ratios are apparent for the two sexes considered separately. This results from the inheritance characteristics of sex-linked genes stemming from the lack of homologous genes on the Y chromosome.

Turning now to the reciprocal cross, it can be analysed in the same way as shown in Fig. 14.8. As can be seen, in this case neither sex shows a $9:3:3:1$ ratio.

CROSSING-OVER AND GENETIC RECOMBINATION

Genetic crosses can be used to show recombination of genes from generation to generation. As has been stated, this genetic recombination results from a physical exchange of parts of homologous chromosomes by a process called crossing-over. Crossing-over occurs in the first division of meiosis during

Cross B   P $xx\,aa$ ♀ × $x^+⌐\,a^+a^+$ ♂

F1 $x^+x\,a^+a$ ♀   × $x⌐\,a^+a$ ♂
              wild-type x, a^+ phenotype

Considering the X-linked gene alone:

F1 × F1 x^+x ♀   × $x⌐$ ♂

F2   ¼ x^+x , ¼ xx , ¼ $x^+⌐$, ¼ $x⌐$
          wild ♀ x ♀   wild ♂ x ♂

Combining this with the autosomal ratio in the same way as for Cross A, we have:

F2 phenotypic ratios

X character	Autosomal character	Probability of occurrence

♀ {
  ¼ x^+ (x^+x) — ¾ a^+ — $\tfrac{3}{16}$ x^+a^+
                  — ¼ a — $\tfrac{1}{16}$ x^+a

  ¼ x (xx) — ¾ a^+ — $\tfrac{3}{16}$ $x\,a^+$
              — ¼ a — $\tfrac{1}{16}$ $x\,a$
}

♂ {
  ¼ x^+ $(x^+⌐)$ — ¾ a^+ — $\tfrac{3}{16}$ x^+a^+
                  — ¼ a — $\tfrac{1}{16}$ x^+a

  ¼ x $(x⌐)$ — ¾ a^+ — $\tfrac{3}{16}$ $x\,a^+$
              — ¼ a — $\tfrac{1}{16}$ $x\,a$
}

And, summarising as for Cross A:

Phenotype ratios

	x^+a^+		x^+a		$x\,a^+$		$x\,a$
♀	3	:	1	:	3	:	1
♂	3	:	1	:	3	:	1
♀ and ♂	6	:	2	:	6	:	2

Fig. 14.8 Genotypes and phenotypes of the parental, F1 and F2 generations of a cross of a female homozygous for a sex-linked recessive and an autosomal recessive mutation, with a wild-type male.

the time when four chromatids are present for each pair of chromosomes (the four-strand stage). Proof for this will be given when meiotic genetic analysis of fungi is discussed.

That genetic recombination does result from a physical exchange of parts of homologous chromosomes was demonstrated by C. Stern in 1931. He overcame the difficulties of distinguishing the members of a pair of homologous chromosomes by using 'abnormal' chromosomes that could be distinguished cytologically. The organism he used was *Drosophila* and the cross he used was as shown in Fig. 14.9.

females have wild-type eye colour. In the case of dominant mutations, the heterozygote will show the mutant rather than the wild-type phenotype and thus the *B*/+ females here have bar-shaped eyes rather than round eyes. In addition the X chromosome with the *car* and *B* mutations is shorter than normal since part of it has broken off and is attached to chromosome 4.

In gamete formation, only two classes of sperm are produced: the Y-bearing and the X-bearing which carries the *car* allele and the wild-type allele of *B*. In producing the eggs, four gamete classes are produced: two of these result from

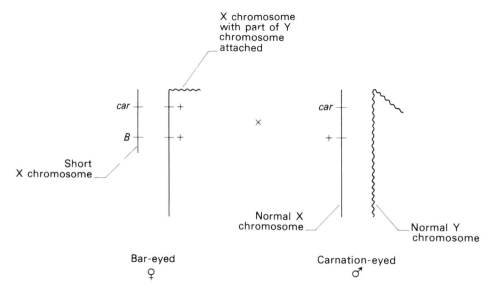

Fig. 14.9 Diagrammatic representation of the chromosomes in the male and female *Drosophila* that were crossed to show that genetic recombination results from a physical exchange of chromosomes.

The male parent carries normal X and Y chromosomes and on the X is the mutant *car* (carnation) and the wild-type allele of the *B* (bar) gene. The male therefore has carnation-coloured eyes. The female parent has two abnormal X chromosomes. One carries the wild-type alleles of both the *car* and *B* genes and in addition has a part of the Y chromosome attached to it. The other X carries the recessive mutation *car* and the dominant mutation *B*. Since the *car* mutation is heterozygous, the

meioses in which no cross-overs occurred between the *car* and *B* loci, and the other two are recombinant gametes produced by such a cross-over. The gamete types and the diploid progeny are shown in Fig. 14.10.

Examination of the chromosomes of the progeny whose phenotypes indicated that genetic recombination had occurred showed that in every case the recombination event was accompanied by an exchange of identifiable chromosome segments.

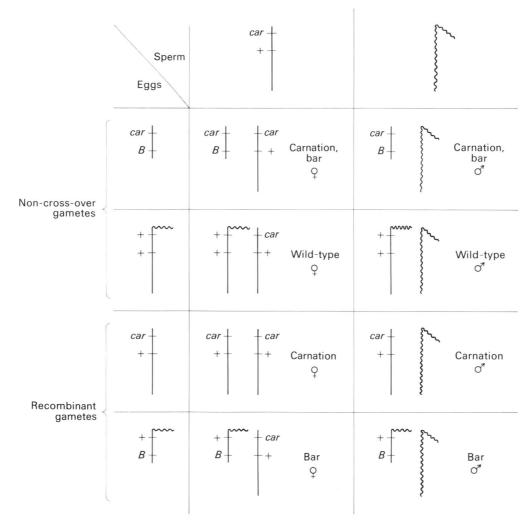

Fig. 14.10 Progeny of cross shown in Fig. 14.9 showing that genetic recombination results from crossing over.

GENE MAPPING IN DIPLOIDS

T.H. Morgan pioneered gene mapping with his experiments using *Drosophila*. He crossed a female homozygous for the sex-linked genes white-eyed (*w*) and miniature-winged (*m*) with a wild-type male and analysed the F1 and F2 progeny (Fig. 14.11).

The results show that 37.6% of the progeny are recombinant and it is expected that these will be distributed equally between females and males. The recombinant types of course result from crossing-over between the *w* and *m* loci and the same frequency of recombination would be found if the genes were in repulsion. Indeed, the recombination frequency between particular pairs of linked genes occurs at quite stable and characteristic frequencies in all organisms. The frequency found obviously depends on the two loci involved and it can be used to describe the *genetic* distance between the two loci on the linkage map. In our

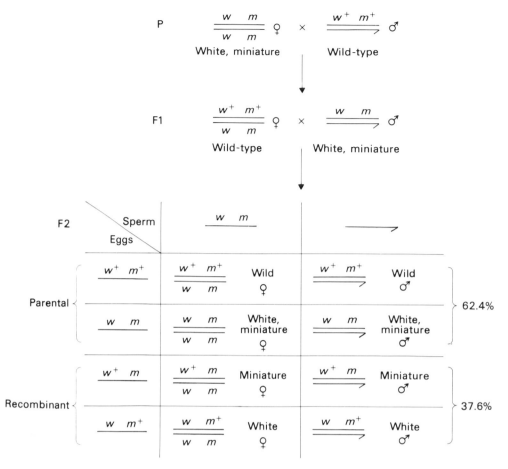

Fig. 14.11 Experimental crosses of *Drosophila* performed by T. H. Morgan to determine the map distance between the sex-linked genes, *white* and *miniature*.

example, *w* and *m* are 37.6 map units apart, meaning that a crossing-over event will occur between the two loci in 37.6% of meioses.

The case described was an example of two-point mapping analysis where the distance between two genes is determined. By doing a series of two-point crosses, the linkage relationships of a number of sex-linked genes were determined and, in 1911, A. Sturtevant realised that the data were compatible with a linear linkage map, and the first such map for the X chromosome of *Drosophila* was constructed. This was done by analysing the recombination frequencies for all possible pairwise

crosses. For example, if *a* and *b* are 5 map units apart, *b* and *c* are 7 map units apart and *a* and *c* are 2 map units apart, then we can determine the linear relationship of the three genes (Fig. 14.12).

In the mapping of recessive sex-linked genes, the crosses are set up so the F1 female is heterozygous for the two genes and the F1 male is hemizygous for both mutant alleles. Since the Y carries no homologous genes to the ones on the X, the F1 × F1 cross is essentially a testcross which is the ideal cross for mapping analysis as we have discussed. A testcross is used also for mapping of recessive autosomal genes, and the usual sequence of crosses

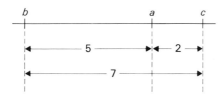

Fig. 14.12 Linear relationship of genes *a*, *b*, *c* as determined by all possible pairwise crosses and recombination analysis.

(here with genes in coupling) for a diploid organism would be as shown in Fig. 14.13.

In the case of mapping dominant mutations, the testcross is also used and here that means crossing the double heterozygote to a strain carrying the wild-type alleles of the mutant genes, which are, of course, recessive (Fig. 14.14). By convention the dominant mutation is indicated by a capital letter, e.g. *Cy* for the curly wing mutation in *Drosophila*. For this example, the wild-type allele would be signified by *Cy*$^+$.

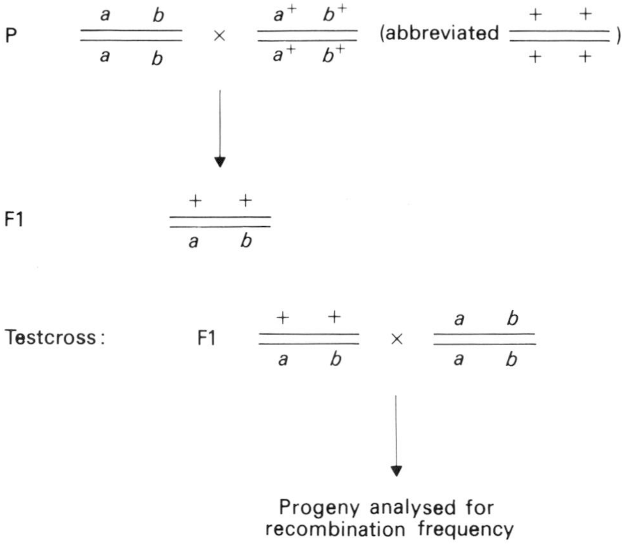

P $\dfrac{a \quad b}{a \quad b}$ × $\dfrac{a^+ \quad b^+}{a^+ \quad b^+}$ (abbreviated $\dfrac{+ \quad +}{+ \quad +}$)

F1 $\dfrac{+ \quad +}{a \quad b}$

Testcross: F1 $\dfrac{+ \quad +}{a \quad b}$ × $\dfrac{a \quad b}{a \quad b}$

Progeny analysed for recombination frequency

Fig. 14.13 Testcross procedure for mapping autosomal recessive mutant genes.

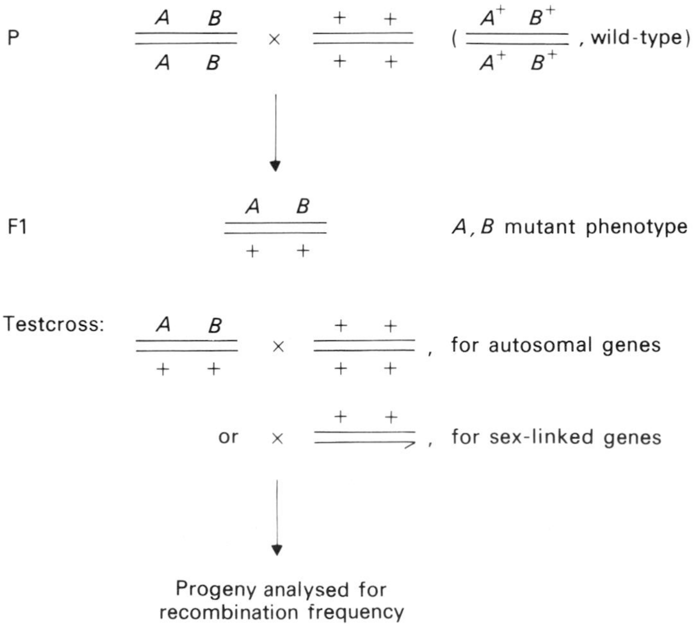

P $\dfrac{A \quad B}{A \quad B}$ × $\dfrac{+ \quad +}{+ \quad +}$ ($\dfrac{A^+ \quad B^+}{A^+ \quad B^+}$, wild-type)

F1 $\dfrac{A \quad B}{+ \quad +}$ *A*, *B* mutant phenotype

Testcross: $\dfrac{A \quad B}{+ \quad +}$ × $\dfrac{+ \quad +}{+ \quad +}$, for autosomal genes

or × $\dfrac{+ \quad +}{}$, for sex-linked genes

Progeny analysed for recombination frequency

Fig. 14.14 Testcross procedure for mapping autosomal (or sex-linked) dominant mutant genes.

THREE-POINT TESTCROSS
AND GENE MAPPING

We have indicated that genetic recombinants result from crossing-over, the interchange of segments between homologous chromosomes. Testcross data, then, provide recombination frequencies as well as giving an estimate of crossing-over frequency. The latter may not be accurate owing to multiple cross-overs occurring between genes, and this can complicate the computation of map distance. If we assume that crossing-over occurs randomly along the length of a chromosome, then the cross-over frequency reflects the distance apart of any two genes. However, the only way that cross-overs can be counted is by looking at recombinant progeny that must have resulted from a cross-over. Therefore, especially if genes are a reasonable distance apart, double- and other even-numbered cross-overs can occur between them, but they will not be counted since the resulting progeny will not be recombinant (Fig. 14.15).

This phenomenon is most significant for double cross-overs as the frequency of higher number cross-overs between two genes is relatively smaller. They do contribute to the map distance calculations since odd-numbered cross-overs generate recombinant gametes and even-numbered cross-overs generate parental gametes. The frequency of occurrence of multiple cross-overs between genes can be computed easily. For example, if the probability of a single cross-over occurring is 10%, then the probability of two cross-overs is $10\% \times 10\% = 1\%$, the probability of three cross-overs is $10\% \times 10\% \times 10\% = 0.1\%$, and so on. For map distances under 5 map units, double crossing-over occurs only infrequently and thus an accurate linkage map can best be constructed by a series of two-point test-crosses where each pair of genes is closely linked, so that recombination frequencies give an accurate measure of cross-over frequencies.

One way to obtain good recombination data is to use three-point testcrosses involving three genes located within a relatively short segment of a chromosome. If we put a gene c between a and b in the above example, we can see one of the advantages of such a testcross in that the double cross-over would be recognisable since it generates recombinant progeny (Fig. 14.16). Obviously double cross-overs in the a–c and b–c region would not be detected amongst the progeny.

To illustrate how a three-point testcross can be used to map genes, we shall discuss data obtained in laboratory experiments with the fruit fly *Drosophila melanogaster*. The data presented are

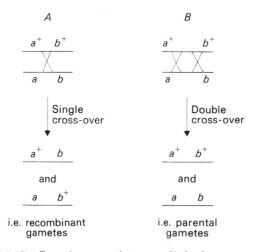

Fig. 14.15 Crossing over between linked genes a and b: (*A*) A single cross-over generates recombinant gametes, (*B*) A double cross-over generates parental gametes.

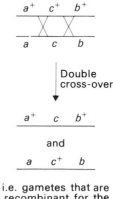

Fig. 14.16 The presence of a third gene, c, between genes a and b allows certain double cross-overs between a and b to be detected as recombinants for c.

testcross values involving the segregation of three mutant alleles. These are: Gl = glued eye (smooth eye surface), a dominant mutation; Sb = stubble bristles (shorter than wild type), a dominant mutation; and ri = radius incomplete (a longitudinal wing vein has a gap in it and henceforth will be called 'gap'), a recessive mutation. The normal allele of each of these genes will be indicated by a '+'. The three genes are known to be linked on chromosome 3 and the following testcross shown in Fig. 14.17 was performed to obtain data for map distance calculations. In this, the order of the genes is arbitrarily given at the moment.

Each progeny type has the phenotype dictated by the genotype of the gamete produced by the heterozygous parent since that always pairs with a gamete from the other parent that carries all recessive alleles. Thus, class 1 is $\dfrac{Gl\,Sb\,+}{+\,+\,ri}$ in genotype and this has a glued, stubble phenotype.

Turning now to the analysis of the data. The heterozygous parent in the cross was $\dfrac{Gl\,Sb\,+}{+\,+\,ri}$ and the gametes produced from this from meioses with no cross-overs are $Gl\,Sb\,+$ and $+\,+\,ri$ which give rise to classes 1 and 2 of the progeny. These are the parental-type progeny and these classes have the largest numbers of individuals, and with about equal numbers in each class as would be expected from such a reciprocal event. In general, even if the parental genotype was not known, the parental-type progeny can be identified by the fact that they have the largest numbers.

The double cross-over progeny can also be picked out by inspection. A double cross-over involves the simultaneous occurrence of two events, each of which alone has a relatively low probability of occurrence. Thus the double cross-over gametes are the least frequent reciprocal pair, in this case classes 5 and 6, $Gl\,Sb\,ri$ and $+\,+\,+$, respectively.

Testcross:

$$\frac{Gl\quad Sb\quad +}{+\quad +\quad ri} \times \frac{+\quad +\quad ri}{+\quad +\quad ri}$$

(glued, stubble,   (gap phenotype)
phenotype)

Progeny:

Class	Phenotypes of testcross progeny	Number of individuals	Genotypes of gamete from heterozygous parent responsible for phenotype		
1	glued, stubble	433	Gl	Sb	$+$
2	gap	456	$+$	$+$	ri
3	glued, gap	29	Gl	$+$	ri
4	stubble	38	$+$	Sb	$+$
5	glued, stubble, gap	2	Gl	Sb	ri
6	wild-type	3	$+$	$+$	$+$
7	glued	48	Gl	$+$	$+$
8	stubble, gap	61	$+$	Sb	ri

Total progeny: 1070

Fig. 14.17 Three-point mapping analysis showing the testcross used and the resultant progeny.

Parental
gametes

Gl   Sb   +
_____

and

+   +   ri
_____

Double cross-over
gametes

Gl   ri   Sb
_____

and

+   +   +
_____

Fig. 14.18 Comparison of the parental and double cross-over gamete types from the testcross progeny data of Fig. 14.17.

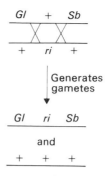

Fig. 14.19 Arrangement of the three genes on the chromosome is *Gl ri Sb*. This figure shows how the two cross-overs in the parent generates the double cross-over recombinant types.

Given that the parental and double cross-over progeny have been established, the order of the genes in the chromosome can be determined. In the diagram of the double cross-over presented before, recall that the double cross-over changed the arrangement of the central gene with respect to the two other genes. To aid the analysis of the data, the parental and double cross-over gamete types can be illustrated (Fig. 14.18). The only arrangement of the three genes that is compatible with the data is *Gl ri Sb* and the heterozygous parent is

$\dfrac{Gl + Sb}{+ \; ri \; +}$. If we illustrate the two cross-overs, the double cross-over types are generated as shown in Fig. 14.19.

Now the data can be rewritten, and for convenience we will call the region between *Gl* and *ri*, region I and the region between *ri* and *Sb*, region II (Fig. 14.20).

Testcross: $\dfrac{Gl \quad + \quad Sb}{+ \uparrow ri \uparrow +}$ × $\dfrac{+ \quad ri \quad +}{+ \quad ri \quad +}$

Region   Region
I   II

Progeny:

Class	Gametic type	Number of individuals	Type
1	*Gl + Sb*	433	Parental types; no cross-over
2	*+ ri +*	456	
3	*Gl ri +*	29	Recombinant — single cross-over (sco) region I
4	*+ + Sb*	38	
5	*Gl + +*	48	Recombinant — sco region II
6	*+ ri Sb*	61	
7	*Gl ri Sb*	2	Recombinant — double cross-over regions I and II
8	*+ + +*	3	

Total progeny = 1070

Fig. 14.20 A rewritten form of the cross of Fig. 14.17 based on the newly determined gene order.

Map distances can then be calculated as before, that is by computing the frequency of crossing-over between two genes. In the example, the crossing-over between the *Gl* and *ri* loci (i.e. in region I) produces classes 3 and 4 (single cross-over, region I) and classes 7 and 8 (double cross-over, regions I and II) above. Out of the 1070 progeny produced then, 72 represent cross-overs between the two loci. This is 6.7% of the progeny, and thus *Gl* and *ri* are 6.7 map units apart. In other words, the *Gl–ri* map distance

$$= \frac{\text{frequency of sco, region I} + \text{frequency of dco}}{\text{Total}}$$
$$\times 100$$
$$= \frac{67 + 5}{1070} \times 100$$
$$= 6.7\%$$

Similarly, the *ri–Sb* map distance involves crossovers in region II and may be calculated from

$$\frac{\text{frequency of sco, region II} + \text{frequency of dco}}{\text{Total}}$$
$$\times 100$$
$$= \frac{(48 + 61) + (2 + 3)}{1070} \times 100$$
$$= \frac{114}{1070} \times 100$$
$$= 10.7\%$$

Therefore the *ri–Sb* distance is 10.7 map units. From these data, the segment of the chromosome involving the three genes is as depicted in Fig. 14.21.

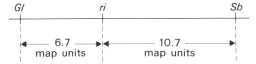

Fig. 14.21 Genetic map of the *Gl-ri-Sb* region of the chromosome determined from the data presented in Fig. 14.20.

Note that in calculating map distance from three-point testcross data, the double cross-over figure must be added to each of the single cross-over figures since in each case a double cross-over represents single crossing-over in *both* regions I and II.

INTERFERENCE AND COINCIDENCE

The map distances obtained from crosses such as those described enable the investigator to gauge whether or not the expected number of double cross-overs occur or whether the occurrence of one cross-over diminishes the probability of a second cross-over occurring close by. In our example, the *Gl–ri* distance of 6.7 map units means that 6.7% of the gametes are expected to reflect crossing-over between those two loci. Similar arguments can be made for the *ri–Sb* gene loci. Now, if we assume for the moment that a cross-over event in region I is independent of a cross-over event in region II, then the probability of cross-overs occurring simultaneously in both regions in a meiosis is equal to the product of the probabilities of the two cross-overs occurring independently. In this case, this would be $0.067 \times 0.107 = 0.0072$ or, in other words, 0.72% double cross-over progeny would be expected in the cross we analysed. Actually, only $\frac{5}{1070}$ or 0.47% double cross-overs occurred in the cross. Indeed, it is quite characteristic of three-point testcross data that the observed number of double cross-over progeny is lower than the expected number of double cross-over progeny. This signifies that the occurrence of one cross-over, perhaps by physical perturbations of the paired homologues, reduces the probability of a second cross-over occurring nearby. The phenomenon is called chromosome interference and this may vary throughout the genome. H. Müller called the ratio of

$$\frac{\text{observed double cross-over frequency}}{\text{expected double cross-over frequency}}$$

the coefficient of coincidence. For the *Gl–Sb* region of the *Drosophila* genome that we have been discussing, the coincidence is $\frac{0.47}{0.72} = 0.65$. Coincidence values normally vary from 0 to 1 and they vary inversely with interference values. Thus a coincidence of 0 means complete interference — two cross-overs did not occur simultaneously at all in that case. Conversely, a coincidence of 1 would indicate absolutely no interference. In our example, the interference value is 0.35, or in other words, only 65% of the expected double cross-overs took place in the region being studied.

SUMMARY STATEMENTS

Firstly, the most accurate mapping analysis is done when genes involved are closely linked. Owing to multiple cross-overs, and particularly double cross-overs, the distances obtained for genes far apart are usually underestimated. Note that the highest recombination frequency that can be obtained for any two genes located far apart on the same chromosome is 50% since the number of even-numbered cross-overs (producing parental progeny) between them will equal the number of odd-numbered cross-overs (producing recombinant progeny). This, of course, signifies that the genes are unlinked, yet they are on the same chromosome. Mapping them to a number of other genes spaced along the length of the chromosome would confirm that fact.

Secondly, the map distances obtained are genetic distances and reflect the probability of occurrence of crossing-over for the region being mapped. Although it is common to assume that crossing-over is random throughout the genome, it is almost certain that this is not true. For example, there is good evidence that crossing-over occurs infrequently near centromeres. Genes located in those areas, then, would appear to be closely linked genetically, whereas physically they could be far apart. Therefore, while the order of the genes is equivalent on the genetic map and on the chromosome, the genetic distance may or may not be an accurate reflection of physical distance depending on what part of the genome one is studying.

Thirdly, the genetic crosses described allow one to construct linkage maps for genes in any organism which provides useful information about the distribution of genes with related function throughout the genome.

REFERENCES

Belling J. 1933. Crossing over and gene rearrangement in flowering plants. *Genetics,* **18**: 388–413.

Bridges C.B. 1916. Nondisjunction as a proof of the chromosome theory of heredity. *Genetics,* **1**: 1–52, 107–163.

Creighton H.S. & B. McClintock. 1931. A correlation of cytological and genetical crossing over in *Zea mays. Proc. Natl. Acad. Sci. USA,* **17**: 492–497.

Gillies C.B. 1975. Synaptonemal complex and chromosome structure. *Annu. Rev. Genet.* **9**: 91–109.

Levine R.P. 1955. Chromosome structure and the mechanism of crossing over. *Proc. Natl. Acad. Sci. USA,* **41**: 727–730.

McClung C.E. 1902. The accessory chromosome — sex determinant? *Biol. Bull.* **3**: 43–84.

McKusick V.A. & F.H. Ruddle. 1977. The status of the gene map of the human chromosomes. *Science,* **196**: 390–405.

Morgan T.H. 1910. Sex-limited inheritance in *Drosophila. Science* **32**: 120–122. (In *Classic Papers In Genetics,* J.A. Peters, (ed.). Prentice-Hall, Englewood Cliffs, New Jersey.)

Morgan T.H. 1911. An attempt to analyze the constitution of the chromosomes on the basis of sex-limited inheritance in *Drosophila. J. Exp. Zool.* **11**: 365–414.

Muller H.J. 1916. The mechanism of crossing over. II. *Am. Nat.* **50**: 284–305.

Roth R. 1976. Temperature-sensitive yeast mutants defective in meiotic recombination and replication. *Genetics,* **88**: 675–686.

Stern C. 1931. Zytologisch-genetische Untersuchungen als Beweise für die Morgansche Theorie des Faktorenaustauschs. *Biol. Zbl.* **51**: 547–587.

Sturtevant A.H. 1913. The linear arrangement of six sex-linked factors in *Drosophila,* as shown by their mode of association. *J. Exp. Zool.* **14**: 43–59.

Sutton W.S. 1903. The chromosomes in heredity. *Biol. Bull.* **4**: 213–251. (In *Classic Papers In Genetics,* J.A. Peters, (ed.). Prentice-Hall, Englewood Cliffs, New Jersey.)

Westergaard M. & D. von Wettstein. 1972. The synaptonemal complex. *Annu. Rev. Genet.* **6**: 74–110.

Wilson E.B. 1905. The chromosomes in relation to the determination of sex in insects. *Science,* **22**: 500–502.

Topic 15
Eukaryotic Genetics: Fungal Genetics

OUTLINE

Fungal life cycles
    Yeast
    Neurospora crassa
    Aspergillus nidulans
Meiotic genetic analysis of yeast and *Neurospora*
    random spore analysis to determine map distance
    tetrad analysis
        gene centromere distance
        map distance between genes
    tests for gene linkage using tetrad analysis
Mitotic genetic analysis in *Aspergillus nidulans*
    mechanism of mitotic crossing-over
    formation of diploid strains
    locating genes to linkage groups by
        haploidisation
    gene mapping by mitotic recombination.

FUNGAL LIFE CYCLES

The advantages of fungi for genetic analysis are that many are haploid and some have a life cycle which enables each of the four products of a single meiosis to be analysed. The latter is called tetrad analysis. The life cycles of two fungi — *Saccharomyces cerevisiae* (a budding yeast) and *Neurospora crassa* (a mycelial-form fungus) — that can be used for tetrad analysis (as well as for molecular experiments) will be described.

Yeast life cycle

In yeast there are two mating types, α and a. The haploid vegetative cells propagate by budding. Fusion of haploid a and α cells produces a diploid cell which is stable and is propagated by budding. If the diploid a/α strain is put into conditions of nitrogen starvation, sporulation is induced which leads to meiosis. The four haploid meiotic products (the ascospores) are contained within an ascus. Two of these are a and two are α mating type. When

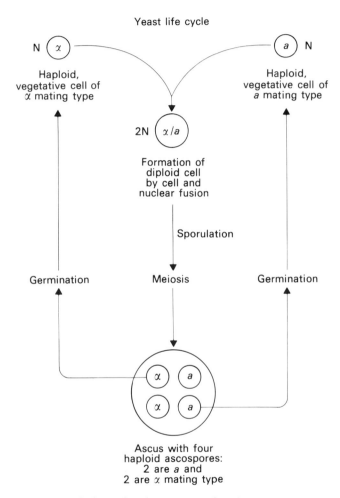

Fig. 15.1 Life cycle of the yeast, *Saccharomyces cerevisiae*.

the ascospores are released and germinate, they produce haploid vegetative cells. In this organism the ascospores are organised randomly within the ascus and thus only unordered tetrads can be isolated.

Life cycle of *Neurospora crassa*

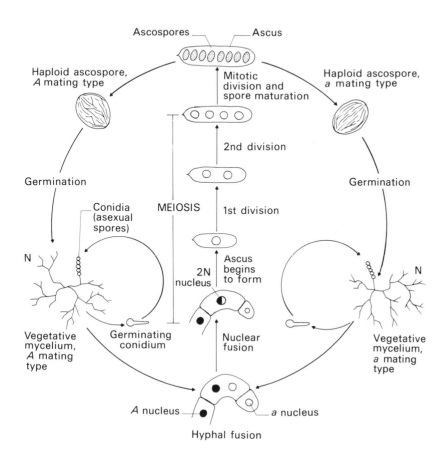

Ascospores — Ascus

Haploid ascospore,
A mating type

Mitotic
division and
spore maturation

Haploid ascospore,
a mating type

Germination

2nd division

Germination

Conidia
(asexual
spores)

MEIOSIS

1st division

N

N

Vegetative
mycelium,
A mating
type

Germinating
conidium

2N
nucleus

Ascus
begins
to form

Nuclear
fusion

Vegetative
mycelium,
a mating
type

A nucleus — — *a* nucleus

Hyphal fusion

Fig. 15.2 Life cycle of *Neurospora crassa*.

Life cycle of *Neurospora crassa*

Neurospora crassa is a haploid organism that grows vegetatively by the formation of a web-like, branching growth called a mycelium. Asexual reproduction in this organism occurs by means of spores called conidia or by propagation through fragments of the mycelium. The mycelium itself consists of cellular compartments separated by a septum with a central hole — this permits circulation of cell contents, including nuclei, throughout the mycelium.

N. crassa has two mating types, *A* and *a*, and these are determined by members of an allelic pair. Sexual reproduction occurs by fusion of nuclei of the opposite mating types to form a diploid zygote which has only a transient existence in the life cycle of this organism. The zygote undergoes meiosis in a linear tubular structure called an ascus. The two divisions occur in tandem within the ascus and then a mitotic division occurs to produce 8 ascospores arranged linearly within the ascus. These eight spores represent the four meiotic products (each doubled) arranged as the chromatids were arranged in the zygote stage, that is it is an ordered tetrad from which the spores can be manually removed in order. All of the ascospores are haploid and half are of the *A* and half are of the *a* mating type. When these germinate, they give rise to the vegetative mycelium.

The asci themselves are formed within a fruiting body called a perithecium (Fig. 15.3). When the asci are 'ripe' the ascospores are ejected through

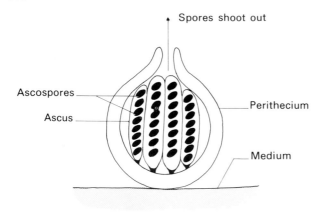

Fig. 15.3 Diagram of a cross-section through the perithecium (fruiting body) of *Neurospora* showing the sets of ascospores arranged linearly within asci.

the neck of the perithecium and may be collected for random spore analysis.

A third fungus that has been and is being studied by geneticists is the mycelial-form fungus *Aspergillus nidulans*. In this organism crossing-over can occur during mitosis and this, in concert with another phenomenon (haploidisation), can be used to map genes to particular chromosomes (see p. 174).

Life cycle of *Aspergillus nidulans*

This fungus is a haploid organism and has a colourless, multinucleate mycelium. The asexual spores, the conidia, bud off from specialised conidiophores, are uninucleate and dark green in colour in the wild type. When these spores germinate, a mycelium is produced and the vegetative cycle is completed. *Aspergillus* has a sexual cycle in which

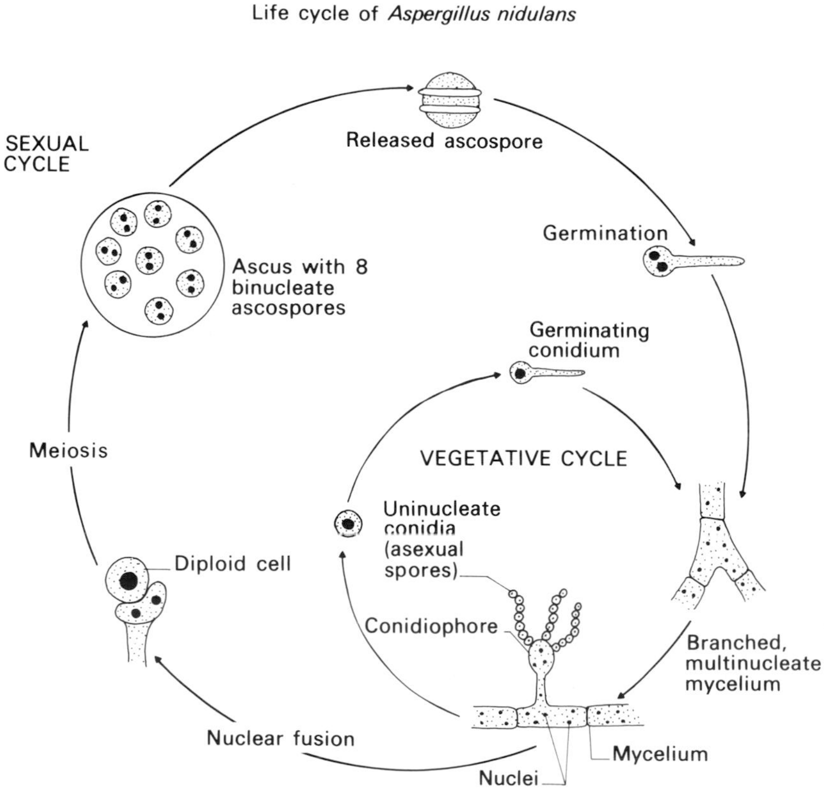

Life cycle of *Aspergillus nidulans*

SEXUAL CYCLE

Released ascospore

Germination

Ascus with 8 binucleate ascospores

Germinating conidium

Meiosis

VEGETATIVE CYCLE

Diploid cell

Uninucleate conidia (asexual spores)

Conidiophore

Branched, multinucleate mycelium

Nuclear fusion

Mycelium

Nuclei

Fig. 15.4 Life cycle of *Aspergillus nidulans*.

asci are produced with eight ascospores which, when they germinate, give rise to the vegetative mycelium. Unlike *Neurospora* which requires fusion of nuclei of two different mating types to instigate the sexual cycle, *Aspergillus* is homothallic — meaning that it can, and does, self-cross. Two nuclei from the same culture can fuse to produce a diploid nucleus which then undergoes meiosis to produce the ascospores. Since this makes the setting up of controlled genetic crosses difficult, meiotic genetic analysis is a problem in this organism. Mitotic genetic analysis, however, is possible in this fungus and this will be described after we have discussed the mechanism of mitotic recombination.

Now that the life cycles of three fungi have been described, we can turn our attention to genetic analysis in these organisms. Firstly, we will discuss meiotic genetic analysis, using *Neurospora* and yeast as examples, and then move on to mitotic genetic analysis in *Aspergillus*.

MEIOTIC GENETIC ANALYSIS OF YEAST AND *NEUROSPORA*

Random spore analysis to determine map distance

In both yeast and *Neurospora,* for example, the ascospores that are released can be collected, induced to germinate and the resulting culture analysed for phenotypic characteristics. The ascospores, then, are equivalent to progeny of the crosses described in the topic on diploid genetic analysis, and indeed two-point and three-point crosses are routinely done in these organisms to map the genes. Their haploid nature, in fact, simplifies the analysis to some extent. In the example shown in Fig. 15.5, the map distances of three linked genes *a*, *b* and *c* can be determined by counting the relative numbers of haploid parental and recombinant progeny.

If this were *N. crassa*, the two haploid parental strains would be of opposite mating type and under nitrogen starvation conditions the diploid zygote will be formed. When the zygote undergoes meiosis, the two parental types (+ + + and *abc*) are produced if no cross-overs occur in that region, and

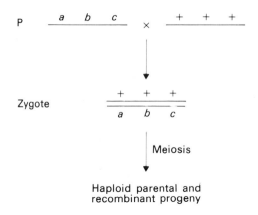

Fig. 15.5 Type of genetic cross that can be used for mapping three genes in a haploid organism, such as yeast or *Neurospora*.

six recombinant types are produced when single or double cross-overs occur in the region. Here the results are progeny, not gametes as was the case in the diploid analysis, and since they are haploid the genotype dictates the phenotype. Note that in this mapping analysis, the multiple heterozygote was constructed in the cross, and this is the configuration we have become used to in thinking about gene mapping. Thus, these procedures allow investigators to determine linkage relationships and construct linkage maps in haploid fungi (and in other haploid organisms amenable to genetic analysis) in just the same way as has been shown for diploid organisms.

Tetrad analysis

The ability to isolate ordered tetrads (the four products of meiosis) from certain fungi (such as yeast and *Neurospora*) allows the investigator to determine the distance between a gene and the centromere (which is usually not possible in random progeny analysis). In addition the isolation of ordered or unordered tetrads provides a different means of establishing the map distance between two or more genes.

1. Gene–centromere distance determination
This requires the isolation of ordered tetrads such as are produced by *N. crassa*. In the example shown in Fig. 15.6 we will consider the mating type alleles

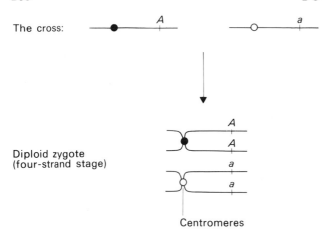

The cross:

Diploid zygote
(four-strand stage)

Centromeres

Fig. 15.6 Determination of the gene–centromere distance of the mating-type locus of *Neurospora*: Formation of diploid zygote by crossing a strain of mating-type *A* with a mating-type *a* strain.

are physically alike but it will be useful to distinguish the two for the sake of discussion. Note that at this diploid zygote stage the chromosomes have divided but the centromeres have not.

Now we can consider two situations. The first is when no cross-over occurs between the mating-type locus and the centromere; the meiotic divisions and resulting ascus in this case are shown in Fig. 15.7 (for simplification the final mitotic division, which merely replicates each of the four products, is omitted).

As can be seen the resulting four ascospores directly reflect the arrangement of the four chromatids in the zygote. Since the centromeres do not divide until just before the second meiotic division, there will be a segregation of 2 : 2 (4 : 4 if we consider the mitotic division that produces 8 ascospores) of the centromere from one parent (●) to the centromere of the other parent (○). We say, then, that the centromeres always segregate to different nuclear areas at the first meiotic division, that is they show *first division segregation*. If no cross-over occurs between the gene and its centromere, that gene will also show first division segregation. Here, after the first division, the two chromatids with the *A* alleles had migrated to a different region of the ascus from the two chromatids carrying the *a* alleles. Also, since it is equally probably that the four chromatids in the zygote are inverted, we would expect to find equal numbers of these two types of asci shown in Fig. 15.8.

A and *a* which are found in linkage group I. At the zygote stage, each chromosome has doubled and the resulting four chromatid (tetrad) stage for linkage group I is shown diagrammatically in the figure. Obviously this stage is in a more compact three-dimensional state in the cell than is shown in the diagram, but for our purposes the two-dimensional representation is adequate. In this and the other figures that will be discussed in this example, we shall use ● to indicate the centromere derived from the *A* parent and ○ to indicate the centromere of the *a* parent. Clearly the centromeres

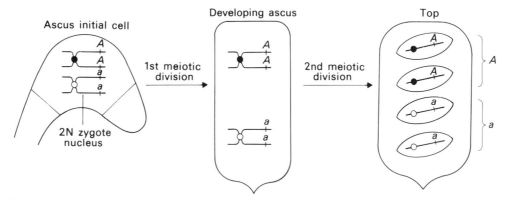

Fig. 15.7 Determination of the gene–centromere distance of the mating-type locus of *Neurospora*: Development of ascus from diploid zygote in which no

cross-over occurred between the centromere and the mating-type locus. The ascus shows first division segregation for the mating-type alleles.

Top   Top

● A   ○ a

● A   ○ a

○ a   ● A

○ a   ● A

1st division segregation

Fig. 15.8 The two possible orientations of mating-type alleles and centromeres (● and ○) in first division segregation asci. The two types occur with equal frequency.

The second situation is if a single cross-over occurs between a gene and its centromere. In this case the centromeres will show first division segregation as before (and thus in a sense we can think of them as being 'gene' markers which always segregate at the first division) and the A/a alleles will show second division segregation. That is, the separation of the allelic genes is delayed until the second meiotic division because of the cross-over event. Since the three-dimensional arrangement of the two non-cross-over and the two cross-over chromatids varies, four second division segregation asci result with equal frequencies (Fig. 15.9). Again the final mitotic division is omitted for the purpose of simplification.

Therefore, by the analysis of ordered tetrads, one can determine the percentage of asci that show second division segregation for a particular allelic pair. For the mating type locus, this is 14%. How do we convert that to map units? As we have learned before, the distance between two genes in map units comes directly from the percentage of

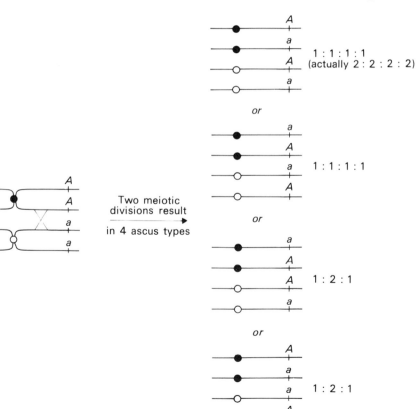

Fig. 15.9 Determination of the gene–centromere distance of the mating-type locus of *Neurospora*: development of asci from diploid zygotes in which a single cross-over has occurred between the centromere and the mating-type locus. The asci produced show second division segregation for the mating-type alleles. The four types of asci are produced in equal proportions.

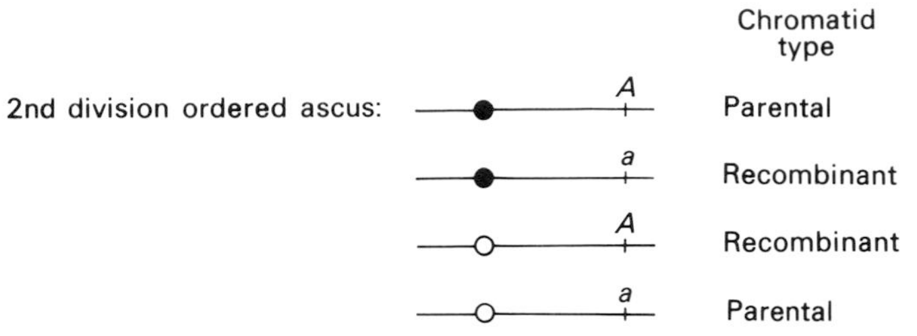

Chromatid
type

2nd division ordered ascus: ──●──────^{A}── Parental

──●──────^{a}── Recombinant

──○──────^{A}── Recombinant

──○──────^{a}── Parental

Fig. 15.10 Parental and recombinant configurations of the centromeres and mating-type alleles in a second-division segration ascus.

recombinant progeny from a particular cross. In the *Neurospora* cross we can consider the centromere to be a chromosome marker with parentals of ──●──────^{A} and ──○──────^{a}. If we now look at a second division ascus, we can examine it for parental and recombinant configurations of the centromere and mating type markers (Fig. 15.10).

As can be seen, half of the resulting spores will be parental (──●──────^{A} and ──○──────^{a}) and half will be recombinant (──●──────^{a} and ──○──────^{A}). Therefore gene–centromere distance is computed by dividing the percentage of second division asci by two. Thus, the mating type locus is $\dfrac{14\%}{2} = 7$ map units from the centromere.

Although we shall not do so in the following examples, if ordered tetrads are isolated, it is possible to combine an analysis of gene–centromere distance with an analysis of map distance between two or more genes.

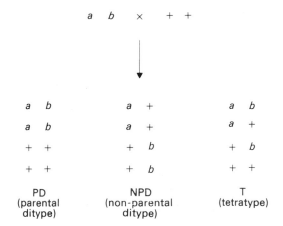

Fig. 15.11 The three types of tetrads that can be produced from a cross $ab \times + +$.

2. Map distance between two genes

With two heterozygous genes in a diploid cell at meiosis there are three possible segregation patterns from a cross. For example, for $ab \times + +$, the three types of tetrads are depicted in Fig. 15.11.

The parental ditype has two types of spores, ab and $+ +$, which are both parental. The NPD has $a+$ and $+b$ spores, both of which are recombinant (non-parental). The tetratype ascus has two parental (ab, $+ +$) and two recombinant ($a+$, $+b$) spores, hence four types of spores. The existence of tetratype asci is prima-facie evidence that crossing-over occurs at the four-strand stage of meiosis.

By making a cross such as $ab \times + +$, we can obtain information about the linkage relationships between the two genes by analysing the progeny derived from unordered or ordered tetrads. Again there are two situations: the first is genes a and b being on different chromosomes. In this case PDs and NPDs arise with equal frequency depending on the arrangement of the two sets of four chromatids at metaphase I, and the Ts arise if a single cross-over occurs between one or other of the genes and its centromere (Fig. 15.12). The proportion of the asci that will be tetratype will depend on how far the two genes are from their centromeres.

The second situation is when two genes a and b are linked on the same chromosome (Fig. 15.13). Ordered tetrads are drawn in the diagrams, but the analysis is the same with unordered tetrads. The possibilities we shall consider are no (Fig. 15.13a), single (Fig. 15.13b), and double cross-overs (Fig. 15.13c).

If no cross-over occurs between the two genes (Fig. 15.13a), all of the progeny are parental. If a single cross-over takes place between the two loci (Fig. 15.13b), half of the progeny are parental and

Two genes on different chromosomes

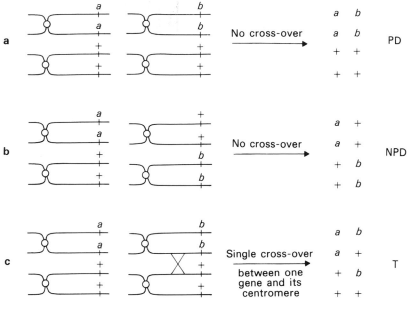

Fig. 15.12 Tetrad analysis for a cross $ab \times + +$ where genes a and b are located in different chromosomes. (a) and (b) show the types of tetrads produced by random orientation of the two sets of four chromatids in meiosis when no cross-over occurs. (c) shows the tetrad type produced when there is a cross-over between one gene and its centromere.

half are recombinant. Now, when we consider double cross-overs (Fig. 15.13c), since there are four chromatids in the zygote, it is necessary to consider three types: two-strand, three-strand and four-strand double cross-overs. There are two ways of having 3-strand double cross-overs and so the relative proportion of two : three : four-strand double cross-overs is 1 : 2 : 1, respectively.

Now the distance between any two genes is computed from the formula: $\dfrac{\text{no. recombinants}}{\text{total progeny}} \times$ 100. Examining the tetrads, we see that NPD asci contain 4 recombinant spores and T asci contain 2 parental and 2 recombinant spores. Therefore converting the 'general' mapping formula into tetrad type terms, the a–b distance $= \dfrac{\frac{1}{2}\text{T} + \text{NPD}}{\text{total}} \times$ 100. Thus if there were 1000 asci, with 900 PD, 96 T and 4 NPD, the a–b distance would be $\dfrac{\frac{1}{2}(96) + 4}{1000}$ $\times 100 = 5.2$ map units.

However, the formula we have derived calculates map distance on the basis of recombination frequency rather than on the basis of cross-over frequency. A correction can be made for this.

In the theory that was presented, three types of double cross-overs were shown to occur; namely, two-strand, three-strand, and four-strand doubles in a ratio of 1 : 2 : 1, respectively. The four-strand double cross-over frequency is directly determined from the number of NPD asci. In the case of two-strand double cross-overs the result is a PD ascus, but here the two cross-overs were not counted (see Fig. 15.13c). From the ratio of double cross-overs, the number of PD asci that result from double cross-overs should equal the number of NPD asci.

Turning to the three-strand double cross-overs, we have shown that there are two types and each gives rise to a T ascus, just as does a single cross-over. Thus, for each three-strand double cross-over, only one of the cross-overs is really being considered in our original formula, and between the two 3-strand doubles, the equivalent of two cross-overs, or one double cross-over is not being included in the calculations. As we have seen, the NPD frequency is equivalent to a double cross-over frequency. The formula can then be modified as in Fig. 15.14.

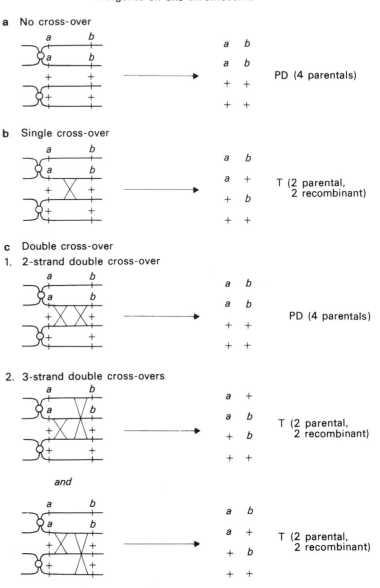

Fig. 15.13 Tetrad analysis for a cross $ab \times + +$ where both genes a and b are located in the same chromosome. (a) No cross-over results in a PD ascus, (b) single cross-over between the two genes results in a T ascus, (c) double cross-overs between the two genes result in PD, T or NPD asci depending on the number of chromatids involved in the exchanges.

Recombinants = ½ T + NPD

+ NPD (to add 2-strand double cross-overs which give PD's with frequency equivalent to NPD's)

+ NPD (to add the equivalent of a double cross-over not counted for the two 3-strand double cross-overs — again equivalent to NPD frequency)

Total = ½ T + 3 NPD

Therefore, map distance based on cross-over frequencies, as manifested by the tetrad types, is given by the formula: $\frac{\frac{1}{2}T + 3NPD}{total} \times 100$. For the numbers given before, this would be: $\frac{\frac{1}{2}(96) + 3(4)}{1000}$ $\times 100 = 6.0$ map units which compares with 5.2 map units for the 'old' recombination frequency formula.

The ascus type formula, then, can be used to calculate the map distance between any two genes. In crosses where more than two genes are segregating, the data should be analysed by considering two genes at a time and sorting the data into PD, NPD, and T ascus types for those two genes.

Tests for gene linkage using tetrad analysis

Instead of random spore analysis, tetrad analysis can be used to determine whether two genes are linked or unlinked. If two genes are unlinked by being on different chromosomes, then the PD frequency will equal the NPD frequency. PD asci contain only parental-type spores and NPD asci contain only recombinant type spores. T asci contain half parental and half recombinant spores. All-in-all then, with PD = NPD — whatever the T value is — the recombination frequency (RF) is 50%, which of course signifies no linkage between the two genes. The frequency of T asci might provide information, however, about whether two unlinked genes are on different chromosomes or far apart on the same chromosome. In the former case, the T asci arise as a result of a single cross-over between one or other of the two genes and their respective centromeres and thus the frequency of T asci will depend on how far away

Fig. 15.14 Origin of formula for determining the distance between two genes by tetrad analysis.

from the centromeres the genes are. In this case the T frequency can vary between 0 and the limiting frequency of 66.7% (see later). On the other hand, if two genes are very far apart on the same chromosome, there will be a large number of cross-overs, even-numbered ones occurring with about the same frequency as odd-numbered ones, thus giving rise to equal numbers of PD and NPD asci respectively. In this situation the T frequency will be 66.7% of all asci, and this will now be explained for a cross $ab \times ++$. If the two loci are far apart, there will be multiple cross-overs between them (Fig. 15.15).

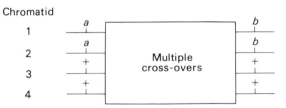

Fig. 15.15 Diagram of the four-chromatid stage of meiosis showing multiple cross-overs between the a and b loci. The chromatids are numbered for the text discussion. (With permission from *An Introduction to Genetic Analysis*, David T. Suzuki and Anthony J. F. Griffiths. Copyright © 1976, W. H. Freeman and Company.)

If we consider chromatid 1 with the a allele, since there are multiple cross-overs between the two loci, there is an equal probability that this chromatid will end up being b or b^+. That is the probability it will be b, $p(b) = 1/2$ and $p(b^+) = 1/2$. If we now consider chromatid 2 and compute the probabilities of the genotypes, this will then determine the ascus type. Thus, if chromatid 1 carries b, then chromatid 2 has a probability of 1/3 to carry b and a probability of 2/3 to carry b^+. If the former occurs, we will have

Table 15.1 Relative frequencies of PD, NPD and T asci when two genes are far apart on the same chromosome*. (With permission, from *An Introduction to Genetic Analysis* David T. Suzuki and Anthony J.F. Griffiths. Copyright © 1976 W.H. Freeman and Company.)

Chromatid 1	Chromatid 2	Total probability	Tetrad genotype	Ascus type
p(b) = 1/2 ⟨ p(b) = 1/3		1/6	ab ab ++ ++	PD
p(b⁺) = 2/3		2/6	ab a+ +b ++	T
p(b⁺) = 1/2 ⟨ p(b⁺) = 1/3		1/6	a+ a+ +b +b	NPD
p(b) = 2/3		2/6	a+ ab +b ++	T

Therefore, frequency of T = 2/6 + 2/6 = 66.66%
PD =   1/6   = 16.66%
NPD =   1/6   = 16.66%

* The cross here is *a b* × + + and, for the purposes of determining tetrad genotype, the two copies of allele *a* are on chromatids 1 and 2. Multiple cross-overs occur between *a* and *b* such that the two genes effectively segregate independently.

a PD and if the latter is the case, we will have a T asci (Table 15.1).

In summary, when two genes are unlinked and very far apart on the same chromosome, the results will be a ratio of 1 PD : 1 NPD : 4 T asci. For two genes on different chromosomes, PD = NPD, and the relative number of T will depend on the distance the two genes are from their centromeres. A low T frequency compared with PD and NPD would certainly indicate that two genes in question are on different chromosomes.

If the two genes *are* linked, NPD asci can only arise as a result of a four-strand double cross-over which is a very rare event. Therefore in this case the frequency of PD asci will greatly exceed the frequency of NPD asci (i.e. PD ≫ NPD). The T asci result from single and three-strand double cross-overs and these will occur with a frequency intermediate between those for PD and NPD.

MITOTIC GENETIC ANALYSIS IN *ASPERGILLUS NIDULANS*

Up to now our discussion of crossing-over has been limited to meiosis. In 1936, C. Stern discovered that crossing-over could take place in the somatic tissue of *Drosophila melanogaster,* that is during mitosis. Here we will discuss mitotic crossing-over in the mycelial-form fungus *Aspergillus nidulans* and we will show how this, in concert with haploidisation, can be used to map genes to particular linkage groups.

Mechanism of mitotic crossing-over

In mitosis each pair of homologous chromosomes replicates, the two pairs of chromatids come independently to the metaphase plate and then segregate to the two daughter cells which thus have the same genotype as the parental cell. This is shown in Fig. 15.16 for a theoretical cell with all genes heterozygous. Thus, normally the maternally and paternally-derived chromatids do not pair together during mitosis. Very rarely, once each chromosome has replicated, the two pairs of chromatids do come together to form a transient tetrad equivalent to the four-strand stage of meiosis. As in the latter, crossing-over can take place during this tetrad stage after the chromatid pairs separate and align independently in the metaphase plate (Fig. 15.16b). The result of this is that some of the progeny cells will be homozygous for one or more

a Normal mitosis

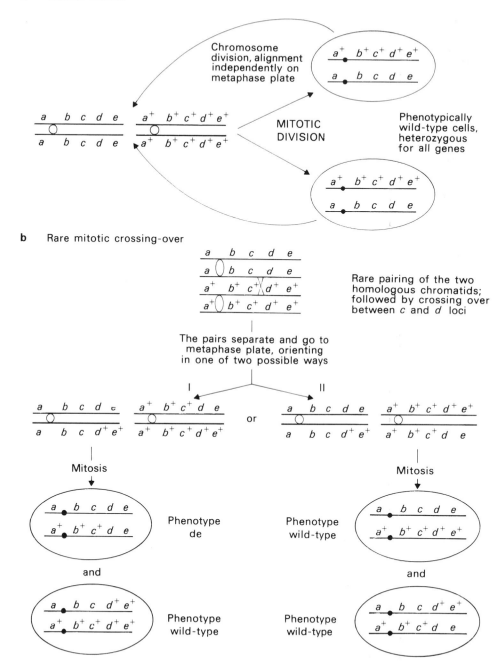

b Rare mitotic crossing-over

Fig. 15.16 Diagrammatic representation of (a) normal mitosis and (b) a mitosis involving a rare crossing-over event.

of the genes. When it is a mutant gene that becomes homozygous, then the cell will now have a mutant phenotype where the parental cell was wild type.

Like meiotic recombination, mitotic recombination involves only two of the four strands. Because of the randomness with which the two chromatid pairs become oriented at metaphase, in only 50% of the time does a recombinant phenotype appear in one of the two progeny cells. Note that, when a mitotic cross-over occurs, the recombinant progeny will be homozygous for *all* gene markers distal to the cross-over site. In addition, since mitotic recombination is a very rare event, for the purposes of discussion double- and other higher-order cross-overs can be ignored.

Formation of diploid strains
Now we turn to the method of genetic analysis in *Aspergillus*.

The first step in mitotic genetic analysis in *Aspergillus* is the formation of stable diploid strains. This can be accomplished by combining complementary auxotrophic strains on a minimal medium (Fig. 15.17). In the example, each strain carries a conidial colour mutation. The only colonies that will grow (barring reversion of a mutation) will be those that result by hyphal fusion of the two strains. The resulting *heterokaryon* contains both types of nuclei in a common cyto-

plasm and since the two auxotrophic mutations will complement under these circumstances, growth will occur. When the heterokaryotic strain produces conidia, most of the asexual spores will be uninucleate, carrying one or other of the parental nuclei. When plated on minimal medium these will not survive. Rarely, nuclear fusion will occur in the heterokaryon and this can lead to uninucleate diploid conidia of the genotype $y\ w^+\ ad\ thi^+/$ $y^+\ w\ ad^+\ thi$ and, owing to the dominance of y^+ and w^+, these conidia will be dark green instead of yellow or white as the haploid conidia will be. When the diploid conidia germinate, they are able to grow on minimal medium and can be used for mitotic genetic analysis.

Locating genes to linkage groups by haploidisation
The diploid that is formed is relatively stable but has a tendency to break down to haploid segregants. This is called haploidisation. We can follow this, for example, if we have a diploid that is $+/y$ and $+/w$ for conidial colour since some of the haploids that result will have white or yellow conidia and will be detected as different-coloured sectors in dark-green colonies. The significant point about haploidisation is that which chromosome of a pair will end up in the haploid segregant is a random event. If the chromosomes are genet-

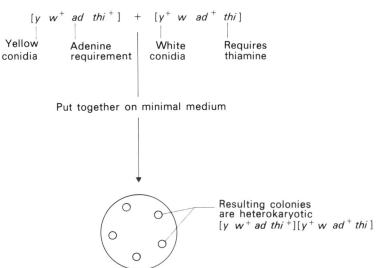

Fig. 15.17 Formation of a diploid strain of *Aspergillus* by combining two strains carrying complementary auxotrophic and conidial colour markers.

Parental diploid
(dark green)

$$\frac{y \quad +}{+ \quad bi} \qquad \frac{+ \quad + \quad +}{w \quad pu \quad ad} \qquad \frac{sm \quad phe}{+ \quad +}$$

Haploidisation to
produce yellow or
white segregants

Genotype							Number obtained
y	bi	w	pu	ad	sm	phe	
y	+	w	pu	ad	sm	phe	7
y	+	w	pu	ad	+	+	11
+	bi	w	pu	ad	+	+	1
y	+	+	+	+	sm	phe	26
y	+	+	+	+	+	+	7

$\underbrace{\qquad}$ $y +/+ bi$ segregation $\underbrace{\qquad}$ $w\ pu\ ad/+ + +$ segregation $\underbrace{\qquad}$ $sm\ phe/+ +$ segregation

Fig. 15.18 Example of how
haploidisation of a diploid strain
can localise genes to chromosomes.
For details, see text.

ically marked we can detect this by segregants
having or not having particular sets of genes. Each
set, of course, will segregate independently of other
sets of genes representing other chromosomes. This
will be made apparent by considering an example
(Fig. 15.18). At the outset we are giving the conclusion of the experiment by showing that there are
three linkage groups. This will be confirmed by
analysing the data.

In the experiment, white or yellow haploid segregants were obtained and analysed for genotype by
further testing. It should be pointed out that the
numbers of each class in the data may not be
significant owing to the effects of the markers on
viability. In the haploid segregants recombination
has not occurred between markers on the same
chromosome, but 'recombination' has occurred between chromosomes. For example, the segregants
are either $y+$ or $+bi$, either $w\ pu\ ad$ or $+ + +$, and
either $sm\ phe$ or $+ +$. Thus, without assuming
gene order, we can designate which markers are on
different chromosomes by analysing which 'blocks'
of genes 'recombine' (actually 'assort') independently of other blocks. This leads to the assignment
of the three linkage groups with particular genotypes shown at the start of this example. Even if a

little mitotic recombination occurs in the diploid
before haploidisation, the same conclusions could
be drawn since it would be such a rare event.

Gene mapping by mitotic recombination

Once genes are assigned to linkage groups, their
order and map locations can be determined by
analysis of segregants that have arisen as a result
of mitotic recombination. We have illustrated
mitotic cross-overs earlier in the Topic — remember
that such an event leads to homozygosity for all
markers distal to the cross-over, and this will be
detected in only 50% of the cases because of the
randomness with which the homologous chromatid
pairs align at metaphase.

The parental diploid and segregant data we will
discuss are presented in Fig. 15.19.

As in the other cases, the diploid was selected
by appropriate markers on this and other chromosomes. The diploid is dark green because it is
$+/y$ and this means we can select for yellow segregants. The diploid is also homozygous for an
adenine auxotrophic mutation and heterozygous
for a recessive suppressor of adenine. If the suppressor is homozygous, the ad/ad strain would no

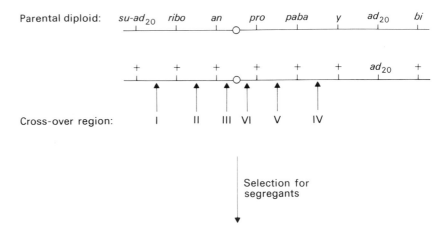

Parental diploid: $su\text{-}ad_{20}$ $ribo$ an pro $paba$ y ad_{20} bi

+ + + + + + ad_{20} +

Cross-over region: I   II   III VI   V   IV

Selection for segregants

Class	Segregant selected	Phenotype	Number
1		+	24
2	adenine-independence	$ribo$	9
3	($su\text{-}ad_{20}\text{----}ad_{20}$)	$ribo\ an$	62
4		$ribo\ an\ pro\ paba\ y\quad bi$	7
5		$y\ ad\ bi$	15
6	Yellow	$paba\ y\ ad\ bi$	42
7		$pro\ paba\ y\ ad\ bi$	12
8		$ribo\ an\ pro\ paba\ y\quad bi$	3

Fig. 15.19 Example of how mitotic crossing-over can be used to determine gene order on chromosome arms. For details, see text.

longer require adenine for growth and this means that we can also select for adenine independence. The resulting segregants can be analysed for their phenotypes with respect to the other markers and thus the gene order can be determined.

As was made apparent, a mitotic cross-over will produce homozygosis for markers distal on the same chromosome arm. It is not possible to make the whole chromosome homozygous by a single cross-over, and a double cross-over (one cross-over on each side of a centromere) is so rare as to be ignored. Thus classes 4 and 8 in Fig. 15.19 must have arisen by haploidisation — note that both are adenine-independent as a result of the recessive suppressor being present in the haploid with the mutant gene it suppresses. These classes will be ignored in further discussions.

Considering the adenine-independent diploid segregant classes 1, 2 and 3 first. Homozygosity for $su\text{-}ad_{20}$ will result following any cross-over between the centromere and the $su\text{-}ad_{20}$ locus, that is in regions I, II or III on the parental diploid diagram. A cross-over in I will only give homozygosity for $su\text{-}ad_{20}$ and the $ribo$ and an genes will remain heterozygous with their respective wild-type alleles. A cross-over in II will give homozygosity for $ribo$ and $su\text{-}ad_{20}$, but never for an (class 2). A cross-over in III will give homozygosity for an, $ribo$ and $su\text{-}ad_{20}$ (class 3). Thus, reading from the distal to the proximal, the gene order must be $su\text{-}ad_{20}\text{--}ribo\text{--}an\text{--}$centromere.

Applying similar logic to the other chromosome arm, a cross-over in IV will give a segregant that is y, ad_{20} and bi in phenotype (class 5) and, from the data presented, the order of the genes distal to *y* cannot be determined.

Class 6 results from a cross-over in region V and class 7 results from a cross-over in region VI, and thus the order must be $\left.\begin{array}{c} bi \\ ad_{20} \end{array}\right\}$ *y–paba–pro–*centromere. Haploidisation showed that all the genes we have discussed are on the same chromosome (see classes 4 and 8) and thus the data allow us to illustrate the gene order as we did in the parental diploid.

We can extend the analysis one step further by calculating the (mitotic) map distances between genes on the chromosome. Here we shall analyse the data for the adenine-independent segregants of Fig. 15.19. These data are presented in a different way in Fig. 15.20.

The 'other markers segregating' are the recombinant types resulting from single cross-overs in one of the three regions. The frequency of segregants resulting from a cross-over in region I, by analogy with meiotic mapping analysis, is a function of the genetic distance between the *ribo* and *su-ad_{20}* loci. The value here is 25.3% of the total, indicating a map distance of 25.3 map units. Similar analyses indicate a *ribo–an* distance (cross-overs in II) of 9.5 map units, and an *an–*centromere distance (cross-overs in I) of 65.3 map units. Obviously these are relative distances and they may or may not tally with meiotic map distances. The gene order would be the same in each case, of course.

In conclusion, fungal genetics is very similar, in many respects, to the formal genetics of other eukaryotic organisms. Tetrad analysis in fungi such as yeast and *Neurospora* permit particular questions about the recombination mechanisms to be asked that are virtually impossible to ask in other eukaryotes. In addition, *Aspergillus* allows the process of genetic recombination to be probed in detail.

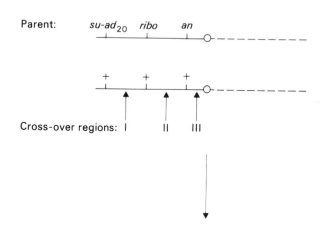

Segregant selected	Other markers segregating	Cross-over region	Number	% of total
$\dfrac{su}{su}$ (adenine-independence)	none	I	24	25.2
	ribo	II	9	9.5
	ribo an	III	62	65.3
		Total	95	

Fig. 15.20 Example of how distances between genes can be determined by mitotic crossing-over. For details, see text.

REFERENCES

Barratt R.W., D. Newmeyer, D.D. Perkins & L. Garnjobst. 1954. Map construction in *Neurospora crassa*. *Adv. Genet.* **6**: 1–93.

Davis R.H. & F.J. deSerres. 1970. Genetics and microbiological research techniques for *Neurospora crassa*. In *Methods in Enzymology*, S. Colowick and N.O. Kaplan (eds.), vol. 17A. pp. 80–143. Academic Press, New York.

Emerson S. 1967. Fungal genetics. *Annu. Rev. Genet.* **1**: 201–220.

Esser K. & R. Kuenen. 1967. *Genetics of Fungi*. Springer Verlag, New York.

Fincham J.R.S. 1970. Fungal genetics. *Annu. Rev. Genet.* **4**: 347–372.

Fincham J.R.S. & P.R. Day. 1971. *Fungal Genetics*, 3rd edn. Blackwell Scientific Publications, Oxford.

Fink G.R. 1970. The biochemical genetics of yeast. In *Methods in Enzymology*, S. Colowick and N.O. Kaplan (eds.), vol. 17A, pp. 59–78. Academic Press, New York.

Houlahan M.B., G.W. Beadle & H.G. Calhoun. 1949. Linkage studies with biochemical mutants of *Neurospora crassa*. *Genetics*, **34**: 493–507.

Kafer E. 1958. An 8-chromosome map of *Aspergillus nidulans*. *Adv. Genet.* **9**: 105–145.

Mortimer R.K. & D.C. Hawthorne. 1966. Yeast genetics. *Annu. Rev. Microbiol.* **20**: 151–168.

Pontecorvo G. 1956. The parasexual cycle in fungi. *Annu. Rev. Microbiol.* **10**: 393–400.

Pontecorvo G. & E. Kafer. 1958. Genetic analysis by means of mitotic recombination. *Adv. Genet.* **9**: 71–104.

Pontecorvo G., J.A. Roper & E. Forbes, 1953. Genetic recombination without sexual reproduction in *Aspergillus niger*. *J. Gen. Microbiol.* **8**: 198–210.

Pritchard R.H. 1955. The linear arrangement of a series of alleles of *Aspergillus nidulans*. *Heredity*, **9**: 343–371.

Roper J.A. 1968. The parasexual cycle. In *The Fungi*, G.C. Ainsworth and A.S. Sussmann (eds.), vol. 2, pp. 589–617. Academic Press, New York.

Stern C. 1936. Somatic crossing-over and segregation in *Drosophila melanogaster*. *Genetics*, **21**: 625–730.

Topic 16
Eukaryotic Genetics:
an overview of human genetics

OUTLINE

Pedigree analysis
   principles of pedigree analysis
   examples of human pedigrees
Chromosomal aberrancies and human diseases
   types of chromosomal aberrancies
   Down's Syndrome
   anomalies of the sex chromosomes
   the Lyon hypothesis.

The study of human genetics is complicated by the fact that, unlike other species of animals and plants, man is not bred experimentally. Therefore we cannot apply the types of genetic analysis that have been discussed in previous Topics. Nonetheless we have discovered that many human traits are due to single pairs of segregating genes. In this Topic we will discuss briefly how family studies can be used to discover the genetic basis of human traits. This involves pedigree analysis where the phenotypic records of families extending over several generations are compiled so that gene segregation patterns can be hypothesised. Naturally the more complete the pedigree, the more accurate the genetic analysis. As we shall see, pedigree analysis is most useful for traits that are the result of a simple gene difference such as an autosomal or sex-linked dominant or recessive mutation. Practically speaking the most interesting traits are those that cause disease and examples of the genetic basis of some diseases will be given later. Also in this Topic we shall discuss some human traits that result from chromosomal aberrancies, such as Down's Syndrome and Turner's Syndrome.

PEDIGREE ANALYSIS

Fig. 16.1 shows some of the symbols used in human pedigrees. There are other specialised symbols, but

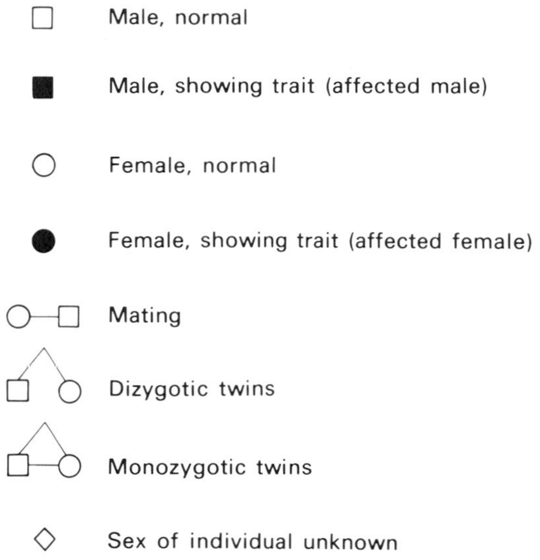

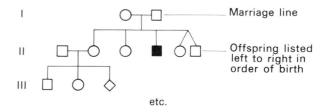

Fig. 16.1 Examples of symbols used in human pedigrees.

Fig. 16.2 An hypothetical human pedigree showing the typical features of pedigrees.

they do not concern us here. The symbols are put together in the pedigree as shown in Fig. 16.2.

Principles of pedigree analysis

As we know, there are 46 chromosomes in man, that is 22 pairs of autosomes and a pair of sex chromosomes. For genes with a simple mode of inheritance, there are four different ways they can be inherited; namely, sex-linked recessive, sex-linked dominant, autosomal recessive or autosomal dominant. In the following we will consider some theoretical pedigrees and analyse them to see what type of inheritance is compatible with the data. As we shall see, this is not the exacting type of analysis we have applied for other eukaryotes.

One trait that is controlled by a single gene pair is the ability to taste phenylthiocarbamide (PTC). When tested, people will either screw up their faces and declare it is bitter (they are 'tasters') or they will detect no taste at all (they are 'nontasters'). Let us suppose we have a small family and have tested them, the results being as shown in Fig. 16.3.

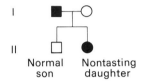

Fig. 16.3 An hypothetical pedigree for the 'trait', PTC nontasting. Persons with the trait are shaded.

In this case we will define the nontasters as having the trait and those are the shaded people in the pedigree. Now we can determine which of the four modes of inheritance apply:

Sex-linked recessive. Nontasting could be the result of a sex-linked recessive gene if we make the assumption that the mother is heterozygous for the gene. That is not unreasonable in this case since nontasting is a fairly common trait affecting about 30% of the population. The father of course is hemizygous for the mutant gene and the 'affected' daughter is homozygous for the mutant gene. Assigning t for the sex-linked recessive mutant gene, and t^+ for the wild-type allele, the pedigree can be presented genotypically as in Fig. 16.4.

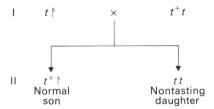

Fig. 16.4 Deduced genotypes for the individuals in the pedigree of Fig. 16.3 if the trait is caused by a sex-linked recessive mutation.

Sex-linked dominant. This could also explain the inheritance of the nontasting trait in this particular pedigree — the husband would be hemizygous for the dominant allele T and the mother would be homozygous for the wild-type allele T^+ (Fig. 16.5).

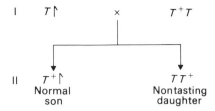

Fig. 16.5 Deduced genotypes for the individuals of the Fig. 16.3 pedigree if the trait is caused by a sex-linked dominant mutation.

Autosomal recessive. This could also be the case here if the father is homozygous and the mother is heterozygous (Fig. 16.6).

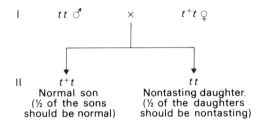

Fig. 16.6 Deduced genotypes for the individuals of the Fig. 16.3 pedigree if the trait is caused by an autosomal recessive mutation.

Autosomal dominant. Again this could be the case if the father was heterozygous and the mother homozygous wild type (Fig. 16.7). Then half of the sons and half of the daughters should be nontasting.

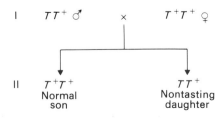

Fig. 16.7 Deduced genotypes for the individuals of the Fig. 16.3 pedigree if the trait is caused by an autosomal dominant mutation.

Thus, we cannot discern the mode of inheritance for PTC nontasting from the pedigree presented. So we need to develop a larger pedigree and, two children later, the pedigree becomes as shown in Fig. 16.8.

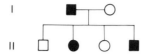

Fig. 16.8 A second hypothetical pedigree for the PTC nontasting trait.

Applying the same type of arguments as before, sex-linked recessive, autosomal recessive and autosomal dominant are all possible modes of inheritance for the trait (Fig. 16.9).

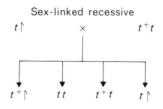

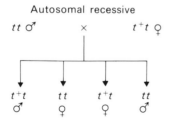

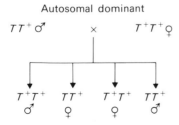

Fig. 16.9 Deduced genotypes for the individuals of the Fig. 16.8 pedigree if the trait is caused by a sex-linked recessive, autosomal recessive, or autosomal dominant mutation.

However, the data do preclude the possibility that the trait results from a sex-linked dominant gene. If this were the case the husband would be $T\Upsilon$ and the wife would be wild type ($T^+ T^+$). From a pairing of this kind, all sons would be $T^+\Upsilon$, that is tasters, and *all* daughters would be heterozygous and hence nontasters. The pedigree shows that these did not occur.

To pin down the mode of inheritance here, more pedigrees would have to be analysed. The two discussed next would strongly indicate that nontasting is the result of an autosomal recessive mutation.

The pedigree shown in Fig. 16.10 rules out a dominant mutation since neither parent showed the trait. It could be argued, though, that the mutation occurred in the germline of one or other of the parents – this is always something that must be considered.

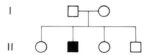

Fig. 16.10 A third hypothetical pedigree for the PTC nontasting trait. This pedigree rules out a dominant mutation as the basis for the trait.

In the pedigree shown in Fig. 16.11, a dominant mutation is again ruled out. Also, sex-linked recessive inheritance is not possible since the affected daughters must be tt and in this case their father would be $t\Upsilon$ and a nontaster – this was not the case.

In conclusion, pedigree analysis is a rather tedious, and sometimes tenuous procedure, especially of small families. The gene mutations causing diseases do not always act in a simple way, thereby leading to variations in the resulting phenotypes and naturally this complicates the analysis. In a

Fig. 16.11 A fourth hypothetical pedigree for the PTC nontasting trait. This pedigree rules out either a dominant mutation or a sex-linked recessive mutation as the basis for the trait.

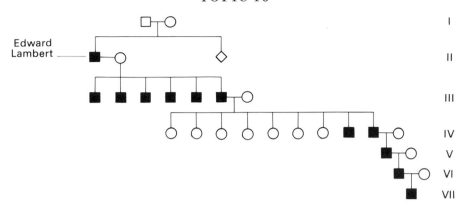

Fig. 16.12 A pedigree for a Y-linked trait: ichthyosis hystrix gravior, the bristly syndrome.

practical sense, it is the job of the genetic counsellor to analyse pedigree data in order to provide estimates of the probability that the children of a particular couple will have a genetic defect. This is a difficult task in view of the complexities of gene expression.

Examples of human pedigrees

1. The pedigree shown in Fig. 16.12 illustrates the inheritance of a *Y-linked (holandric) trait* that presumably is the result of mutation on the Y chromosome. The trait here (according to data 300 years old) is ichthyosis hystrix gravior which arose in England in the Lambert family. Edward Lambert grew bristles all over his body (except the palms of his hands and the bottoms of his feet) and he was

called the 'Porcupine man'. Reportedly the trait followed the Y chromosome among the progeny and hence all sons were bristly.

There is only one trait that is presently considered to be Y-linked in humans and that is hypertrichosis of the ear. The phenotype here is the presence of relatively long hairs on the pinnae of the ears. Indeed, it is intriguing that in both examples the phenotypic expression of the Y-linked trait involves hairs, but obviously it cannot be determined whether the same gene is involved in both cases.

2. An example of a *sex-linked recessive trait* is haemophilia, a defect in blood coagulation, and the classic pedigree is that of Queen Victoria. Part of her pedigree is presented in Fig. 16.13. The

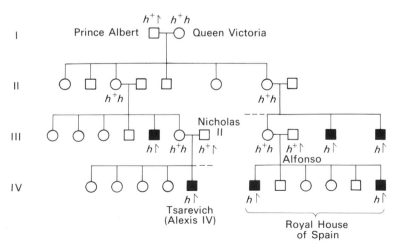

Fig. 16.13 A pedigree for a sex-linked recessive trait: part of Queen Victoria's pedigree showing the inheritance of haemophilia in the Royal families.

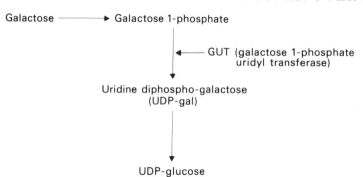

Fig. 16.14 Part of the pathway for the metabolism of galactose showing the reaction catalysed by the enzyme 'GUT' (galactose 1-phosphate uridyl transferase). 'GUT' is non-functional in galactosaemic individuals.

presumed genotypes, where *h* is the recessive mutant gene, are given next to the people.

It seems that the *h* mutation arose in the germline of Queen Victoria. Note that only males show the trait since they are hemizygous *h*↑. Females would have to be *hh* to have haemophilia and, since it is a rare gene, this is very unlikely as it requires the pairing of a haemophiliac male with a carrier (heterozygous) female.

3. An example of an *autosomal dominant trait* is Huntington's chorea, a disease which results in involuntary movements, progressive central nervous system degeneration and, eventually, death. The American folk singer Woody Guthrie was afflicted by this disease.

4. An example of an *autosomal recessive trait* is galactosaemia. Homozygosity for the recessive mutant gene must occur for such diseases to be manifest and this is usually brought about in one fourth of the children when both parents are carriers (heterozygous a^+a). For many such diseases the biochemical defect is known and some of these defects can be tested for by culturing cells taken from the amnion (the sac surrounding the developing foetus) by a procedure called amniocentesis. If familial pedigrees suggest that both parents are carriers, then this could be a worthwhile procedure to consider.

Galactosaemia is a disease where the biochemical defect is known. The disease is manifested in *gg* babies when they are fed milk — they do not grow well and develop permanent brain damage as well as other problems. If they are removed from a milk diet, the children can develop normally. The defect here is that an enzyme, galactose 1-phosphate uridyl transferase ('GUT'), is nonfunctional. This enzyme is essential for the conversion of galactose to UDP-glucose (Fig. 16.14). In the absence of the 'GUT' reaction, galactose 1-phosphate accumulates and this damages cells.

CHROMOSOMAL ABERRANCIES AND HUMAN DISEASES

In this section we shall describe the types of chromosomal aberrancies that can occur in eukaryotic organisms and we shall discuss some human traits that have their bases in chromosomal aberrancies.

Types of chromosomal aberrancies

There are two types of chromosomal aberrancies. The first involves changes in the entire set of chromosomes. A eukaryote, be it haploid or diploid, is considered to be a *euploid*. Occasionally, by a basic breakdown in spindle fibre formation during meiosis, for example, gametes may be produced which have a diploid chromosome content while others will have no chromosomes. Fusion of these gametes with a normal gamete will produce triploid and monoploid progeny, respectively. In other words they have three sets or one set of chromosomes instead of the normal two. These individuals, which have alterations in the number of sets of chromosomes are still considered to be euploids (Table 16.1). In humans the development and the functioning of the adult organism is dependent upon the correct gene dosage, that is, the diploid state. Thus, in general, the presence of fewer or

Table 16.1 Terminology for variations in chromosome number.

Chromosome complement	Shorthand formula based on diploid cell	
monosomic	2N − 1 ⎫	
trisomic	2N + 1 ⎬	Aneuploidy
tetrasomic	2N + 2 ⎭	
monoploid	N ⎫	
diploid	2N ⎬	Euploidy
triploid	3N ⎬	
tetraploid	4N ⎭	

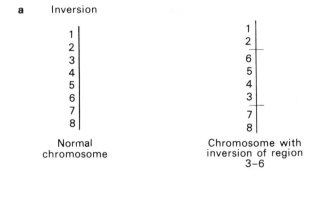

a   Inversion

Normal chromosome

Chromosome with inversion of region 3–6

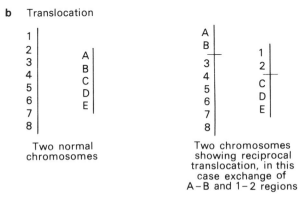

b   Translocation

Two normal chromosomes

Two chromosomes showing reciprocal translocation, in this case exchange of A–B and 1–2 regions

Fig. 16.15 Diagrammatic representations of (a) a chromosomal inversion, and (b) a translocation — in this case a reciprocal translocation involving two chromosomes.

extra sets of chromosomes results in a drastic perturbation of gene activities and development is aberrant. To this end, monoploid, triploid and polyploid (many sets of chromosomes) individuals are found only among aborted foetuses.

By contrast chromosome anomalies of this kind can be, and are, tolerated by lower eukaryotes and by higher plants. Since no human diseases are caused by monoploidy, triploidy or polyploidy, this type of chromosomal aberrancy will not be considered further.

The second type of chromosomal aberrancy affects individual chromosomes rather than the complete set of chromosomes. The aberrancies here may affect an entire chromosome or part of one. If the number of a particular chromosome present in an organism is abnormal, the individual is said to be aneuploid (Table 16.1). Thus a diploid individual such as man might have only one of a particular chromosome (monosomy) or he might have three copies (trisomy) instead of two. Again, since gene dosage is important in humans, monosomic and trisomic individuals, if they survive until birth, generally have serious defects. Of less serious consequence to the human individual are the chromosomal aberrancies in which the basic diploid set of genes is maintained but their arrangement in the genome is altered. Examples of this are inversions and translocations and these are shown in Fig. 16.15. They can be induced in organisms by treatment with ionizing radiation. These aberrations do not usually have drastic effects unless the genes at the break points that gave rise to the altered chromosomes have critical functions.

Serious consequences do result, however, when organisms carrying inversions or translocations are crossed with normal individuals. Fig. 16.16 illustrates this for an inversion in which the centromere is *not* included in the inverted chromosomal segment (a paracentric inversion), and Fig. 16.17 shows this for an inversion in which the centromere *is* included in the inverted segment (a pericentric inversion). In both cases, normal gametes are produced by meiosis if no cross-overs occur in the inverted segment. However, a single cross-over in the inverted segment results in the formation of some gametes that have extra copies of genes (duplications) or which have some genes missing (deficiencies). Fusion with a normal

Meiosis in paracentric inversion heterozygote

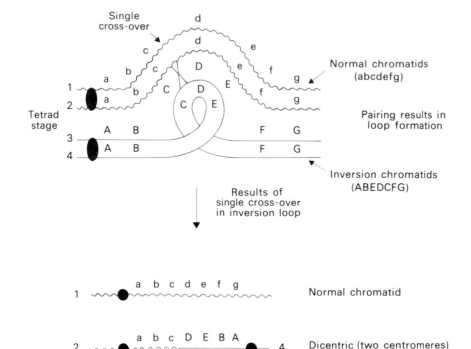

Fig. 16.16 Production of gametes when a single cross-over takes place in the inversion loop in a heterozygote involving a normal and a paracentric inversion chromosome. Half of the gametes are defective.

gamete often results in inviable zygotes in higher organisms owing to the gene dosage problem mentioned earlier. Thus, only normal gametes are generally viable in these crosses. A somewhat similar situation pertains to translocations, although this will not be illustrated here.

We now turn to a consideration of human traits that are the result of chromosomal aberrancies.

Down's Syndrome

Individuals with Down's Syndrome (mongolism) are characterised by a number of abnormal attributes including a very low IQ, epicanthal folds, a protruding, furrowed tongue, short, broad hands with incurving of the fifth finger, and below average

stature. In the most common instances of Down's Syndrome, these abnormalities are the result of an extra chromosome 21. In other words the individuals are aneuploid or more specifically they are trisomic for chromosome 21, hence the more formal name for the condition of the aberrant individuals, *trisomy-21*. That they survive at all is presumably related to the small size of that particular chromosome and the roles of the particular genes it contains.

As we indicated earlier in this section, trisomic individuals can arise by fusion of a gamete with two of a particular chromosome and a gamete with one of that chromosome. The abnormal gamete occurs very rarely as a result of non-disjunction of

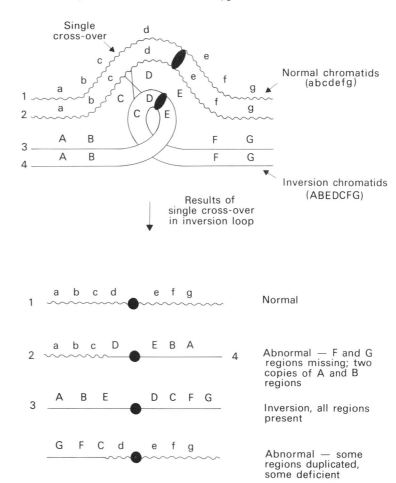

Meiosis in pericentric inversion heterozygote

Fig. 16.17 Production of gametes when a single cross-over takes place in the inversion loop in a heterozygote involving a normal and a pericentric inversion chromosome. Half of the gametes are defective.

chromosome pairs during either of the two meiotic divisions as shown in Fig. 16.19. The molecular basis for the abnormal meiosis is not known, but is presumably a consequence of spindle fibre failure. As the figure illustrates, there should be equal numbers of gametes produced which lack a particular chromosome as those which have the abnormal complement of two. Since no individuals have been born with monosomy-21, we can conclude that this condition is lethal.

In mature females the ovaries contain the primary oöcytes that are the progenitors of all the eggs that will be released by the ovary throughout the lifetime of the woman. The primary oöcyte is a cell that is arrested in the first division of meiosis. Each month, starting at or near the onset of menstruation, one (or, at least, usually one) primary oöcyte completes meiosis and the resulting egg is released into the fallopian tube. One might expect, then, that the probability of non-disjunction occurring during egg maturation would increase with age since spindle fibre malfunctions would be more likely to arise as the cell aged. In fact there is a strong correlation between the mother's age and the incidence of trisomy-21 in the children (Table 16.2), and the data presented in the table can be used by genetic counsellors to assess the potential risk of older women having trisomy-21

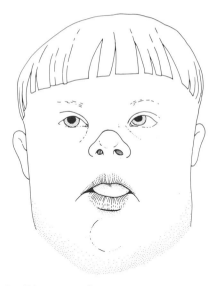

Fig. 16.18 Diagram of an individual with Down's Syndrome (trisomy-21).

Table 16.2 The relationship between the age of the mother and the risk of a trisomy-21 child.

Age of mother	Risk of trisomy-21 in child
<29	1/3000
30–34	1/600
35–39	1/280
40–44	1/70
45–49	1/40
All mothers combined	1/665

children. This, and other chromosomal aberrancies, can be detected prior to birth in karyotypes prepared following amniocentesis.

Down's Syndrome individuals can also result from a different sort of chromosomal aberrancy. This came to light in studies of certain families which had a high percentage of Down's Syndrome children (33%) even when the mothers were young. Karyotype studies showed that the mothers had a reciprocal translocation (exchange of chromosome parts) involving chromosomes 15 and 21 (Fig. 16.20). The mothers have a normal phenotype since all of the genes are present in normal dosages. When these females produce gametes, many have abnormal numbers of chromosomes owing to random segregation of the translocation chromosome and normal chromosome in various combinations to the gametes. This process, and the types of cells produced by fusion with a normal gamete, are shown in Fig. 16.21. Of the living children that are produced, 1/3 have Down's Syndrome (since they have three copies of chromosome 21 genes), 1/3 carry the same translocation as the mother and are phenotypically normal and 1/3 are normal both phenotypically and karyotypically. Thus in this case the possibility of conceiving trisomy-21 children can be inherited.

There are a number of other examples of surviving trisomic individuals in humans, and these generally involve the small chromosomes. Many other human traits also have their basis in chromosome aberrancies such as partial deletions of chromosomes, rearrangements of chromosomes, etc. More drastic changes in chromosome complement such as polyploidy, monosomy and trisomy of the larger chromosomes have been detected in aborted foetuses, thus reinforcing the notion that correct gene dosage is very important to human development and function. Aberrancies of the X chromosome (which is a large chromosome in humans) are an exception to this and will be discussed next.

Anomalies of the sex chromosomes

In humans, sex determination is based on the presence (male) or absence (female) of the Y chromosome. Normally males are XY and females are XX and, since there are few, if any, genes on the Y that are homologous to those on the X, there is a difference in gene dosage for X chromosome genes between males and females. However, microscopic examination of nuclei of human cells (and of the cells of many mammalian species) reveals a condensed mass of chromatin in females that is not apparent in males. This heterochromatinised material is called a Barr body after its discoverer Murray Barr. The Barr body is associated with the X chromosomes in that individuals who have extra X chromosomes also have extra Barr bodies (Table 16.3). A general formula is: number of Barr bodies = number of X chromosomes minus one.

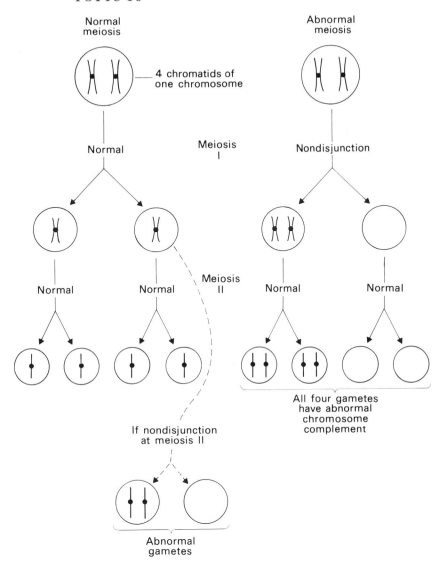

Fig. 16.19 Production of gametes with abnormal numbers of chromosomes when non-disjunction occurs at either the first division (right-hand diagram) or second division (left-hand diagram) of meiosis.

What the Barr bodies represent is all but one of the X chromosomes in a genetically inactive or virtually inactive state. This means that adult human males and females both have only one active copy of most X chromosome genes. (Recent evidence suggests that not all of the X chromosome in a Barr body is inactivated.) This point will be elaborated later in the discussion of the sex chromosome anomalies that will now be described.

Turner's Syndrome (XO)

Individuals with Turner's Syndrome have a 45-chromosome karyotype with only one sex chromosome, an X. They lack the Barr body and they arise by a non-disjunction mechanism with an occurrence of approximately one per 3000 live births. Since there is no Y chromosome, they are female. Turner's Syndrome females have few major defects up until puberty but then they do not

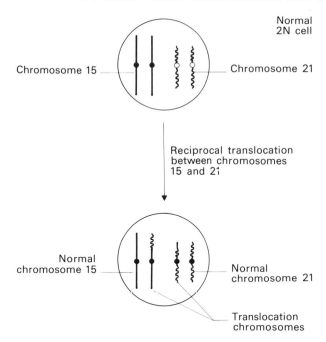

Normal
2N cell

Chromosome 15

Chromosome 21

Reciprocal translocation
between chromosomes
15 and 21

Normal
chromosome 15

Normal
chromosome 21

Translocation
chromosomes

Fig. 16.20 Diagram showing a cell with the normal complement of chromosomes and a cell with a reciprocal translocation involving chromosomes 15 and 21.

Table 16.3 Relationship of number of Barr bodies to X chromosome complement.

Chromosomal constitution	No. of Barr bodies
XO	0
XY	0
XX	1
XXY	1
XXX	2
XXXX	3
XXXXY	3

develop secondary sexual characteristics. Generally speaking they are short, have a web-like neck, poorly developed breasts and immature internal sexual organs. Sometimes they exhibit mental deficiencies, and very occasionally they may be fertile.

Superfemale (XXX)

These individuals have a 47-chromosome karyotype with three X chromosomes. Two Barr bodies can be seen in many cells. Superfemales are sometimes mentally deficient or infertile hence their detection in hospitals for the retarded and in infertility clinics. Phenotypically most superfemales are reasonably normal with perhaps poorly developed secondary sexual characteristics.

Klinefelter's Syndrome (XXY)

These individuals have an extra X chromosome and are male owing to the presence of the Y chromosome. The incidence of Klinefelter's Syndrome is approximately one in 400 live births and again non-disjunction is the culprit. Phenotypically, many of the Klinefelter males are mentally deficient, they have poorly developed testes and many are tall. These individuals have one Barr body in their cells. In more extreme cases of Klinefelter's Syndrome, the chromosomal constitution is more irregular, e.g. XXXY, XXXYY, XXXXY, XXYY. In these cases the number of Barr bodies follows the general formula mentioned previously.

Other anomalies than the ones just described have been described in the literature and are beyond the scope of this text. In general for individuals showing anomalies of the sex chromosomes, an individual with a Y chromosome has a male phenotype and an individual without a Y chromosome has a female phenotype. However, the degree of maleness or femaleness a particular individual has varies greatly but basically the more abnormal the chromosome complement, the more aberrant the phenotype.

The Lyon hypothesis

The basic question to ask at this point is why can multiple X chromosomes be tolerated by humans with relatively minor consequences while extra copies of all but the smallest chromosome lead to lethality? The answer to this question has already been alluded to; the Barr body represents an inactive or mostly inactive X chromosome. In 1961 Mary Lyon formulated a hypothesis to explain, among other things, the survival of individuals with X chromosome anomalies. This has become

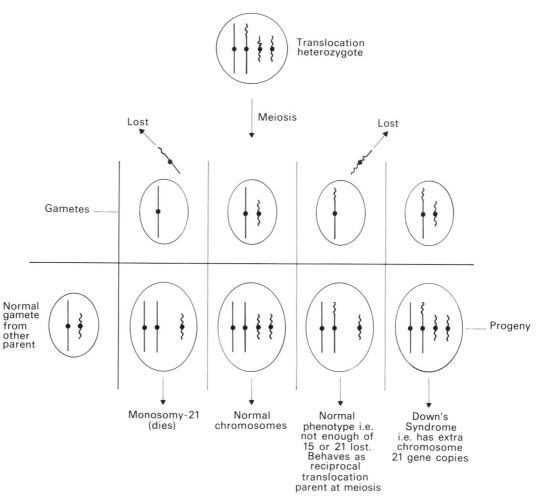

Fig. 16.21 Production of gametes by random chromosome segregation in meiosis of the cell with the reciprocal translocation shown in Fig. 16.20, and the chromosome constitutions and phenotypes of the individuals produced following fusion with normal gametes.

known as the Lyon hypothesis which states that:
1. The Barr body is a genetically inactivated X chromosome.
2. The inactivated X chromosome can be either of paternal or maternal origin in different cells of the same individual.
3. X chromosome inactivation occurs early in embryonic life (the 16th day following fertilisation in humans).

We now know that X inactivation is random and that, once inactivated in a cell, the descendants of that cell have the same inactive X, be it maternal or paternal.

Thus we can see that the Lyon hypothesis affords a simple explanation of why normal XY and XX have the same single set of active X chromosome genes. This is essentially a gene dosage compensation mechanism. Similarly the Lyon hypothesis can explain why, for example, XXY or XXX individuals are not so drastically different from normal individuals. That is to say, inactivation of the extra X chromosomes (Lyonisation) results

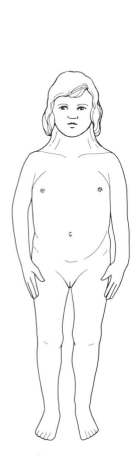

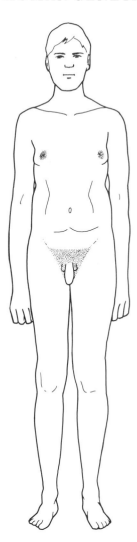

Fig. 16.22 Diagram of an individual with Turner's Syndrome (XO).

Fig. 16.23 Diagram of an individual with Klinefelter's Syndrome (XXY).

in gene dosage compensation so that only one set of X chromosome genes is ultimately active no matter how many X chromosomes are present in the cell. The phenotypic differences from normality that the individuals with the anomalous sex chromosome have, are presumably the result of the activity of the extra chromosomes during the first 16 days before chromosome inactivation is initiated.

In conclusion, this Topic has presented an over-view of two areas of human genetics, namely pedigree analysis and the occurrences of chromosome abnormalities. Present-day research is directed towards investigating whether particular human diseases have a genetic basis and, if so, what biochemical or chromosomal alterations are present. Once that is known, physicians and genetic counsellors can advise the prospective parents as to the potential risks of the woman giving birth to a child with the genetic defect.

REFERENCES

Barr M.L. & E.G. Bertram. 1949. A morphological distinction between neurones of the male and female, and the behavior of the nucleolar satellite during accelerated nucleoprotein synthesis. *Nature,* **163**: 676–677.

Bloom A.D. 1972. Induced chromosome aberrations in man. *Adv. Human Genet.* **3**: 99–153.

Bodmer W.F. & L.L. Cavalli-Sforza. 1976. *Genetics, Evolution, and Man.* W.H. Freeman, San Francisco.

Boyer S.H. (ed.) 1963. *Papers on Human Genetics.* Prentice-Hall, Englewood Cliffs, New Jersey.

Dice L.R. 1946. Symbols for human pedigree charts. *J. Hered.* **37**: 11–15.

Fraser F.C. 1974. Current issues in medical genetics: Genetic counseling. *Amer. J. Hum. Genet.* **6**: 636–659.

Garrod A.E. 1909. *Inborn Errors of Metabolism.* Frowde, Hodder and Stoughton, London.

Harris H. 1962. *Human Biochemical Genetics.* Cambridge University Press.

Levitan M. & A. Montagu. 1971. *Textbook of Human Genetics.* Oxford University Press, London.

Lyon M.F. 1961. Gene action in the X-chromosomes of the mouse (*Mus musculus L*). *Nature,* **190**: 372–373.

Lyon M.F. 1962. Sex chromatin and gene action in the mammalian X-chromosome. *Am. J. Hum. Genet.* **14**: 135–148.

McKusick V.A. 1965. The Royal hemophilia. *Sc. Amer.* **213**: 88–95.

McKusick V.A. & R. Claiborne (eds.). 1973. *Medical Genetics.* H.P. Publishing Co., New York.

Penrose L.S. 1933. The relative effects of paternal and maternal age in mongolism. *J. Genet.* **27**: 219–224.

Penrose L.S. & G.F. Smith. 1966. *Down's Anomaly.* Little Brown, Boston.

Penrose L.S. & C. Stern. 1958. Reconsideration of the Lambert pedigree (ichthyosis hystrix gravior). *Ann. Hum. Genet.* **22**: 258–283.

Shaw M.W. 1962. Familial mongolism. *Cytogenetics,* **1**: 141–179.

Stanbury J.B., J.B. Wyngaarden & D.S. Fredrickson. 1972. *The Metabolic Basis of Inherited Disease.* 3rd edn. McGraw-Hill, New York.

Stern C. 1973. *Principles of Human Genetics,* 3rd edn. W.H. Freeman, San Francisco.

Thompson J.A. & M.W. Thompson. 1966. *Genetics in Medicine.* W.B. Saunders, Philadelphia.

Topic 17
Extrachromosomal Genetics

OUTLINE

Mitochondria and chloroplasts
  the mitochondrial genetic system
Characteristics of extrachromosomal inheritance
Examples of extrachromosomal inheritance
  The iojap trait in maize
  Respiratory deficiencies of fungi
    the *poky* mutant of *Neurospora*
    petite strains of yeast.

Up to now we have considered the inheritance of traits that are under the control of the nuclear genome. In these cases the transmission pattern of the traits can be predicted from the known patterns of chromosome segregation and assortment. There are other traits whose inheritance does not follow these rules; these exhibit extrachromosomal inheritance. In the Topics on bacterial genetics and recombinant DNA, we discussed episomes and plasmids and these are examples of extrachromosomal genetic elements in prokaryotes. In eukaryotes there are many examples of traits showing extrachromosomal inheritance and many of these have their basis in the genetic material found in the cell organelles, namely the mitochondria and chloroplasts. This Topic will concentrate on examples of extrachromosomal inheritance in eukaryotes in order to differentiate clearly this mode of inheritance from that shown by traits coded by the nuclear genome. In so doing we'will present only a relatively small fraction of the information known about extrachromosomal elements that have genetic continuity.

MITOCHONDRIA AND CHLOROPLASTS

Mitochondria are essential constituents of all aerobic animal and plant cells. They are relative complex organelles with a double membrane and they contain the enzymes of the electron transport chain (the cytochromes) which are involved with the generation of ATP by oxidative phosphorylation.

Chloroplasts are found only in plant cells and they are the site of photosynthesis, the transfer of light energy into chemical energy. Within the chloroplast is a series of flattened sacs called thylakoids whose membranes contain chlorophyll and the other pigments involved in the photosynthetic process. An average leaf cell of a higher plant contains 40–50 chloroplasts.

Both mitochondria and chloroplasts contain genetic material in the form of circular, naked, double-stranded DNA. In this regard, then, the organelle genomes resemble those of the bacteria. Often this DNA has a different buoyant density from that of nuclear DNA which facilitates its isolation by CsCl density gradient centrifugation. Further, these organelles contain ribosomes and carry out protein synthesis, although not all of the proteins found in the organelles are made *in situ*. To illustrate this point, we will consider the coding capacity of the mitochondrial genome.

The mitochondrial genetic system

Mitochondrial ribosomes show many similarities to bacterial ribosomes, particularly with regard to sedimentation coefficient, ribosomal protein number and function. Protein synthesis in mitochondria requires the initiation codon AUG and initiation involves a specific fmet-tRNA·f molecule. The RNAs of the mitochondrial ribosomes are coded for by single-copy genes in the mitochondrial genome. The genes in *Neurospora* mitochondria, for example, code for the 25S and 19S rRNAs found in the large and small ribosomal subunits respectively. What about the ribosomal proteins of the mitochondrial ribosomes? Protein synthesis by mitochondrial ribosomes can be inhibited by the antibiotic chloramphenicol (to which cytoplasmic ribosomes are resistant) but not by the antibiotic cycloheximide (to which cytoplasmic ribosomes are sensitive). By using these inhibitors it has been shown that most of the ribosomal proteins of mitochondrial ribosomes are made on the cytoplasmic ribosomes and thus these must be encoded by nuclear genes. Assembly of the ribosomes occurs within the mitochondria and thus the completed ribosomal proteins must be transported across the mitochondrial membranes.

The mitochondria contain several important respiratory enzymes called cytochromes. The cytochromes are proteins, each of which has an atom of iron held within a haem group. Their synthesis has been examined in studies involving the use of the previously mentioned antibiotics with the following interesting results:

a. Cytochrome c. The protein part of the molecule is synthesised entirely on cytoplasmic ribosomes, the haem group is attached and then the whole molecule is transferred into the mitochondria. Thus, apparently only nuclear genes are involved in its synthesis.
b. Cytochromes $a + a_3$. The protein part is synthesised on cytoplasmic ribosomes while the attachment of the haem group apparently occurs within the mitochondria.
c. Cytochrome oxidase. The assembly of this complex molecule requires some polypeptides synthesised on mitochondrial ribosomes and some synthesised on cytoplasmic ribosomes.

In summary, mitochondrial DNA contains genes for a number of mitochondrial components including rRNA, possibly a few ribosomal proteins, some tRNAs, some cytochrome components and a few other constituents. The other essential components are encoded by nuclear genes. Clearly, then, mutations in the organellar genes (among other things) would lead to phenotypes that would show extrachromosomal inheritance and this is defined in detail in the next section.

CHARACTERISTICS OF EXTRACHROMOSOMAL INHERITANCE

Certain predictions can be made for a trait that shows extrachromosomal inheritance.

1. Differences should be observed in the progeny of reciprocal crosses. A characteristic form of this is *maternal inheritance* where progeny show the phenotypes of the female parent. This is the case since in many organisms the female gamete provides vastly more cytoplasm to the zygote than does the male gamete and thus there is transmission through the cytoplasm. As we have shown before (with the exception of sex-linked genes), the results of reciprocal crosses are identical for nuclear gene mutations.
2. A second prediction is that the extrachromosomal mutation should be non-mappable. Specifically, in organisms that have well-mapped linkage groups, it should not be possible to find linkage of the mutation to any of the chromosomal genes.
3. A third prediction is that the presumed extrachromosomally-coded characteristics should persist when the nuclei in the cells are substituted with nuclei of a different genetic constitution.

Clearly, then, for a trait to show extrachromosomal inheritance, the gene or genes involved must be non-nuclear in order for the above criteria to be met. Thus extrachromosomal inheritance should be distinguished from the phenomenon of maternal effect where the phenotype of the offspring is determined by the mother's nuclear genotype. The determination of coiling direction of the shell in the snail, *Limnaea peregra,* provides an illustration of maternal effect as we shall see.

In the snail the direction of coiling is determined by a single pair of alleles, *D* for dextral coiling (i.e. to the right) and *d* for sinistral coiling (i.e. to the left). Genetic crosses have shown that *D* is dominant to *d* and that the direction of coiling is always determined by the genotype of the mother (Fig. 17.1). Specifically, reciprocal crosses can be performed between a homozygous *DD*, right-

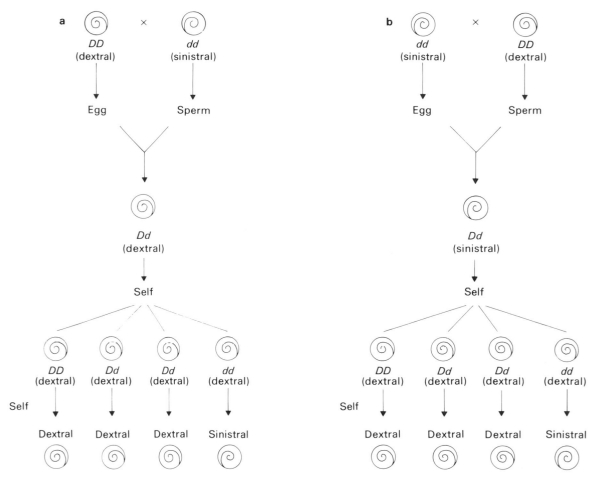

Fig. 17.1 An example of maternal effect: the direction of shell coiling in the snail *Limnaea peregra*.
(a) Inheritance pattern when two homozygous individuals are crossed where a dextral shell snail is used as the female and successive generations are selfed; (b) Inheritance pattern for the reciprocal cross of a.

handed (R) snail and a homozygous *dd*, left-handed (L) snail. When the R snail is the source of the egg (Fig. 17.1a) the F1 is *Dd* in genotype and the snail has an R phenotype. Selfing of these snails produces a 1 : 2 : 1 ratio of *DD* : *Dd* : *dd* snail genotypes but *all* of the snails, even the *dd*, are right handed. When these snails are, in turn, selfed, the *DD* and *Dd* produce *all* R progeny whereas the right-handed *dd* snail gives rise to left-handed progeny. (Clearly this is the evidence that the parent was *dd* here and illustrates nicely the maternal effect concept.) Conversely if the left-handed (L) *dd* snail is used as the maternal parent (Fig. 17.1b) then the F1 is *Dd* as in the reciprocal cross. However, here the snails are L since the mother was L. Selfing of these snails results in the same distribution of genotypes as in the reciprocal case and, because the parent here was *genotypically Dd*, all of the progeny are right-handed snails.

In general, then, maternal effects last only one generation and have, as their basis, extranuclear components that are encoded by nuclear genes. In

the particular example given, the coiling phenotypes are dependent upon the spiralling cleavage
patterns of the cells in the first few divisions after
eggs are fertilised. However, the precise factors
responsible for left- and right-handedness are not
known.

EXAMPLES OF EXTRACHROMOSOMAL INHERITANCE

The following examples are taken from the
numerous, well-studied instances of extrachromosomal inheritance in eukaryotes and illustrate the
essential criteria of that mode of inheritance. As we
shall see, unlike maternal effect, the differences in
the results of reciprocal crosses do not disappear
after one generation but rather occur as long as the
extrachromosomal factor is maintained.

The iojap trait in maize

In maize, *Zea mays*, there is a leaf striping trait
called 'iojap' that is initiated by a nuclear chromosomal mutation but which then shows a non-
Mendelian form of inheritance. The name 'iojap'
comes from the origin of the strain of maize used,
Iowa, and the name of a similar striped variety of
maize, 'japonica'. The gene mutation responsible
for the iojap trait, *ij*, is recessive, and, when it
is homozygous the result is white-leaved plants.
The genetic properties of the *ij* mutation will now
be discussed.

In 1924, M. Jenkins pollinated a homozygous
wild-type plant with pollen from a homozygous
ij/ij plant (Fig. 17.2). The heterozygous F1 plants all
had normal green leaves. When these F1 plants
were self-fertilised, the F2 plants consisted of 2498
green and 782 iojap (striped) plants. Twelve of the
latter had white leaves. These results conform to
the 3 : 1 ratio expected for Mendelian segregation.
Further, the testcrosses of representative green F2
plants confirmed the expected 1 : 2 ratio of homozygotes to heterozygotes. Thus, Jenkins concluded
that the iojap trait shows Mendelian inheritance
when it is used as the male parent. However, when
he performed the reciprocal cross in which the

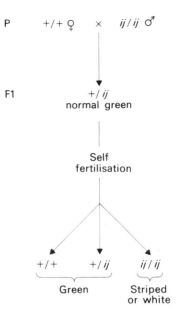

Fig. 17.2 Mendelian inheritance of the iojap trait in
maize when *ij/ij* is used as the male parent.

ij/ij was the female parent, the F1 plants often had
striped, typical iojap leaves in spite of their being
+/*ij* in genotype where the + allele should be
dominant. This phenomenon was examined in
detail by M. Rhoades in 1943 with the results shown
in Fig. 17.3. When he crossed female *ij/ij* plants with

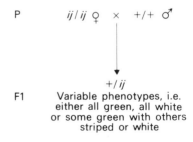

Fig. 17-3 Non-Mendelian inheritance of the maize
iojap trait when *ij/ij* is used as the female parent.
The F1s in this case show varying phenotypes.

male +/+ plants, he found that the F1 plants
showed a lot of variation depending on the experiment. That is to say they were either all green, all
white, or some were green with the others striped
(iojap) or white. To explain the results, Rhoades

proposed that when the nuclear gene ij/ij is homozygous, the striping phenotype is initiated but then the trait is inherited through the cytoplasm and thus only through the egg cell. In other words, iojap shows maternal inheritance. This hypothesis was confirmed by using an F1 striped plant as the female parent in a cross with an unrelated homozygous $+/+$ male (Fig. 17.4). In terms of gene

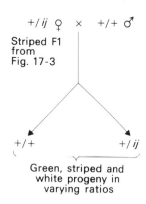

Fig. 17.4 Evidence for maternal inheritance of the iojap trait in maize: here a striped F1 plant from the cross shown in Fig. 17.3 is pollinated with pollen from an unrelated $+/+$ plant. The resulting progeny show phenotypes like those of the female parent.

segregation, half of the progeny were $+/+$ and half were $+/ij$. Phenotypically, however, the resulting plants consisted of green, striped and white individuals in varying ratios. Some individuals were entirely green and others were entirely white. Clearly, then, the phenotypes observed did not follow classic Mendelian segregation patterns.

All in all, the simplest interpretation of iojap is that the phenotypes observed are related to the

chloroplasts which are under some nuclear control but which are autonomously self-replicating structures. The striped and white plants, then, would be expected to have defective chloroplasts in the white or yellow areas and indeed colourless, abnormal chloroplasts can be found there. However, the exact relationship between the chloroplast structure and the iojap mutation is not known.

Respiratory deficiencies of fungi

1. The poky mutant of Neurospora.

In *Neurospora* aerobic respiration is essential for the organism to survive. This aerobic respiration is a property of the mitochondria. Some slow-growing strains of this fungus has been characterised which have defects in respiration. For example the slow-growing *poky* mutant lacks cytochromes $a + a_3$ and b, and also has an excess of cytochrome c (Fig. 17.5). The cytochrome defect is manifested in all slow-growing progeny of a cross between *poky* and wild type, thus attesting to the genetic continuity of the mitochondria. In the following, genetic evidence will be presented which shows that *poky* is the result of a mutation in the mitochondrial genome, that is it shows extrachromosomal inheritance.

When we discussed the life cycle of *Neurospora crassa* in an earlier Topic, we showed that the sexual cycle is initiated under nitrogen starvation conditions when there is fusion of nuclei from *A* and *a* strains. One added fact needs to be mentioned. If we inoculate the starvation medium with only one strain, it will produce a pre-fruiting structure called a protoperithecium. At this point, addition of conidia or mycelial fragments of a strain of the

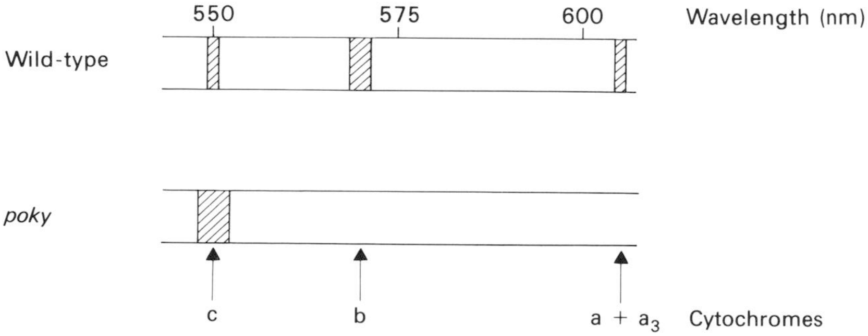

Wild-type

550   575   600   Wavelength (nm)

poky

c   b $a + a_3$ Cytochromes

Fig. 17.5 Cytochrome spectrum of wild-type and *poky* strains of *Neurospora* showing the cytochrome aberrancies characteristic of the latter.

opposite mating type will lead to nuclear fusion and the formation of an *A/a* diploid zygote. The protoperithecial parent (female parent), then, is analogous to the female gamete in that it contributes the bulk of the cytoplasm to the cross. The conidial parent is equivalent to the male gamete. This makes it possible to make reciprocal crosses in *Neurospora* to examine the contribution of the cytoplasm to the trait under study.

The results of reciprocal crosses between wild type and *poky* are shown in Table 17.1. In these crosses the mating-type genes and any other nuclear marker genes introduced into the cross, segregate 4 : 4 as is characteristic of nuclear genes.

Table 17.1 Pattern of extrachromosomal inheritance of the slow-growing *poky* mutant of *Neurospora* as shown by reciprocal crosses with the wild type.

Female (protoperithecial) parent	Male (conidial) parent	Segregation pattern of progeny tetrads
+	×	[*poky*]	8 + : 0 [*poky*]
[*poky*]	×	+	0 + : 8 [*poky*]

By contrast the transmission of *poky* follows the cytoplasmic line in that the progeny *all* have growth rate phenotypes like that of the female parent in the cross. This, then, is an example of extrachromosomal inheritance and, owing to the cytochrome deficiencies in *poky*, the simplest hypothesis is that the mutant strain carries a mutation in the mitochondrial genome. Support for this hypothesis has come from microinjection experiments in which a filament of a wild-type strain was converted to a slow-growing state with mitochondria purified from the *poky* strain. More recently it has been shown that *poky* mitochondria have ribosomal subunits which are present in disproportionate amounts compared with the wild type owing to a virtual lack of the small ribosomal subunit. This leads to a slower rate of mitochondrial protein synthesis compared with the wild type. Since, as we have discussed, cytochromes or parts of cytochromes are made on the mitochon-

drial ribosomes, the abnormal ribosomal subunit complement in *poky* is presumably responsible for the cytochrome deficiencies and thus the slow-growth phenotype. There is some evidence that the abnormal ribosome phenotype of *poky* is the result of an altered or deficient ribosomal protein that is coded by the mitochondrial genome. (Most of the proteins of mitochondrial ribosomes are coded by the nuclear genome.)

2. Petite strains of yeast

When cells of the yeast, *Saccharomyces cerevisiae*, are plated onto a glucose-containing medium, a small percentage (up to 10%) of the colonies are petite (small) in that they are 1/3–1/2 the diameter of the other (normal) colonies. When cells from a normal-sized colony are then plated, up to 1% of those colonies will be *petite*. By contrast, *petite* cells breed true in that cells of *petite* colonies give rise to *petite* colonies only. The cells that make up the *petite* colonies are the same size as the normal cells. The reason for the small colony size is that the colonies developed from a *petite* mutant cell which lacked cytochromes b, c, a + a_3 and cytochrome oxidase — all of which are enzymes of the inner mitochondrial membrane. Unlike wild-type, the *petite* mutants, then, cannot carry out oxidative phosphorylation for energy production (a mitochondrial function) and hence their growth rate and cell division rate are lower since their energy comes principally from glycolysis. Normal cells can be induced to grow at the same low rate if oxygen is removed from the environment.

The *petite* strains can mate with normal strains and analysis of the phenotypes of the progeny of such matings allow us to distinguish three types of *petite* strains, two of which show the characteristics of extrachromosomal inheritance.

a. Nuclear *petites*. The cytochrome deficiency (and hence slow-growth phenotypes) of these strains is the result of a nuclear gene mutation, *pet⁻*. As illustrated in Fig. 17.6, when a *pet⁺/pet⁻* diploid is formed and sporulated, the diploid is normal and the resulting four ascospores show a 2 : 2 segregation, of normal (*pet⁺*) to slow-growing (*pet⁻*) colonies. This is indicative of Mendelian inheritance.

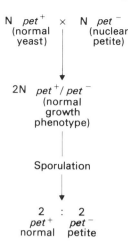

Fig. 17.6 Mendelian inheritance pattern for a nuclear *petite* mutant of yeast.

b. Neutral extrachromosomal *petites*. This class of *petites* shows extrachromosomal (non-Mendelian) inheritance for the slow-growing and cytochrome deficiency phenotypes. The extrachromosomal factor in this case is called $[\text{rho}_{(\overline{N})}]$ where the normal (wild-type) factor is called $[\text{rho}^+]$. Fig. 17.7 shows the segregation properties of a neutral *petite* mutation. When a diploid is formed with a normal strain, the diploid grows normally as was the case with the nuclear *petites*. Sporulation of the $[\text{rho}^+]/[\text{rho}_{(\overline{N})}]$ diploid results in four ascospores all of which germinate to produce normally growing cells that

have no cytochrome deficiencies. The *petite* phenotypes do not appear in future generations either. Nuclear gene markers segregating in the same cross all segregate $2 : 2$. Thus the neutral *petites* show extra-chromosomal inheritance. Since the two yeast cells that fuse to produce the diploid are the same size, it is not possible here to determine if maternal inheritance would occur with this type of *petites*.

At the molecular level, the mitochondria of the neutral petites have no DNA. Thus clearly they are unable to produce the cytochromes coded by the mitochondrial genome and consequently aerobic respiration is non-existent and the strains grow slowly.

c. Suppressive extrachromosomal *petites*. Like the neutrals, the suppressive *petites* show extrachromosomal inheritance for the slow-growth and respiration-deficient phenotypes. The extrachromosomal factor in this case is called $[\text{rho}_{(\overline{S})}]$. As their name indicates, however, they can suppress normal

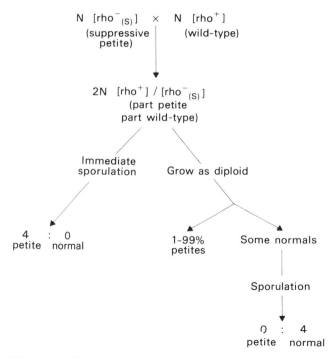

Fig. 17.8 Extrachromosomal inheritance pattern for a suppressive *petite* mutant of yeast.

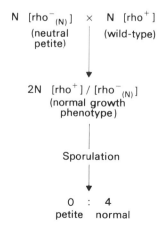

Fig. 17.7 Extrachromosomal inheritance pattern for a neutral *petite* mutant of yeast.

aerobic respiration when in the presence of normal cytoplasm (and thus normal mitochondria). The properties of suppressive *petites* are shown in Fig. 17.8. When a [rho$^+$]/[rho$_{\overline{(S)}}$] diploid is formed, it has respiratory properties intermediate between those of normal and *petite* strains. In the presentation of the yeast life cycle in Topic 15, it was pointed out that diploid formation and meiosis (sporulation) can be separated temporally. The [rho$^+$]/[rho$_{\overline{(S)}}$] diploid can be sporulated immediately it is formed and the result of this is that the ascospores show 4 : 0 segregation of *petite* to normal phenotypes while nuclear marker genes segregate 2 : 2. In other words, the normal cytoplasm has been suppressed by the *petite* cytoplasm.

If, on the other hand, the [rho$^+$]/[rho$_{\overline{(S)}}$] diploid is grown vegetatively for a number of generations, then they will produce diploid progeny in which the proportion of *petites* is between 1 and 99% depending upon the particular strain. Sporulation of the normal diploid cells produced by vegetative reproduction gives 4 : 0 normal to *petite* phenotypes.

At the molecular level, the suppressive petites have altered mitochondrial DNA. The more suppressive the *petite* is, the more mitochondrial DNA is altered in terms of its buoyant density. Thus, the simplest explanation is that the suppressive *petites*

show extrachromosomal inheritance owing to a defect in the mitochondrial genome.

Further evidence that the non-nuclear *petites* are the result of mutations residing in the mitochondria has come from work with two chemicals. As was pointed out earlier the spontaneous rate of formation of extrachromosomal *petites* is approximately 1%. If replicating wild-type cells are treated with 10^{-6}M acriflavin, almost 100% of the cells are converted to the *petite* state. A similar result is obtained for non-growing or growing cells if they are treated with ethidium bromide. The site of action of both of these chemicals has been shown to be the mitochondrial DNA, and the resulting *petites* have mitochondrial DNA that has altered buoyant density or no DNA depending on the type of *petite* induced.

In summary then, there are distinctive properties of traits that show extrachromosomal inheritance and a few examples of mutants that show non-Mendelian inheritance have been discussed. Numerous examples of extrachromosomal inheritance have been found in many other organisms. Studies of these characters are leading to an understanding of the structure and function of the mitochondrial and chloroplast genomes.

REFERENCES

Aloni Y. & G. Attardi. 1971. Expression of the mitochondrial genome in HeLa cells. II. Evidence for complete transcription of mitochondrial DNA. *J. Mol. Biol.* **55**: 251–270.

Ashwell M. & T.S. Work. 1970. The biogenesis of mitochondria. *Annu. Rev. Biochem.* **39**: 251–290.

Attardi B. & G. Attardi. 1971. Expression of the mitochondrial genome in HeLa cells. I. Properties of the discrete RNA components from the mitochondrial fraction. *J. Mol. Biol.* **55**: 231–249.

Beale G.H., A. Jurand & J.R. Preer. 1969. The classes of endosymbionts of *Paramecium aurelia. J. Cell Sci.* **5**: 69–91.

Beisson J., A. Sainsard, A. Adoutte, G.B. Beale, J. Knowles & A. Tait. 1974. Genetic control of mitochondria in *Paramecium. Genetics,* **78**: 403–413.

Boardman N.K., A.W. Linnane & R.M. Smillie (eds.). 1971. *Autonomy and Biogenesis of Mitochondria and Chloroplasts.* North Holland, Amsterdam.

Borst P. 1972. Mitochondrial nucleic acids. *Annu. Rev. Biochem.* **41**: 333–376.

Ephrussi B. 1953. *Nucleo-cytoplasmic Relations in Microorganisms.* Oxford University Press, New York.

Galper J.B. & J.E. Darnell. 1971. Mitochondrial protein synthesis in HeLa cells. *J. Mol. Biol.* **57**: 363–367.

Gillham N.W. 1974. Genetic control of the chloroplast and mitochondrial genomes. *Annu. Rev. Genet.* **8**: 347–392.

Gillham N.W. 1978. *Organelle Heredity.* Raven Press, New York.

Gillham N.W., J.E. Boynton & R.W. Lee. 1974. Segregation and recombination of non-Mendelian genes in *Chlamydomonas. Genetics,* **78**: 439–457.

Jinks J.L. 1964. *Extrachromosomal Inheritance.* Prentice-Hall, Englewood Cliffs, New Jersey.

Kirk J.T.O. 1971. Chloroplast structure and biogenesis. *Annu. Rev. Biochem.* **40**: 161–196.

Lambowitz A.M. & D.J.L. Luck. 1976. Studies on the *poky* mutant of *Neurospora crassa. J. Biol. Chem.* **251**: 3081–3095.

Lambowitz A.M., N.H. Chua & D.J.L. Luck. 1976. Mitochondrial ribosome assembly in *Neurospora*. Preparation of mitochondrial ribosomal precursor particles, site of synthesis of mitochondrial ribosomal proteins and studies on the *poky* mutant. *J. Mol. Biol.* **107**: 223–253.

Perlman P.S. & C.W. Birky. 1974. Mitochondrial genetics in Baker's yeast: a molecular mechanism for recombinational polarity and suppressiveness. *Proc. Natl. Acad. Sci. USA*, **71**: 4612–4616.

Preer J.R. 1971. Extrachromosomal inheritance: hereditary symbionts, mitochondria, chloroplasts. *Annu. Rev. Genet.* **5**: 361–496.

Preer, J.R., L.B. Preer and A. Jurand. 1974. Kappa and other endosymbionts in *Paramecium*. *Bacteriol. Rev.* **38**: 113–163.

Rifkin M.R. & D.J.L. Luck. 1971. Defective production of mitochondrial ribosomes in the *poky* mutant of *Neurospora crassa*. *Proc. Natl. Acad. Sci. USA*, **68**: 257–290.

Saccone C. & A.M. Kroon (eds.). 1976. *The Genetic Function of Mitochondrial DNA*. North Holland, Amsterdam.

Sager R. 1972. *Cytoplasmic Genes and Organelles*. Academic Press, New York.

Topic 18
Biochemical Genetics (Gene Function)

OUTLINE
Genetic control of metabolism in man
Genetic control of *Drosophila* eye pigments
Biochemical mutants of *Neurospora*
Colinearity
Summary.

Up to this point in the book, we have discussed genes as individual entities. We have described their structure in terms of nucleotide sequence, their transcription into an RNA copy (where the gene is DNA) and the translation of the RNA into the amino acid sequence of a polypeptide chain where the gene product is a protein. In the cell the polypeptide chains have either structural or enzymatic roles. However, the cell is very complex and its function depends upon the integration of the activities of many genes. In the remainder of this book we will consider how genes interact within the organism. In this Topic we shall describe the historical accumulation of evidence for the relationship between genes and enzymes, and the involvement of enzymes in biochemical pathways. Then in subsequent Topics we shall describe the regulation of expression of genes that have related functions in prokaryotes and in eukaryotes, and the role of gene interactions as they pertain to population genetics.

GENETIC CONTROL OF METABOLISM IN MAN

Early evidence for the relationship between genes and enzymes came from the work of the physician Archibald Garrod in 1909, the results of which were published as a book called *Inborn Errors of Metabolism*. Dr Garrod was interested in human diseases that had an apparent genetic basis and among those

he studied was *alkaptonuria*. This is a rare disease characterised by a number of symptoms including a hardening and blackening of the cartilage and a blackening of the urine when it is exposed to the air. The symptoms are the result of the accumulation of high amounts of homogentisic acid which is not a usual component of the cartilage or urine. Dr Garrod showed that the amount of homogentisic acid excreted was increased when patients with alkaptonuria were fed diets high in phenylalanine or tyrosine. Thus he deduced that homogentisic acid is one of the intermediates in the degradation of those two amino acids, and he surmised that alkaptonuriacs probably lacked an enzyme activity that in normal people breaks down homogentisic acid. In concomitant work, pedigree analysis by A. Garrod and W. Bateson showed that alkaptonuria is caused by a recessive mutation. Thus Dr Garrod had obtained the first suggestion of a link between genes and enzymes in that a hereditary disorder lacked an enzyme activity. He argued that this sort of causal link probably explained other 'inborn errors of metabolism'.

Garrod's hypothesis was proved correct in 1958 when the biochemical pathways for phenylalanine and tyrosine metabolism were elucidated. Part of these pathways is shown in Fig. 18.1. Each reaction in the pathways is catalysed by an enzyme and each enzyme is encoded by a gene. Thus for normal metabolism of the two amino acids, the coordinated effort of a large number of gene products is required. To return to the subject of Garrod's studies, alkaptonuriacs cannot convert homogentisic acid to maleylacetoacetic acid since the enzyme homogentisic oxidase is non-functional in these individuals. This pathway also provides examples of other metabolic diseases such as phenylketonuria, albinism and tyrosinosis and the positions of the enzyme deficiencies in these diseases are shown in Fig. 18.1.

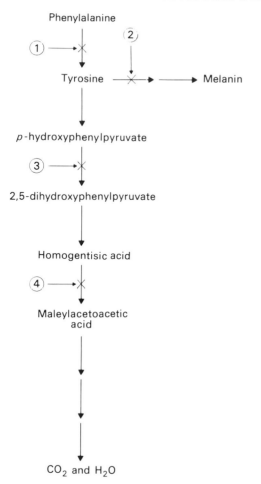

Phenylalanine

Tyrosine — X → → Melanin

p-hydroxyphenylpyruvate

2,5-dihydroxyphenylpyruvate

Homogentisic acid

Maleylacetoacetic acid

CO_2 and H_2O

Fig. 18.1 Part of the biochemical pathway for the metabolism of phenylalanine and tyrosine. The sites of metabolic blocks associated with human diseases are indicated by crosses: (1) phenylketonuria (2) albinism, (3) tyrosinosis, (4) alkaptonuria.

GENETIC CONTROL OF *DROSOPHILA* EYE PIGMENTS

As described in the preceding section, Garrod's pioneering work provided the first clue to the link between Mendelian factors and proteins. In 1935 G. Beadle and B. Ephrussi obtained more evidence for gene-enzyme relationships in a biochemical pathway in their studies of the synthesis of the eye pigments in the fly, *Drosophila melanogaster*.

There are two pigment types in the *Drosophila* eye, the bright red pterins and the brown ommo-chromes. As we now know, these two pigments are made in two, multistep, biochemical pathways, each reaction of which is catalysed by an enzyme. The bright red and brown pigments which are the end products of these pathways become attached to protein granules and are deposited in the eye cells. The combination of the two pigments results in the dull red eye colour characteristic of the wild type. Further, in *Drosophila*, the larval stages contain groups of cells called discs, each of which develops into a particular adult structure during metamorphosis. There are two discs that are the progenitors of eyes that can be identified. In their initial experiments, Beadle and Ephrussi showed that it is possible to transplant an eye disc from a larva into the abdomen of a second larva and the disc will develop into an identifiable eye which can be found in the abdomen of the adult following metamorphosis. This paved the way for a series of elegant experiments in which eye discs were transplanted between larvae of different genetic constitutions.

At the time of their experiments, three gene loci were known that were implicated in the production of brown pigment, namely *scarlet, cinnabar* and *vermilion*. Mutations in any one of these genes resulted in a bright orange eye. Thus Beadle and Ephrussi first addressed the question of whether or not embryonic eye tissue from the three mutant types *st* (scarlet), *cn* (cinnabar) or *v* (vermilion) transplanted into a wild-type larval host would develop a wild-type eye colour. In the case of the *st* mutant the transplanted disc developed into a scarlet-coloured eye (Fig. 18.2a) and is an example of autonomous development in which the wild-type host was unable to provide substances to the disc to enable the brown pigment to be produced. By contrast, transplanted discs from either *v* or *cn* developed into normal-coloured eyes in the wild-type host and thus showed non-autonomous development (Figs. 18.2b and 18.2c, respectively). The conclusion in this case was that the wild type provided a diffusible substance that the discs used to bypass the genetic block and to make brown pigment. Of more importance, historically, were the results of reciprocal transplant experiments

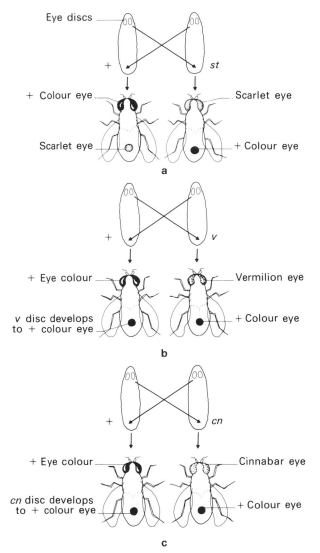

Fig. 18.2 Results of Beadle and Ephrussi's reciprocal transplant experiments with eye discs of *Drosophila*. (a) Reciprocal transplants between + and *st* (scarlet). The *st* disc develops into a scarlet-coloured eye in the wild-type host indicating autonomous development for *st*. (b) Reciprocal transplant between + and *v* (vermilion). The *v* disc develops into a wild-type-coloured eye in the wild-type host indicating non-autonomous development of the *v* disc. (c) Reciprocal transplant between + and *cn* (cinnabar). Like *v*, the *cn* disc shows non-autonomous development in that it produces a wild-type-coloured eye in the wild-type host.

Table 18.1 Results of eye disc transplantation experiments involving eye colour mutants.

Source of eye disc	Host fly	Colour of transplanted eye after metamorphosis
+	*v*	+
v	+	+
+	*cn*	+
cn	+	+
cn	*v*	cinnabar
v	*cn*	+

between *v* and *cn* larvae; these (along with the control results) are shown in Table 18.1. The experiments showed that *cn* discs transplanted into *v* hosts produced cinnabar-coloured eyes whereas *v* discs transplanted into *cn* hosts developed into wild-type coloured eyes. They concluded from these results that the production of the brown pigment involved a biochemical sequence with at least two precursors, the wild type having both, *cn* having one, and *v* having neither, with the gene-reaction step relationships shown in Fig. 18.3. Thus the production of a wild-type coloured eye when a *v* disc is implanted into a *cn* host follows logically. That is to say, the *v* disc cannot make the v^+ substance (the product of the wild-type allele of *v*) whereas the *cn* host can. This v^+ substance diffuses to the *v* disc where it is converted to the cn^+ substance (since genotypically the *v* disc carries the wild-type allele of *cn*) and this substance is then converted to the brown pigment. In other words, the *cn* host makes up for the deficiency of the *v* disc by supplying it with a diffusible substance so that it becomes wild type. In the reciprocal experiment, the *cn* disc cannot convert the v^+ substance to the cn^+ substance and the *v* host is deficient in the production of v^+ substance so no brown pigment can be produced.

At that time, the relationship between genes and enzymes was not known. Beadle and Ephrussi's work was important historically since it indicated a strong link between phenotype (in this case eye colour) and genotype (*v* and *cn*). E. Tatum further investigated this system by making extracts of *v* and *cn* flies by homogenization and by injecting the

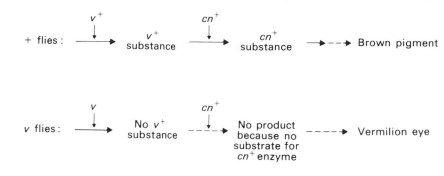

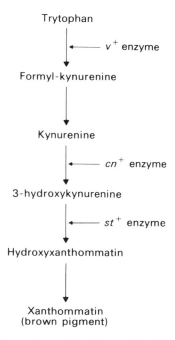

Fig. 18-3 The biochemical sequence and gene-reaction step relationships for $+$, v, and cn as deduced from the results of Beadle and Ephrussi's experiments shown in Table 18.1.

extracts into host larvae. He found that the same results were obtained as for the eye disc transplantation experiments, namely that the injection of a cn extract into v larvae resulted in the production of wild-type-coloured eyes, whereas in the reciprocal experiment the cn host produced cinnabar-coloured eyes. This experiment was the first in a series designed to discover the nature of the v^+ and cn^+ substances. At that time, it was difficult to analyse the extracts he had prepared, and his limited analysis showed that the substances were water-soluble and of low molecular weight. He proposed that the substances were relatives of amino acids and set about trying to supplement the fly food in a defined way to try to bypass the genetic blocks in the brown pigment pathway and therefore to identify the biochemical precursors in the pathway in this way. Several complex media such as peptone resulted in brown pigment formation in adult eyes when fed to v and cn larvae. One exception was gelatin which is deficient in tryptophan and tyrosin. Tatum then injected tryptophan into v and cn larvae and no brown pigment was produced in most experiments. However, in one experiment brown pigment was produced and it turned out that this occurred as a result of a bacterial contamination which resulted in the metabolism of tryptophan into chemicals that are required by v and cn for the production of brown pigment. With the implication of tryptophan metabolism in the production of brown pigment,

it was then a relatively simple matter to work out the biochemical pathway and to localise the steps catalysed by the v^+ and cn^+ gene products (Fig. 18.4). As can be seen the v mutants are unable to

Fig. 18.4 The pathway for the production of the brown eye pigment of *Drosophila* from the amino acid tryptophan. The steps catalysed by the enzymes coded for by the v^+, cn^+, and st^+ genes are indicated on the diagram.

convert tryptophan to formylkynurenine and the *cn* mutants cannot convert formylkynurenine to 3-hydroxykynurenine owing to the fact that the enzymes they have for these respective steps are non-functional.

The pathway shown in Fig. 18.4 can serve to illustrate a general point about the sequence of reactions and the consequences of genetic mutations. In the *v* mutations, for example, the flies accumulate high amounts of tryptophan since it is not converted to the brown pigment in this pathway. This is analogous to the accumulation of homogentisic acid in alkaptonuriacs. Here it provided further evidence that tryptophan metabolism is involved with eye pigment formation. Indeed, if one feeds wild-type larvae with radioactive tryptophan, the brown ommochrome pigment of the adult eye is found to be radioactive.

BIOCHEMICAL MUTANTS OF *NEUROSPORA*

The work of Beadle, Ephrussi and Tatum led directly to studies of biochemical mutants of microorganisms. Of historical importance here is the work of G. Beadle and E. Tatum in the early 1940s on the relationships between genes and enzymes in the fungus *Neurospora crassa*. The switch from *Drosophila* to *Neurospora* for studies of gene function was made since the latter organism is relatively simple, it is haploid so dominance relationships do not have to be considered, and it can be handled like other microorganisms. Further, as we indicated in the Topic on mutation, wild-type *Neurospora* has very simple nutritional requirements, namely, a minimal medium consisting of inorganic salts, a carbon source and the vitamin, biotin. When growing on this medium, *Neurospora* can synthesise all the other necessary components including amino acids, nucleotides, vitamins, etc., and, as we described in the mutation Topic, it is relatively simple to isolate auxotrophic (nutritional) mutants that can no longer grow on the minimal medium but require a particular supplement in order to grow. This is illustrated in Fig. 18.5. The mutants obtained, then, include amino acid auxotrophs, purine auxotrophs, and so on.

Beadle and Tatum made the basic assumption that cells function by the interaction of the products of a large number of genes and that wild-type *Neurospora* converted the constituents of minimal medium into the amino acids, nucleotides, etc. by a series of reactions organised into biochemical pathways. In this way, the synthesis of complex biochemical compounds occurs by a series of small steps each catalysed by its own enzyme; the product of each step is used as the substrate for the next enzyme (Fig. 18.6).

Beadle and Tatum then reasoned that if the enzymes of a biochemical pathway are specified by genes, it should be possible to obtain mutant strains that have a defect in one of the genes so the corresponding enzyme is no longer produced or is no longer active. In the hypothetical pathway shown in Fig. 18.6, a mutation affecting the production or activity of any one of four enzymes, will lead to the common phenotype of a requirement for end-product E. Genetic analysis would confirm that the mutations causing a requirement for E fall into four complementation groups and thus represent four distinct polypeptide products (enzymes in our example). How can we order the biochemical reactions by analysing the gene mutations? We can do this by determining whether the mutant strains can grow if supplemented with any of the postulated intermediates in the pathway. For example, a mutation in gene 4 would result in the inability of the strain to carry out the conversion of D to E. Such a strain would only grow on minimal plus end product E. None of the intermediates A, B, C or D would be a suitable supplement. By contrast a strain carrying a mutation in gene 2 will not be able to convert B to C and this strain will be able to grow when supplemented with any of the compounds in the pathway after the lesion, that is with C, D or E. It will not grow on minimal plus A or minimal plus B, since the lack of enzyme 2 will prevent the B to C conversion and no end product will result. Thus from the pattern of growth that mutant strains exhibit on presumed intermediates in a biochemical pathway, it is possible to order the sequence of reactions in the production of the end product.

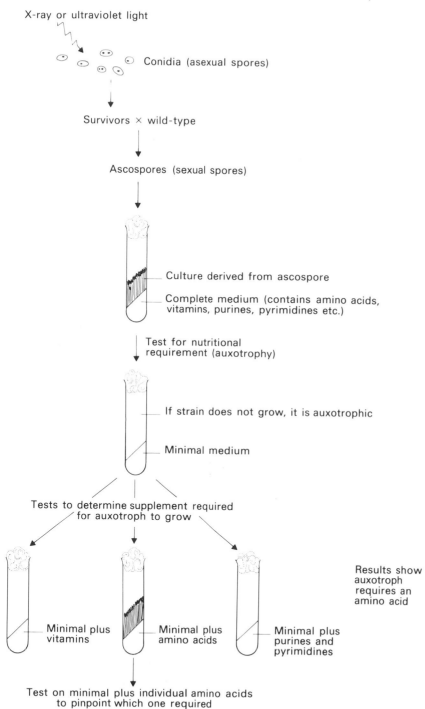

X-ray or ultraviolet light

Conidia (asexual spores)

Survivors × wild-type

Ascospores (sexual spores)

Culture derived from ascospore

Complete medium (contains amino acids, vitamins, purines, pyrimidines etc.)

Test for nutritional requirement (auxotrophy)

If strain does not grow, it is auxotrophic

Minimal medium

Tests to determine supplement required for auxotroph to grow

Minimal plus vitamins

Minimal plus amino acids

Minimal plus purines and pyrimidines

Results show auxotroph requires an amino acid

Test on minimal plus individual amino acids to pinpoint which one required

Fig. 18.5 Procedure used by Beadle and Tatum for the induction, isolation, and classification of nutritional mutants of *Neurospora crassa*. The diagram illustrates the detection of an amino acid auxotroph. Similar steps would be used to detect strains auxotrophic for vitamins, purines and pyrimidines, etc.

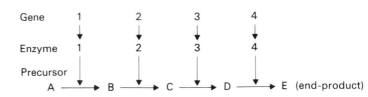

Fig. 18.6 An hypothetical biochemical pathway showing the gene-enzyme relationships.

The principles just described are illustrated by the analysis of arginine auxotrophs of *Neurospora* carried out by A. Srb and N. Horowitz. All of the strains grow on minimal plus arginine, some could grow on minimal plus citrulline, and others could grow on minimal plus ornithine. The patterns of growth are shown in Table 18.2. Based on the arguments already presented, a logical interpretation of the data is that the mutant genes represent sequential blocks in a biochemical pathway with

Table 18.2 Growth responses of arginine auxotrophs of *Neurospora crassa* (After A. Srb and N.H. Horowitz. 1944. *J. Biol. Chem.* **154**: 133).

Mutant strain	Growth response on minimal plus:			
	nothing	ornithine	citrulline	arginine
1	−	+	+	+
2	−	−	+	+
3	−	−	−	+

+ = growth on the medium
− = no growth on the medium

the end product arginine (Fig. 18.7). Thus, strain (3) is blocked in the citrulline to arginine reaction and will grow if supplemented with arginine, but not if supplemented with citrulline or ornithine. Strain (2) is blocked in the ornithine to citrulline reaction and thus can grow on citrulline or arginine. The logic for proposing a pathway, then, is straightforward — the earlier in the pathway a step is blocked, the greater the number of supplements that can be used to allow the strain to grow. The pathway we have discussed here is, in fact, part of a more complex series of reactions. Arginine, for example, can be broken down by the action of the enzyme arginase to ornithine and urea, hence completing the so-called ornithine cycle.

From this kind of work, Beadle and Tatum concluded that there is a very definite, direct relationship between genes and enzymes. Indeed they put forward the 'one gene–one enzyme' hypothesis which, simply stated, said that each biochemical reaction in the cell was catalysed by an enzyme that is coded by one gene in the DNA. This concept is of very limited value now since the idea that each gene codes for one enzyme is too simple; genes also are known to code for single polypeptide chains

Fig. 18.7 Reaction steps in the arginine biosynthesis pathway as deduced from the growth patterns of arginine auxotrophs of *Neurospora* shown in Table 18.2. The steps blocked by the mutant strains 1, 2, and 3 are indicated by arrows.

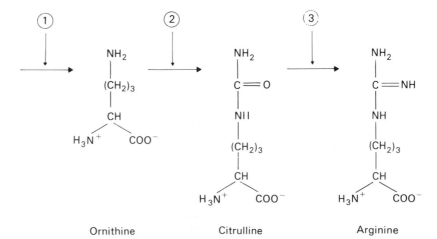

that are parts of complex enzymes, for antibodies, for structural proteins and for various types of non-translated RNA. An updated version of Beadle and Tatum's hypothesis for DNA sequences that produce translatable RNA is the 'one cistron–one polypeptide' where a cistron (defined by the cis-trans test) codes for a single polypeptide chain. Enzymes with multiple, heterogeneous polypeptides, then, would be coded for by multiple cistrons. Nonetheless it is clear that Beadle and Tatum's work laid the foundations in the field of biochemical genetics.

COLINEARITY

A more precise understanding of the relationships between a gene (cistron) and the sequence of amino acids in the polypeptide for which it codes came from the work of C. Yanofsky and his group in 1967. Their studies centered on the enzyme tryptophan synthetase from *E. coli*. This enzyme consists of two copies each of two distinct polypeptides A and B, which are coded for by two adjacent genes. So here is an example of where two genes code for one enzyme — a clear exception to Beadle and Tatum's hypothesis. The reactions that the enzyme carries out are in the biosynthetic pathway for the amino acid tryptophan (Fig. 18.8).

Tryptophan synthetase is easily isolated and the two polypeptides A and B can be purified. Further,

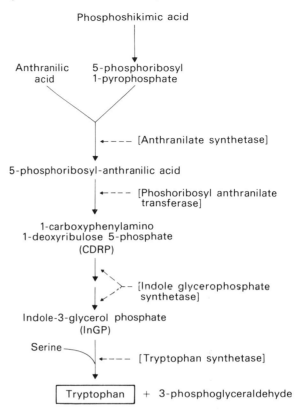

Fig. 18.8 The tryptophan biosynthetic pathway. The enzymes that catalyse the reaction steps are shown in square brackets. Tryptophan synthetase catalyses the last step in the pathway.

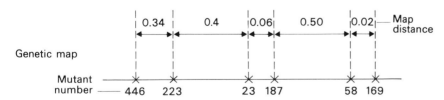

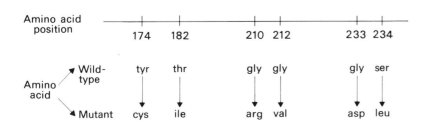

Fig. 18.9 Part of the *trp A* gene of *E. coli* and its product, the tryptophan synthetase A polypeptide, showing the colinearity of mutation positions and amino acid substitutions.

the amino acid sequence of the 267-amino acid long A polypeptide was determined at the outset of their experiments. They then isolated a series of tryptophan auxotrophs and, with further tests, identified those that carried missense mutations in the *trp A* gene which codes for the A polypeptide. The *trp A* mutants cannot carry out the production of tryptophan from indole glycerol phosphate and serine. Fine-structure mapping was used to locate the various mutations within the gene and amino acid sequencing was used to pinpoint the amino acid substitutions in the A polypeptide resulting from the missense mutation. This enabled a comparison to be made of the relative locations of the corresponding amino acid substitutions in the polypeptide. The data showed that there is a complete correspondence between the sequence and relative positioning of the mutations and the amino acid substitutions and this was termed *colinearity* (Fig. 18.9). In addition the data indicated that no single mutation affected more than one amino acid and different, very closely-linked, mutations can cause different amino acid substitutions at the same position in the A polypeptide. All in all, Yanofsky's work was highly significant since it confirmed the hypothesis that genes code for the amino acid sequence in polypeptides.

SUMMARY

1. Many proteins consist of complexes of one or more polypeptide subunits.

2. Each polypeptide is specified by a single cistron.
3. A mutation in a gene coding for an enzyme or one of its subunits can result in a biochemical defect affecting one of the steps in a biosynthetic pathway. The strain carrying the mutations can grow if supplemented with any intermediate in the pathway that is past the block. The strain will also accumulate intermediates in the pathway that are prior to the block.
4. The sequence of mutations within a cistron is colinear with the sequence of amino acid substitutions in the polypeptide for which it codes.

REFERENCES

Beadle G.W. & B. Ephrussi. 1937. Development of eye colors in *Drosophila*: diffusible substances and their interrelationships. *Genetics,* **22**: 76–86.

Beadle G.W. & E.L. Tatum. 1942. Genetic control of biochemical reactions in *Neurospora. Proc. Natl. Acad. Sci. USA,* **27**: 499–506.

Garrod A.E. 1909. *Inborn Errors of Metabolism*, Oxford University Press.

Srb A.M. & N.H. Horowitz. 1944. The ornithine cycle in Neurospora and its genetic control. *J. Biol. Chem.* **154**: 129–139.

Wagner R.P. & H.K. Mitchell. 1964. *Genetics and Metabolism,* 2nd edn. Wiley, New York.

Yanofsky C. 1967. Structural relationships between gene and protein. *Annu. Rev. Genet.* **1**: 117–138.

Yanofsky C., G.R. Drapeau, J.R. Guest & B.C. Carlton. 1967. The complete amino acid sequence of the tryptophan synthetase A protein (or subunit) and its colinear relationship with the genetic map of the A gene. *Proc. Natl. Acad. Sci. USA,* **57**: 296–298.

Topic 19
Gene Regulation in Bacteria

OUTLINE
The lactose operon of *E. coli*
  function in the wild type
  elucidation of the regulation of the lactose
    operon by studies of mutants
  positive control of the lactose operon (catabolite
    repression)
  some recent data on lactose operon regulation
The arabinose operon of *E. coli*
The tryptophan operon of *E. coli*
Summary of operon regulation
Major control of transcription and translation
  stringent control
  control at the translational level.

In the previous Topic we developed the notion that the biosynthesis of a cellular component requires a number of steps in a biochemical pathway. Each step is catalysed by a different enzyme and each enzyme is coded by one or more cistrons in the chromosome. In bacteria and phages the different genes controlling a particular pathway (and hence determining a particular trait) are often found clustered together in a group called an *operon*. As we shall see this facilitates the regulation of expression of these genes as a single unit and contrasts sharply with the situation in eukaryotes where related genes are usually scattered throughout the genome. In this Topic we will discuss some examples of bacterial operons and how their expression is regulated. As a precursor to this, we must remember that bacteria grow and divide rapidly in a nutrient medium and, to survive, they must be capable of making rapid adjustments when the environment changes, for example when components of the medium are altered. To cope with this bacteria have evolved very effective ways of rapidly turning on and off the relevant sets of genes which, when coupled with the short half-lives of the

mRNAs, renders the organisms very efficient in energy utilisation. As we shall see, this regulatory system operates at the transcriptional level.

THE LACTOSE OPERON OF *E. COLI*

Function in the wild type
The sugar lactose cannot be used as an energy source of *E. coli* unless it is first broken down into its glucose and galactose components. This reaction is catalysed by the enzyme β-galactosidase which has a tetrameric structure of identical 135 000-dalton polypeptides (Fig. 19.1).

In a wild-type cell growing in a medium that does not contain lactose there are only a few molecules of β-galactosidase. However, if the cell is growing in a medium containing lactose as the sole carbon and energy source, there are about 3000 copies of the enzyme. In other words, *induction* of enzyme synthesis occurs as a result of the presence of lactose. Recent work has shown that lactose itself is not the actual inducer in this system, but rather induction is brought about by the action of allolactose, which is produced from lactose by the enzyme activity of the few molecules of β-galactosidase present in the uninduced cell (Fig. 19.2). As long as lactose is present, the system will remain induced so that the level of β-galactosidase will remain high, but once the lactose runs out the enzyme level will diminish rapidly. However this is not an all-or-none phenomenon; the amount of enzyme produced is directly proportional to the amount of inducer present (up to the maximum amount found in the cell) (Fig. 19.3).

The addition of lactose to the cell not only brings about the increase in β-galactosidase levels, but also a rapid increase in the synthesis of β-galactoside permease and thiogalactoside transacetylase, two enzymes also needed for lactose breakdown.

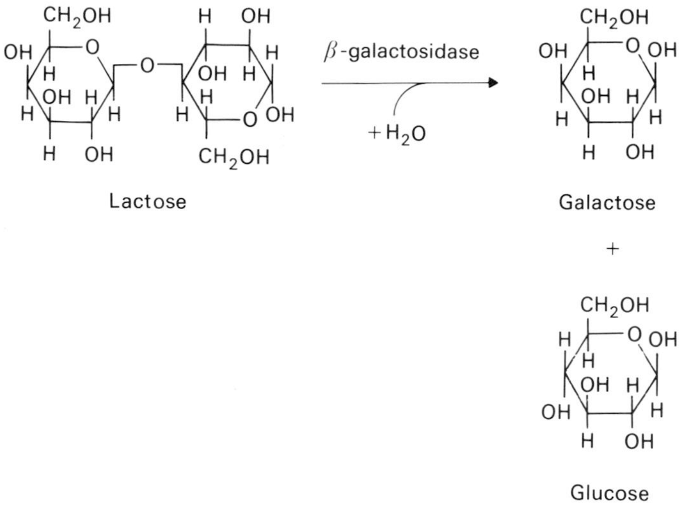

Fig. 19.1 The reaction catalysed by the enzyme β-galactosidase: the cleavage of lactose to galactose and glucose.

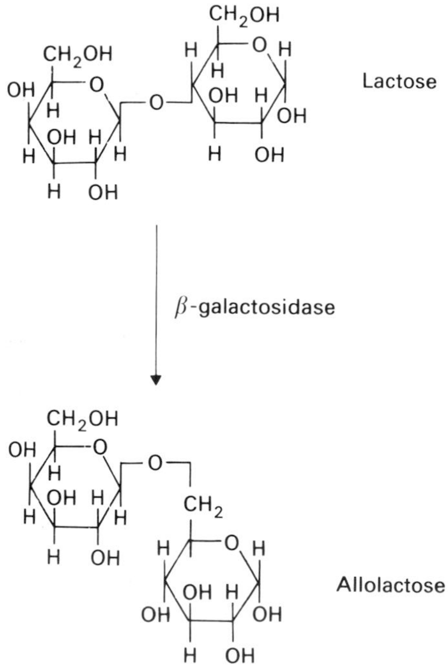

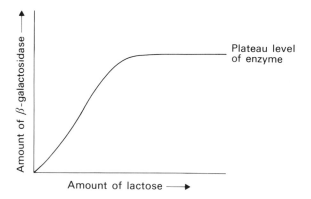

Lactose

β-galactosidase

$+ H_2O$

Galactose

+

Glucose

Lactose

β-galactosidase

Allolactose

Fig. 19.2 The production of the actual inducer of the lactose operon, allolactose, from lactose by the action of β-galactosidase.

Amount of β-galactosidase ⟶

Amount of lactose ⟶

Plateau level of enzyme

Fig. 19.3 Graph showing the direct proportionality between the amount of β-galactosidase produced and the amount of lactose added to the cells. The plateau level of enzyme amount reflects the maximum rate at which β-galactosidase can be synthesised.

Genetic experiments have shown that the genes for all of the three proteins are linked on the chromosome and adjacent to them are two regulatory sites, the operator and the promoter. A short distance away is the i gene which codes for a repressor gene involved in the regulation of the system. As a result of studies of the phenotypic properties of regulatory mutations F. Jacob and J. Monod proposed their classic operon model for the control of gene expression in bacteria, a more up-to-date description of which we will give here. Before proceeding we must define an operon: it is a genetic unit consisting of contiguous genes that function coordinately under the joint control of an operator and a repressor. A diagram of the lactose operon and its regulatory sites and repressor gene in a wild-type strain growing in the absence of lactose is given in Fig. 19.4.

which is located adjacent to the z^+ gene. When this complex is formed, RNA polymerase cannot bind to the structural gene promoter (p^+) region, and thus transcription of the genes is prevented.

If the cell is now placed in a medium containing lactose as the sole carbon source, lactose is transported into the cell by the activity of the few permease molecules present and then it is converted to the inducer, allolactose, by β-galactosidase. The inducer binds to the repressor, one molecule for each polypeptide, and causes a con-

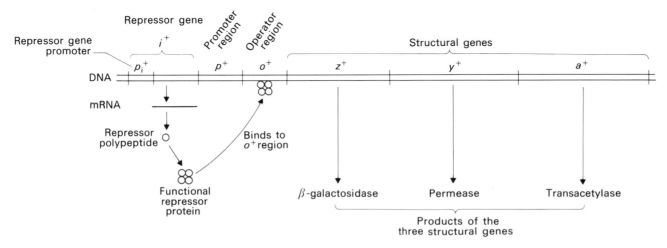

Fig. 19.4 Organisational features of the lactose operon and its regulatory sites in wild-type *E. coli*. The diagram shows the condition when no lactose is present; the repressor protein produced binds to the operator region and translation of the structural genes is prevented.

The z^+, y^+ and a^+ structural genes code for β-galactosidase, the permease and the transacetylase, respectively. Genetic experiments with mutant strains lacking enzyme activity showed that the three genes are adjacent on the chromosome. The i^+ gene (repressor gene) codes for a repressor protein. The expression of this gene is constitutive, that is the product is synthesised all the time. Control of the amount of gene product, then, is a function of the rate at which RNA polymerase can bind to the i gene promoter (p_i^+) and initiate the transcription of the repressor message. Translation of the message produces a polypeptide which aggregates to produce the functional repressor tetramer. If the cell is growing on a medium lacking lactose, the repressor will bind to the operator (o^+) region

formational change in the repressor such that it no longer has affinity for the DNA. As a consequence, the repressor falls off the operator and, once that has occurred, RNA polymerase can bind to the structural gene promoter and initiate transcription of the operon. The lactose operon is a single transcriptional unit so that the RNA polymerase transcribes the z, y and a genes, in that order, onto a single, *polycistronic* mRNA. This mRNA is translated by ribosomes attaching to the 5′ end (z gene end) and moving down the molecule. Thus the β-galactosidase is made first and, after the stop codon of that region, the ribosome continues moving towards the 3′ end of the message. Then after recognising the initiation sequence for the permease gene, the ribosome begins to translate that part of the message to produce permease. The process is repeated at the boundary between the permease and transacetylase gene sequences. Thus the general principle that pertains here is that ribosomes can *only* initiate translation at the 5′ end of a polycistronic mRNA. (This of course makes the coordinate production of proteins of related function easy to control.) It is not possible for ribosomes to bind and initiate translation at the start regions of the permease or transacetylase sequences presumably because the correct 'initiation' sequence is only found at the 5′ end of the message. All of this is summarised in Fig. 19.5. As long as lactose is present in sufficient amounts to bind with the repressor, the operon will be transcribed and the three enzymes will be produced.

partial diploid strains and the construction of these strains will be described before we discuss the properties of the mutants themselves.

In the discussion of *Hfr* strains of *E. coli*, it was shown that *Hfr* strains could revert to the F^+ state if there was a reversal of the process for integration of the circular *F* factor. In most cases the *F* factor is 'outgrated' correctly but occasionally an error is made and part of the bacterial chromosome is looped out. The resulting episome is called an *F′* (F-prime) factor and, by conjugation, the genes picked up by the *F* can be transferred rapidly through an F^- population. Thus, by this process, episomes can be produced that carry the lactose region of the chromosome and these are called *F′ lac* (Fig. 19.6).

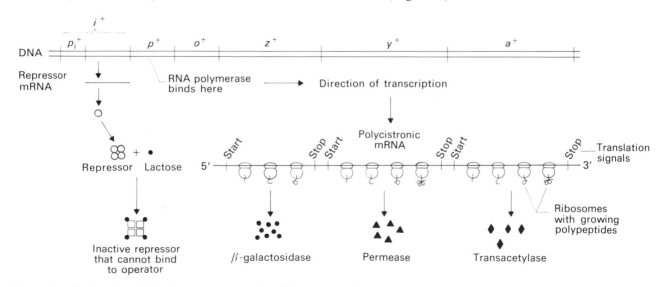

Fig. 19.5 Induction of the lactose operon in wild-type *E. coli* when lactose is present as the sole carbon source. Lactose inactivates the repressor thereby allowing transcription of the structural genes.

Translation of the resulting polycistronic mRNA produces the three enzymes required for lactose utilisation by the cell.

Elucidation of the regulation of lactose operon by studies of mutants

The Jacob-Monod operon model of control of gene expression was proposed by those two gentlemen and was based on their studies of a number of regulatory mutants in which the control of the expression of the lactose operon was abnormal. An important part of their mutant studies involved

Now to return to the studies of mutants of the lactose system.

a. Mutants of the structural genes. Both missense and nonsense mutants have been isolated for the three structural genes. Mapping these mutants provided evidence for the order z–y–a on the chromosome. A missense mutation in one of the genes

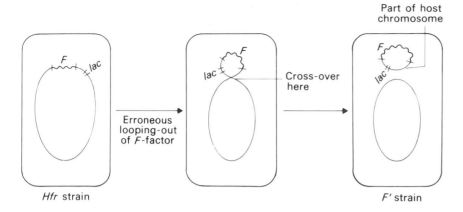

Fig. 19.6 Production of an episome (*F′ lac*) carrying the *lac* region of the *E. coli* chromosome by erroneous looping out of the *F* factor in an *Hfr* strain.

results in the loss of the activity of the enzyme for which the gene codes but does not affect the activities of the other two enzymes in the system. On the other hand the nonsense mutations have some interesting properties. For example, a nonsense mutation in the *z* gene (*z*⁻), providing it is not too near the end of the gene, will not only abolish *z* gene activity but, in addition, it will reduce or abolish the expression of the *y* and *a* genes. This is called the polar effect and thus the nonsense mutations are sometimes called polar mutations. Indeed there is a gradient of polarity of the effect of chain-terminating mutants in the *z* gene on the expression of the other two genes. Specifically, the closer the mutation is to the operator, the more severe the polar effect, that is the less likely permease and transacetylase will be produced. Such mutations in the *y* gene affect the activity of the *y* gene and the *a* gene but not the activity of the *z* gene. The interpretation of these data was that the three genes are transcribed onto a single polycistronic mRNA which is translated with 5′ to 3′ polarity in the order *z–y–a*, as we discussed previously. The polar effect of chain-terminating mutations, then, is the consequence of the distance ribosomes have to travel before recognising a new start codon — the farther it is, the more likely the ribosome will fall off the message.

b. Operator mutants. A series of mutants were isolated which were constitutive for enzyme pro-

duction. In other words they had lost regulatory control such that the enzymes were produced whether or not the inducer was present. When the mutations were mapped it was found that several were clustered next to the *z* gene in a region that is now called the operator. These so-called operator-constitutive (*o*ᶜ) mutants have lost the ability to bind the repressor protein such that transcription cannot be prevented. This latter point is supported by genetic and biochemical evidence. We will discuss the former here.

The effect of *o*ᶜ mutations on the contiguous structural genes and on *z*, *y* and *a* genes located on a different piece of DNA was examined in haploid and partial diploid strains. In these studies, mutant strains were constructed with various combinations of operator and structural gene alleles and the production of the enzymes was monitored in the absence and presence of the inducer, lactose. Representative data are shown in Table 19.1.

In the table, class 1 is the wild type where neither enzyme is produced until lactose is added as the inducer. Class 2 is a haploid strain carrying an *o*ᶜ mutation which leads to constitutive production of the enzymes. Class 3 is a partial diploid carrying *o*ᶜ and *y*⁻ mutations on one chromosome and the *o*⁺ and *z*⁻ alleles on the other chromosome. In this case β-galactosidase (*z* gene product) was produced constitutively but permease (*y* gene product) was produced only in the presence of lactose. Class 4 is a similar case, with *o*ᶜ and *z*⁻ on

Table 19.1 Effect of an o^c mutation in the lactose operon on enzyme production in haploid and partial diploid strains grown in the presence and absence of lactose.

Genotype	No Lactose		Lactose	
	β-galactosidase (z)	Permease (y)	β-galactosidase	Permease
1. $i^+p^+o^+z^+y^+$	−	−	+	+
2. $i^+p^+o^cz^+y^+$	+	+	+	+
3. $i^+p^+o^cz^+y^-/i^+p^+o^+z^-y^+$	+	−	+	+
4. $i^+p^+o^cz^-y^+/i^+p^+o^+z^+y^+$	−	+	+	+

one piece of DNA and the o^+ and y^- alleles on the other DNA. Here the permease is produced constitutively and the β-galactosidase is inducible. The two latter cases show that the effect of the o^c mutation is limited to those genes adjacent to it on the same piece of DNA, that is, it is a cis-dominant mutation. To take class 3 as an example: z^+ is on the same DNA as o^c and hence is transcribed constitutively. The y^- mutation results in a non-functional permease. In the strain, the wild-type y^+ allele is on the chromosome which also carries the o^+ allele. This DNA, then, is under normal regulatory control, and hence the y^+ is inducible. All of these data are consistent with the thesis that the operator region does not produce a diffusible product. Further, biochemical experiments showed that DNA from o^c strains will not bind the repressor thus showing that the operator is the site of repressor action. The operator, then, is a controlling region that in the wild type can bind the repressor which then prevents transcription. In the o^c mutants, the region has been altered (e.g. nucleotide pair change or deletion) so that the required protein–nucleic acid interaction does not occur and transcription cannot be stopped. Finally, since the o^c mutants do not affect the properties of β-galactosidase, it has been concluded that the operator is most probably a region distinct from the z cistron.

c. Repressor gene mutants. Mutations in the repressor (i) gene affect the control of expression of the lactose operon and the study of these types of mutants was also instrumental in the formulation of the operon model for gene regulation.

We know of a number of classes of mutations mapping at the i locus that affect the regulation of the *lac* operon. One class, the i^- mutations, gives rise to a constitutive phenotype. Mapping experiments showed that these mutations are at a locus (the repressor gene locus) distinct from the operator region. Other classes of mutations in the repressor gene include the i^s (super-repressed) and the i^{-d} (trans-dominant) — the properties of which are summarised in Table 19.2.

In the table, class 1 is the wild type as before and class 2 shows the constitutive nature of i^- mutations. Classes 3 and 4 involve partial diploids in which a z^- mutation is on one DNA and a y^- mutation is on the other DNA in both combinations with the i^- mutation. In both cases normal, inducible production of the enzymes is the case showing that the i^+ is dominant to i^- for genes either on the same or a different DNA. This is called trans-dominance and this is interpeted to mean that the i^+ gene codes for a diffusible product which acts to prevent transcription in the absence of lactose. This we now know to be the repressor protein which binds the operator. The i^- mutations produce an inactive repressor which cannot bind to the operator. The i^- mutations produce an inactive repressor which cannot bind to the operator and this results in constitutive enzyme production in haploid cells. In the i^+/i^- partial diploids, there are enough repressor molecules produced from the

Table 19.2 Effect of mutations in the repressor gene (i) of the lactose operon on enzyme production in haploid and partial diploid strains grown in the presence and absence of lactose.

Genotype	No Lactose		Lactose	
	β-galactosidase (z)	Permease (y)	β-galactosidase	Permease
1. $i^+p^+o^+z^+y^+$	−	−	+	+
2. $i^-p^+o^+z^+y^+$	+	+	+	+
3. $i^+p^+o^+z^+y^-/i^-p^+o^+z^-y^+$	−	−	+	+
4. $i^-p^+o^+z^+y^-/i^+p^+o^+z^-y^+$	−	−	+	+
5. $i^sp^+o^+z^+y^+$	−	−	−	−
6. $i^sp^+o^+z^+y^+/i^+p^+o^+z^+y^+$	−	−	−	−
7. $i^{-d}p^+o^+z^+y^+$	+	+	+	+
8. $i^{-d}p^+o^+z^+y^-/i^+p^+o^+z^-y^+$	+	+	+	+
9. $i^{-d}p^+o^+z^-y^+/i^+p^+o^+z^+y^-$	+	+	+	+

i^+ gene to bind to the two operators present and thus normal regulation is in effect.

An i^s mutation results in a completely negative phenotype with respect to *lac* enzyme production (class 5). In partial diploids, i^s is trans-dominant to the i^+ allele (class 6). The interpretation here is that i^s mutations result in the production of an altered repressor molecule (a super-repressor) which binds to the operator normally but is not capable of recognising the inducer molecule, lactose. Hence, transcription cannot be initiated since the defective repressors remain stuck on the operators.

The last class of i mutations we shall discuss are the i^{-d} mutations. Like the i^- mutations, these result in constitutive enzyme production in haploids (class 7) but, unlike the i^- alleles, they are trans-dominant to the wild-type i^+ allele (classes 8 and 9). These i^{-d} mutants are very rare and their phenotype is believed to be related to the tetrameric nature of the repressor. That is to say, the wild type produces a repressor protein of four identical polypeptides. This repressor molecule has only one binding site for the operator. In i^{-d} mutants the hypothesis has been proposed that the subunits do not combine normally so that no operator-specific binding site results. In the $i+/i^{-d}$ strains there is a mixture of normal and defective repressor subunits and the trans-dominance of i^{-d} is thought to be because repressors made from combinations of

the two cannot bind to the operator. Only purely wild-type repressor could bind and, statistically speaking, they would be quite rare. With this in mind then, the i^- mutations discussed previously must either be nonsense mutations which produce short polypeptides or missense mutations which result in polypeptides that do not participate in tetramer formation.

In summary, the i mutations provide evidence that the i gene produces a diffusible product — the repressor — which prevents transcription of the *lac* operon. The site of action of the repressor is the operator region. Also, the i mutations show that the lactose repressor has three recognition reactions coded into its structure:

(i) With the inducer, lactose; this presumably alters its shape causing it to dissociate from the DNA.

(ii) With the operator region.

(iii) With itself in that it acts as a tetramer.

d. Promoter mutants. These mutants map to the 'left' of the operator region and are characterised by the lack of mRNA production in the presence or absence of lactose. The properties of these mutations are shown in Table 19.3. The data show that p^- mutations are cis-dominant to p^+ in that their effect is limited only to the genes on the same piece of DNA. This is shown especially clearly by classes 5 and 6 where inducible enzyme production

Table 19.3 Effect of mutations in the promoter region (p) for the structural genes of the lactose operon on enzyme production in strains grown in the presence or absence of lactose.

Genotype	No Lactose		Lactose	
	β-galactosidase (z)	Permease (y)	β-galactosidase	Permease
1. $i^+p^+o^+z^+y^+$	−	−	+	+
2. $i^+p^-o^+z^+y^+$	−	−	−	−
3. $i^+p^-o^cz^+y^+$	−	−	−	−
4. $i^-p^-o^+z^+y^+$	−	−	−	−
5. $i^+p^-o^+z^+y^-/i^-p^+o^+z^-y^+$	−	−	−	+
6. $i^+p^-o^+z^-y^+/i^-p^+o^+z^+y^-$	−	−	+	−

resulted for the wild-type gene on the p^+ DNA whereas no enzyme was produced from the wild-type structural gene on the p^- DNA. Classes 3 and 4 also indicate that the properties of p^- mutations are not affected by o^c or i^- constitutive mutants. These facts are interpreted if we assume the p^+ region is the binding site for RNA polymerase. If it is altered so that polymerase cannot bind then no transcription will occur.

Positive control of the lactose operon (catabolite repression)

The lactose operon is under negative control in that a specific repressor molecule binds to the DNA to prevent transcription of the structural genes. There is also good evidence that a positive control signal must be present for the operon to function normally. That is to say, in the discussions thus far it has been carefully stated that when lactose was present it was the *sole* carbon and energy source. Now, if both glucose and lactose are present in the growth medium, then the cells will preferentially catabolise glucose and the lactose operon is not transcribed. A similar situation applies to a number of other operons involved in the catabolism of other sugars, for example arabinose and galactose. These are called glucose-sensitive operons, and the phenomenon is commonly called catabolite repression.

The effect of glucose on transcription of the operons in question is the result of the action of a break-down product of that sugar to lower the intracellular amount of *cyclic AMP* (cAMP: 3′,5′-cyclic adenosine monophosphate). This molecule is made from ATP in a reaction catalysed by adenyl-cyclase and is broken down with the aid of the enzyme phosphodiesterase. Thus the (unknown) catabolite of glucose could bring about the decrease in cAMP levels either by inhibiting adenyl-cyclase or by stimulating phosphodiesterase, or perhaps by doing both (Fig. 19.7).

The importance of cAMP to the transcription of the lactose and other glucose-sensitive operons is that a complex of cAMP with a catabolite gene activator (CGA) protein (a dimer of molecular weight 44 000) must bind to the promoter before RNA polymerase can bind and initiate transcription. Schematically, the events shown in Fig. 19.8 are needed for transcription of the operons. Clearly, then, if insufficient cAMP is present to make the complex, transcription is blocked at the various operons.

Some recent data on lactose operon regulation

Recently, information has been obtained about the sequence of nucleotide pairs involved in the promoter and operator regions of the lactose operon. Without going into all of the details of the techniques involved or of the actual sequences found, Fig. 19.9 is a generalised diagram of the results.

The actual sequencing involved nucleic acid fingerprinting techniques. The repressor binding site was determined by sequencing the stretch of DNA protected from DNase digestion when the

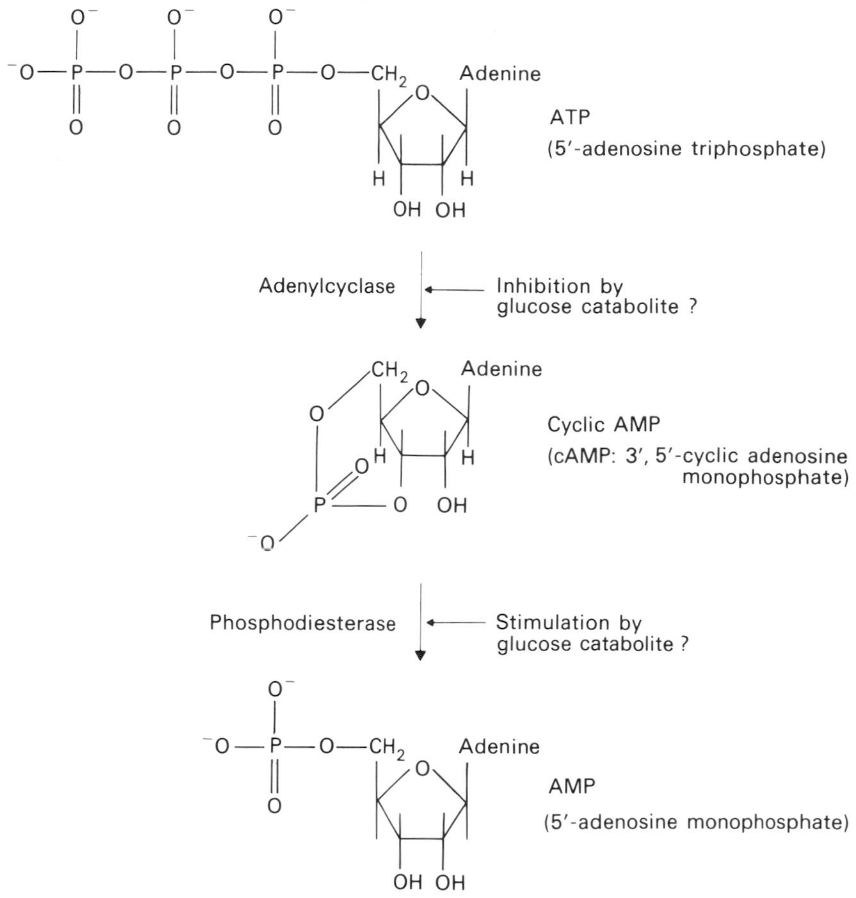

ATP
(5'-adenosine triphosphate)

Adenine

Adenylcyclase ← Inhibition by glucose catabolite ?

CH₂

Adenine

Cyclic AMP
(cAMP: 3', 5'-cyclic adenosine monophosphate)

Phosphodiesterase ← Stimulation by glucose catabolite ?

Adenine

AMP
(5'-adenosine monophosphate)

Fig. 19.7 The biosynthesis of cyclic AMP from ATP and its breakdown to AMP. In catabolite repression, a metabolic product of glucose inhibits adenyl cyclase activity or stimulates phosphodiesterase activity or it does both. The result is a decrease in cAMP level in the cell and this 'turns off' glucose sensitive operons in the cell.

repressor protein is bound to the operator. The exact limits of the cAMP-CGA binding site and RNA polymerase interaction sites are not known, but genetic mutations that affect the level of expression of the lactose operon, and which presumably are in the promoter, show the promoter region to be about 80 nucleotide pairs long. It is noteworthy that the end of the *i* gene is immediately adjacent to the promoter sequence. Comparison of the sequence of the lactose operon mRNA with the DNA sequence shows that transcription begins in the operator region within the region protected by the repressor. This is shown in more detail in Fig. 19.10, which also shows the region of the mRNA protected by the ribosome during the initiation step.

As can be seen, the first 38 bases of the mRNA are not translated, the codon for fmet starting at position 39. The first part of the mRNA is a copy of most of the operator region. The ribosome binding site was determined by sequencing the part of the mRNA remaining after RNase digestion while the ribosome is bound in its initiation configuration. This latter arrangement was achieved *in vitro* by having only fmet-tRNA in the reaction mixture so initiation, but not elongation, could occur. The ribosome binding site of the lactose operon covers 50 nucleotides of the mRNA and includes the codons for the first seven amino acids of the β-galactosidase (only three are given in the diagram). The boxed nucleotides in the diagram indicate sequences that have also been found in

Key: ◯ = cAMP

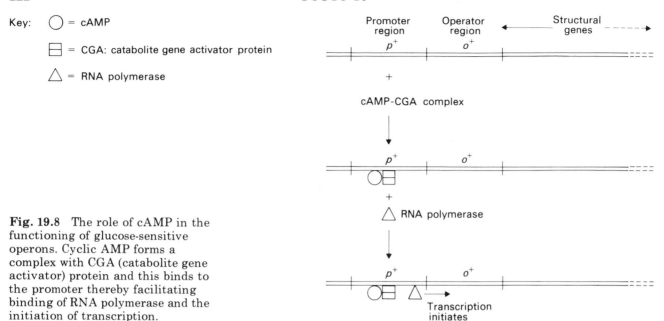

Fig. 19.8 The role of cAMP in the
functioning of glucose-sensitive
operons. Cyclic AMP forms a
complex with CGA (catabolite gene
activator) protein and this binds to
the promoter thereby facilitating
binding of RNA polymerase and the
initiation of transcription.

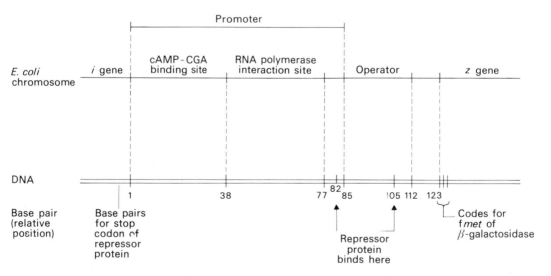

Fig. 19.9 Diagram of the promoter-operator region of
the lactose operon showing the extents and positions
of the binding sites for cAMP-CGA, RNA polymerase
and repressor protein. (After R. C. Dickson et al, 1975.
Science, **187**: 27–35.)

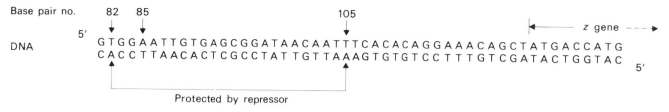

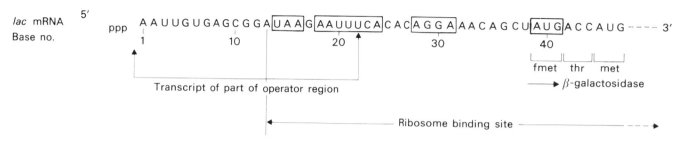

Fig. 19.10 The first 47 bases of the lac mRNA and the DNA sequence corresponding to it. The first 21 bases of the message are transcribed from the operator region. Other features of this region are described in the text. (After R. C. Dickson *et al*, 1975. *Science*, **187**:27–35 and N. Maizels, 1974. *Nature*, **249**:647–649.)

ribosome binding sites of other prokaryotic mRNAs. Specifically there is a nonsense codon (UAA here), a purine-purine-UUU-X-purine (where X is usually a purine), an AGGA sequence, and the start codon AUG.

In summary, the lactose operon has proved to be a model system for understanding gene regulation for a number of operons in prokaryotes and presumably, when a large number of promoter and operator sequences have been obtained, it will be possible to generalise about the molecular bases for the regulatory mechanisms at the DNA and RNA levels.

THE ARABINOSE OPERON OF *E. COLI*

The arabinose operon is another example of glucose-sensitive operon. As with lactose, when arabinose is absent only a few molecules of the enzymes needed for arabinose catabolism are present in the cell. When arabinose is added (provided glucose is absent), there is a very rapid increase in the number of arabinose catabolic enzymes. As

we shall see, this is controlled by a quite different mechanism from that described for the lactose operon.

The genes governing the metabolism of arabinose comprise what is called a *regulon* which is composed of at least three operons. The *araBAD* operon contains the genes for the enzymes involved with the conversion of L-arabinose to D-xylulose 5-phosphate. The controlling sites for this operon are located adjacent to it. There are also two operons that control the transport of arabinose into the cell: *araE*, which is the structural gene for the L-arabinose binding protein. The regulator gene for the system, *araC*, is located between the *araBAD* controlling site region and the *leu* operon. The *araC* gene controls the expression of *araBAD* by its positive and negative action in the controlling site region. Since *araBAD*, *araE*, and *araF* are inducible by L-arabinose and controlled coordinately by *araC*, it is assumed that the structures of the three controlling site regions are similar. In the following we will concentrate our attention on the *araBAD* operon which is shown in Fig. 19.11.

The operon is thought to be controlled as follows. The *araC* gene codes for a P1 protein which has

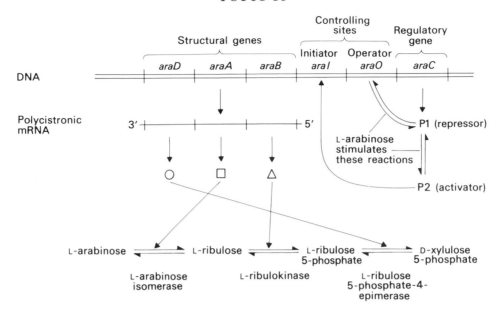

Fig. 19.11 The *araBAD* (arabinose) operon of *E. coli* and the associated controlling sites and regulatory gene. The regulation of this operon is described in the text. (After the work of E. Englesberg.)

repressor function and exerts its effect by binding to the adjacent *araO* (operator) controlling site and preventing the RNA polymerase from binding. Thus the operon is under negative control by P1. When L-arabinose is present, it stimulates the release of P1 from the DNA and the conversion of P1 to P2 which is an activator of the operon. The P2 binds to the *araI* (initiator) site, facilitating RNA polymerase binding and the initiation of transcription. (All this, of course, requires the prior binding of cAMP-CGA complex, and this is thought to occur in the same vicinity.) The three structural genes are transcribed on a single polycistronic mRNA. The operon, then, is under positive control by P2.

Much of this is hypothesis, but there is good evidence for some parts of it. There is genetic evidence for the existence of the *araI* site and it is thought that part of this has promoter activity and another part is involved with cAMP-CGA binding. The function of the *araC* gene has been demonstrated by the study of genetic mutants. Mapping experiments with these mutants have

shown that *araC* consists of only one cistron. Moreover, *araC* nonsense mutants have been shown to have no cis effect on the *araBAD* operon thus indicating that *araC* is not in the *BAD* operon.

Three classes of *araC* alleles are known:

1. $araC^+$ — the wild-type allele renders the operon inducible; that is the three enzymes are induced by L-arabinose.

2. $araC^-$ — these occur with high frequency and result in a pleiotropically L-arabinose-negative phenotype. In other words, the enzymes are not inducible by L-arabinose.

3. $araC^c$ — these are quite rare and give a pleiotropically constitutive phenotype in that the enzymes are produced even in the absence of the inducer.

As with the regulatory mutants of the lactose operon, diploid studies have been used to obtain an understanding of the function of the *araC* gene. From these studies it was shown that $araC^-$ is recessive to C^+ in either the cis or trans arrangement. The C^- alleles are complemented by A^-, B^- and D^- alleles, and thus the pleiotropic negative phenotype is not the result of a polarity effect on the *araBAD* operon. The conclusion from studies of C^- and C^+/C^- strains was that C^+ produces a protein which, in the presence of L-arabinose, is necessary for the expression of the region. This suggested some positive control in the system and contrasts with the lactose operon C^- mutants which are constitutive owing to the loss of negative control.

The C^c alleles are cis- and trans-dominant to C^-, suggesting that they produce the activator P2 in the absence of L-arabinose and this activator is able to turn on the operon both cis and trans to the C^c allele. On the other hand C^+ is dominant to C^c. This suggests that there is some negative control of the operon. Based on the model for regulation of the operon that was presented (which was, of course, proposed on the basis of the data we are now discussing), in the absence of arabinose, P1 acts as a repressor preventing expression of the operon by $araC^c$ activator. That is, when P1 is on the operator, transcription ceases even if P2 is present.

There is some biochemical evidence to support the regulatory model. Studies of heat-sensitive $araC^-$ mutants have shown that *both* the repressor and activator functions are heat-labile thereby indicating that *araC* produces a protein that can serve both functions. More recently the *araC* protein has been purified and it has been shown to have both repressor and activator activity. Indeed there is some evidence that the P2 form of the protein is a dimer of the P1 form. There is also evidence that L-arabinose interacts directly with the *araC* protein to bring about the necessary conversion. Also, there is direct evidence that the activator form of *araC* protein, P2, is required for transcription of the operon. That is, in an in-vitro system, synthesis of *ara* mRNA shows an absolute requirement for *araC* protein.

In conclusion, the L-arabinose regulon is under both positive and negative control, with the *araC* protein playing a pivotal role in the regulatory process. The exact nature of the controlling sites remains to be worked out. In contrast to the lactose operon where inhibition must be removed in order for the genes to be expressed, the arabinose operon requires activation for transcription to begin.

THE TRYPTOPHAN OPERON OF *E. COLI*

The tryptophan operon of *E. coli* is an example of a *repressible* operon which is regulated by many of the basic features of the Jacob-Monod operon model. In this case tryptophan is an amino acid that is needed for protein synthesis. When tryptophan is absent from the medium, the enzymes necessary for tryptophan biosynthesis are produced by the cell and, when tryptophan is present in the medium, the tryptophan genes are repressed. This makes sense from energetic considerations. Most operons controlling the synthesis of a compound are repressible whereas operons involved with the breakdown of something (e.g. a carbon source) are usually inducible. The repressible operons have a regulatory gene that codes for an apo-repressor molecule which cannot bind to the operator. The end-product of the biosynthetic pathway or a derivative of it (e.g. a charged tRNA molecule) acts as a co-repressor (and not as an inducer) in the system and, by binding with the apo-repressor, produces a functional repressor molecule capable of binding to the operator and blocking transcription. The general features of a repressible operon are shown in Fig. 19.12.

As we shall see, this basic type of control is in effect in the tryptophan operon although there are some interesting additional elements in that particular system. Most of the work that will be described is that of C. Yanofsky and his co-workers.

The tryptophan operon of *E. coli* is shown in Fig. 19.13 along with the estimated lengths of the structural genes and other regions. In this operon there are two promoters, *p1* and *p2*, the former is adjacent to the operator (*o*), and the latter is at the end of the *trpD* cistron and acts as a low efficiency internal promoter. The function of the operon is controlled by the apo-repressor protein which is the polypeptide product of the *trpR* gene. When tryptophan is present in the medium, a complex of the repressor with L-tryptophan binds at the operator region and represses transcription by preventing the attachment of RNA polymerase at the *p1* promoter. In the absence of tryptophan, the operon is not repressed and mRNA transcription is ini-

A repressible operon

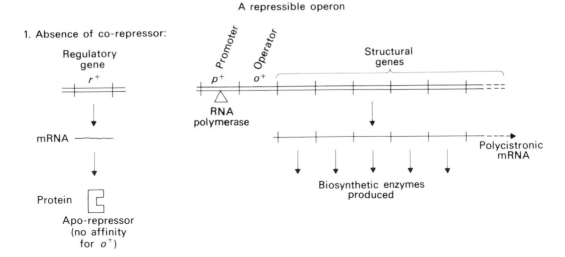

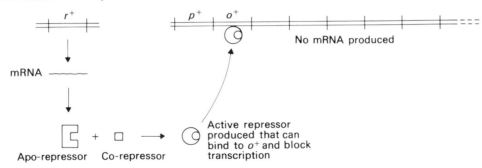

Fig. 19.12 The general functioning of a repressible
operon in bacteria.

tiated. Polycistronic mRNA transcribed from the
operon contains a leader sequence of 162 nucleotides which is coded by the *trpL* region. This leader
sequence precedes the translation initiation codon
for the *trpE* polypeptide. A short transcript is also
produced which contains only the first 142 nucleo
tides of *trp* mRNA. This latter RNA results from
the termination of transcription at the attenuator
(*a* in the figure) which is a regulatory site in the
leader region of the operon. As we shall see, the
attenuator plays an important role in the regulation of expression of the *trp* structural genes.

The existence of the leader sequence was shown
by the sequencing of the 5′ segment of the *trp*

operon mRNA synthesised either *in vivo* or *in vitro*.
Internal deletions of the *trp* operon which have
one end between *p*1 and about nucleotide 130 of
the leader sequence and the other end in one of
the structural genes generally result in an increase
in operon expression. On the other hand, internal
deletions which leave the leader region intact do
not give this result. This indicated that there is a
site called the attenuator, which normally limits
operon expression. (This is not unique to the *trp*
operon but is found in a number of other operons,
for example the histidine operon of *E. coli*.)
Quantitative hybridisation experiments showed
that the ratio of the number of copies of leader

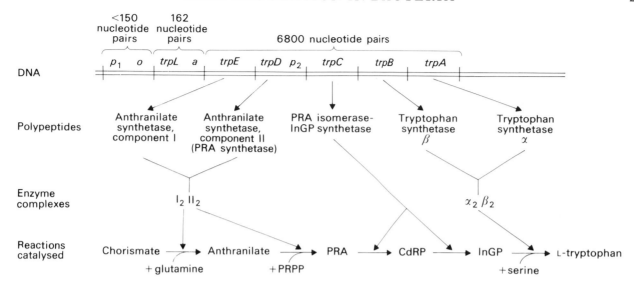

Key: PRPP = phosphoribosyl pyrophosphate
      PRA = phosphoribosyl anthranilate
     CdRP = 1-(o-carboxyphenylamino)-1-deoxyribulose 5-phosphate
     InGP = indole-3-glycerol phosphate

Fig. 19.13 The tryptophan operan of *E. coli* showing the arrangement of structural genes, their polypeptide products and the reactions they catalyse. p_1 is the principal promoter; *o* is the operator; p_2 is an internal promoter; *trpL* is the leader region and *a* is the attenuator. (After the work of C. Yanofsky.)

mRNA to structural gene *trp* mRNA is about 10 : 1, thus supporting the attenuator concept. Since short, 142 nucleotide-long transcripts are found, the attenuator must be at about that position in the leader region of the DNA. Indeed, in an in-vitro transcription system, it has been shown that the RNA polymerase terminates transcription in the attenuator region.

So how does the system function? Under repressing conditions (that is when cells are grown in the presence of excess tryptophan) the attenuator functions maximally to prevent distal transcription thereby amplifying the effect of the repressor-operator interaction perhaps ten-fold. Transcription termination at the attenuator is regulated in response to changes in the number of charged tRNA · trp; the more charged it is, the fewer mRNAs are transcribed through the attenuator. Thus, when tryptophan is limited, repression is lifted and in addition attenuation is relaxed. In extreme starvation conditions for tryptophan, the attenuation is completely relaxed. Thus it appears that the attenuator's function is to extend the range of expression of the operon beyond that possible by operator–repressor interaction alone. The mechanism of attenuator control is not completely understood, although it is thought that the translatability of the leader RNA and the secondary structure of that RNA plays an important role in this process. Some evidence for this has been obtained, for example, from RNA sequencing experiments, from studies with genetic mutants, and from investigations using an in-vitro system. Specifically, it has been found that there are two adjacent tryptophan codons (UGG) quite close to the termination codon (UGA) on the leader mRNA. Between the UGA codon and the attenuator site (at about nucleotide 142 on the leader RNA) the nucleotide sequence has the potential to form various secondary structures as a result of hydrogen bonding between the bases. So, after RNA polymerase has synthesised this sequence, the RNA can fold up. Further, the leader RNA is being translated by ribosomes while transcription is proceeding, and it is proposed that the ribosomes are following the RNA polymerase quite closely. If excess tryptophan is present, then the ribosome can read the two tryptophan codons and will terminate at the UGA codon, which is adjacent to the folded RNA. In this case, the ribosome's position will result in the formation of a particular pairing of nucleotides. This, in turn, will affect the RNA polymerase's association with the DNA, and thus transcription will terminate at the attenuator. When the cells are starved for tryptophan, however, there are no charged tRNA·trp molecules present and hence the ribosome will stall on the adjacent tryptophan codons. In this event, a different nucleotide pairing will be favoured in the folded RNA and this will have no effect on RNA polymerase and, hence, transcription will continue past the attenuator region and into the structural genes. This, then, is a good working model to explain how the continued transcription of a DNA segment can be controlled by the extent to which the RNA product is translated. Apparently this model is generally applicable to all bacterial operons that have an attenuator control region.

Finally, studies with internal deletion mutants showed the presence of an internal promoter, $p2$, within $trpD$, in the operator distal segment of the gene. The efficiency of this promoter is only a few per cent that of the $p1$ promoter and it does not appear to be subject to tryptophan-mediated repression. The physiological role of $p2$ is unknown.

SUMMARY OF OPERON REGULATION

In *E. coli* there are 100–200 operons that have been identified. Operons are ubiquitous throughout the prokaryotes (bacteria and phages) but, with one possible exception (the galactose catabolism genes of yeast), there appear to be no operons in eukaryotes. As discussed earlier in this Topic, and as

exemplified by the three operons described, the organisation of genes and controlling regions into operons affords a very simple and effective way of coordinating the activity of genes with related functions. As we have seen, with the basic operon organisation of repressor gene, and structural genes contiguous with promoter and operator regions, regulation of gene expression can be achieved in an inducible or repressible manner (negatively-controlled operons) or by an activation mechanism (positively-controlled operons). This regulation is at the transcriptional level, although for operons with attenuators the translational system may also play a role. In general, the control of gene expression by an operon system may be considered a fine control system since it involves the regulation of one or a few transcriptional units by specific regulatory signals. This allows the organism to adapt rapidly to the changing environment. By contrast, in the next section we shall see how the overall growth and/or gene expression of a bacterium or phage is regulated by effects on (usually) several transcriptional units.

MAJOR CONTROL OF TRANSCRIPTION AND TRANSLATION

The overall control of bacterial cell growth and division or of phage reproduction is related to macromolecular synthesis, and particularly to RNA and protein synthesis. In this section we shall consider some examples of general regulation at the transcriptional and translational levels in prokaryotes.

Stringent control
Bacteria have evolved a general cellular regulatory system to survive under harsh conditions. In the laboratory such conditions can be imposed, for example, by depriving an amino acid auxotrophic strain of the required amino acid or by moving the cells rapidly from a medium containing high amounts of amino acids to a medium containing no amino acids (a nutritional shift-down). Wild-type cells respond in a stringent way to these condi-

tions and studies of this response have shown that there is a regulatory mechanism which coordinates protein synthesis with the activities of a variety of other cellular activities. In other words, the starvation conditions described bring about a plethora of effects including the inhibition of synthesis of proteins, rRNAs, tRNAs, mRNAs, lipids, carbohydrates and nucleotides; the inhibition of glycolysis; the stimulation of protein degradation processes; and the inhibition of some membrane transport processes. In the absence of these regulatory mechanisms, starved cells would grow in an uncontrolled manner with all components (except proteins) accumulating in large amounts.

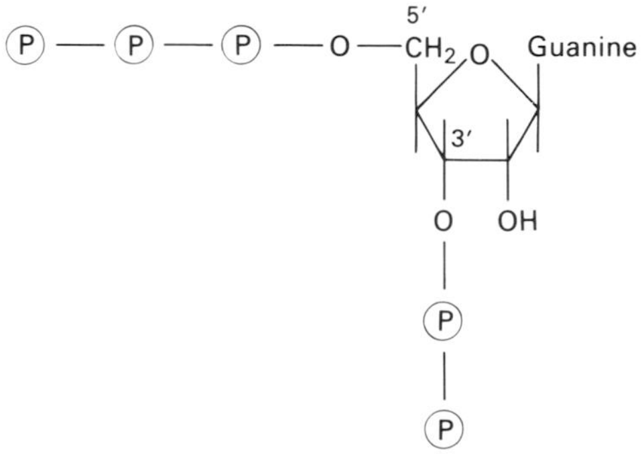

Fig. 19.14 Shorthand formulae for unusual guanine ribonucleotides implicated in the stringent response. (P) represents a phosphate group. (a) Guanosine tetraphosphate (5'-diphospho-guanosine-3' diphosphate), (b) guanosine pentaphosphate (5'-triphospho-guanosine-3'-diphosphate).

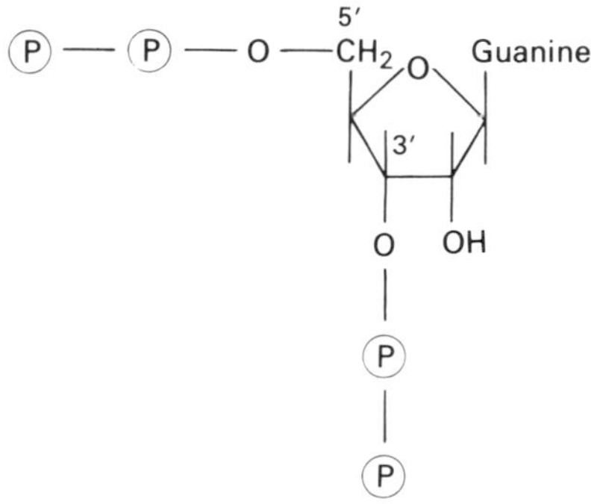

In an attempt to understand the stringent control mechanism, mutants of *E. coli* were isolated that did not show the response to amino acid starvation. The 'relaxed' (*relA*) mutants were compared with the wild-type cells. From this it was discovered that starvation results in the accumulation of two unusual nucleotides, guanosine tetraphosphate and guanosine pentaphosphate (ppGpp and pppGpp; Fig. 19.14) in the wild type, but not in the *relA* strains. The interpretation is that the *rel*+ gene is involved with the production of a *stringent factor* that is responsible for the intracellular accumulation of ppGpp and pppGpp. Further, it has been shown that ppGpp and pppGpp are made on the ribosomes and that this occurs in response to the presence of non-aminoacylated (uncharged) tRNAs that accumulate during amino acid starvation. Thus when uncharged tRNAs interact with ribosomes, the stringent factor can also bind (a protein of molecular weight 77 000 daltons), and this results in the formation of the two unusual nucleotides. How the two nucleotides produce the pleiotropic effects characteristic of the stringent response is not known. Finally, with the exception of yeast, the stringent response is an exclusively prokaryotic phenomenon.

Control at the translational level

There are a number of ways by which gene expression can be modulated at the translational level. For example, one characteristic of the operon organisation is that a polycistronic mRNA is produced when the structural genes are transcribed. For some operons translation of the mRNA results in stoichiometrically-equal amounts of the proteins coded for, whereas for other operons an unequal ratio of protein products is the case. For the lactose operon of *E. coli*, for example, there is a 10 : 5 : 2 ratio of *z*, *y* and *a* gene products. The relative amounts of the products of a polycistronic mRNA in fact depend upon the primary structure of the mRNA. Specifically, at the end of each cistron on the message there is at least one chain terminating codon followed by an intercistron segment prior to the next translation initiation sequence. Thus, the nucleotide sequences of these regions between stop and start codons can govern whether or not a ribosome 'out of gear' will fall off the message before initiating the next polypeptide. Presumably in the lactose operon, approximately half of the ribosomes fall off between z and y and another half fall off between y and a. With the DNA cloning and sequencing techniques now available (see Topic 12), it will be possible to obtain more detailed information about this regulatory control mechanism.

Some other examples of translational control have come from studies of the metabolic changes in a bacterial cell following phage infection. For instance, when T4 infects *E. coli* there is a rapid production of phage-specific mRNA molecules. These are translated selectively on the host-cell ribosomes owing to the fact that the latter become modified by the binding of a phage-specific protein. The exact nature of this functional modification is unknown, but it certainly serves as an effective means of channelling the host's metabolic activities towards phage reproduction.

A similar example of this type of translational channelling has been found in T2-infected *E. coli* cells. Messengers transcribed from the T2 genome contain no CUG (leucine) codons and thus there is no need for the presence of leu-tRNAs with the anticodon for CUG in the host. The T2 phage itself codes for its own leu-tRNAs (for other leucine codons) and also codes for a ribonuclease which acts (with about 50% efficiency) to degrade the host's leu-tRNA with anticodons for CUG. This effectively favours the translation of phage-specific mRNAs over that of host-specific messengers.

In sum, there are a number of ways that gene expression can be regulated in prokaryotes. The regulatory mechanisms can operate at the transcriptional and/or the translational level. The examples presented illustrate the complexity of these control systems and it is anticipated that more molecular details will be obtained when, for example, important controlling regions are cloned and sequenced.

REFERENCES

The lactose operon

Beckwith J.R. 1967. Regulation of the lactose operon. *Science*, **156**: 597–604.

Beckwith J.R. & D. Zipser. 1970. *The Lactose Operon*. Cold Spring Harbor Press, New York.

Beyreuther K. 1978. Revised sequence for the *lac* repressor. *Nature*, **274**: 767.

Beyreuther K., K. Adler, N. Geisler & A. Klemm. 1973. The amino acid sequence of *lac* repressor. *Proc. Natl. Acad. Sci. USA*, **70**: 3576–3580.

Calos M.P. 1978. DNA sequence for a low-level promoter of the *lac* repressor gene and 'up' promoter mutation. *Nature*, **274**: 762–765.

Dickson R.C., J. Abelson, W.M. Barnes & W.S. Reznikoff. 1975. Genetic regulation: the *lac* control region. *Science*, **187**: 27–35.

Edelmann P.L. & G. Edlin. 1974. Regulation of the synthesis of the lactose repressor. *J. Bacteriol.* **120**: 657–665.

Farabaugh P.J. 1978. Sequence of the *lacI* gene. *Nature*, **274**: 765.

Gilbert W., N. Maizels & A. Maxam. 1974. Sequences of controlling regions of the lactose operon. *Cold Spring Harbor Symp. Quant. Biol.* **38**: 845–855.

Gilbert W. & A. Maxam. 1973. The nucleotide sequence of the *lac* operator. *Proc. Natl. Acad. Sci. USA*, **70**: 3581–3584.

Gilbert W. & B. Muller-Hill. 1966. Isolation of the lac repressor. *Proc. Natl. Acad. Sci. USA*, **56**: 1891–1898.

Jacob F. & J. Monod. 1961. Genetic regulatory mechanisms in the synthesis of proteins. *J. Mol. Biol.* **3**: 318–356.

Jacob F. & J. Monod. 1965. Genetic mapping of the elements of the lactose region of *Escherichia coli*. *Biochem. Biophys. Res. Commun.* **18**: 693–701.

Maizels N. 1973. The nucleotide sequence of the lactose messenger ribonucleic acid transcribed from the UV5 promoter mutant of *Escherichia coli*. *Proc. Natl. Acad. Sci. USA*, **70**: 3585–3589.

Maizels N. 1974. *E. coli* lactose operon ribosome binding site. *Nature New Biol.* **249**: 647–649.

Miller J.H., C. Coulondre & P.J. Farabaugh. 1978. Correlation of nonsense sites in the *lacI* gene with specific codons in the nucleotide sequence. *Nature*, **274**: 770–775.

Ptashne M. & W. Gilbert. 1970. Genetic repressors. *Sc. Am.* **222**: 36–44.

Reznikoff W.S. 1972. The operon revisited. *Annu. Rev. Genet.* **6**: 133–156.

The arabinose operon

Englesberg E. 1971. Regulation in the L-Arabinose System. In *Metabolic Pathways*, H. Vogel (ed.), vol. 5, pp. 256–296. Academic Press, New York.

Englesberg E., J. Irr, J. Power & N. Lee. 1965. Positive control of enzyme synthesis of gene *C* in the L-arabinose system. *J. Bacteriol.* **90**: 946–957.

Englesberg E., C. Squires & F. Meronk. 1969. The L-arabinose operon in *Escherichia coli* B/r: a genetic demonstration of two functional states of the product of a regulatory gene. *Proc. Natl. Acad. Sci. USA,* **62**: 1100–1107.

Englesberg E. & G. Wilcox. 1974. Regulation: positive control. *Annu. Rev. Genet.* **8**: 219–242.

Heffernan L., R. Bass & E. Englesberg. 1976. Mutations affecting catabolite repression of the L-arabinose regulon in *Escherichia coli* B/r. *J. Bacteriol.* **126**: 1119–1131.

Heffernan L. & G. Wilcox. 1976. Effect of *araC* gene product on catabolite repression in the L-arabinose regulon. *J. Bacteriol.* **126**: 1132–1135.

Irr J. & E. Englesberg. 1970. Nonsense mutants in the regulator gene *araC* of the L-arabinose system of *Escherichia coli* B/r. *Genetics,* **65**: 27–39.

Lee N., G. Wilcox, W. Gielow, J. Arnold, P. Cleary & E. Englesberg. 1974. *In vitro* activation of the transcription of *araBAD* operon by *araC* activator. *Proc. Natl. Acad. Sci. USA,* **71**: 634–638.

Sheppard D.E. & E. Englesberg. 1967. Further evidence for positive control of the L-arabinose system by gene *araC. J. Mol. Biol.* **24**: 443–454.

Wilcox G., K.J. Clemetson, P. Cleary & E. Englesberg. 1974. Interaction of the regulatory gene product with the operator site in the L-arabinose operon of *Escherichia coli. J. Mol. Biol.* **85**: 589–602.

The tryptophan operon

Bertrand K., L. Korn, F. Lee, T. Platt, C.L. Squires, C. Squires & C. Yanofsky. 1975. New features of the structure and regulation of the tryptophan operon of *Escherichia coli. Science,* **189**: 22–26.

Bertrand K., C. Squires & C. Yanofsky. 1976. Transcription termination *in vivo* in the leader region of the tryptophan operon of *Escherichia coli. J. Mol. Biol.* **103**: 319–337.

Bertrand K. & C. Yanofsky. 1976. Regulation of transcription termination in the leader region of the tryptophan operon of *Escherichia coli* involves tryptophan as its metabolic product. *J. Mol. Biol.* **103**: 339–349.

Jackson E.N. & C. Yanofsky. 1972. Internal promoter of the tryptophan operon of *Escherichia coli* is located in a structural gene. *J. Mol. Biol.* **69**: 307–313.

Jackson E.N. & C. Yanofsky. 1973. The region between the operator and first structural gene of the tryptophan operon of *Escherichia coli* may have a regulatory function. *J. Mol. Biol.* **76**: 89–101.

Lee F. & C. Yanofsky. 1977. Transcription termination at the *trp* operon attenuators of *Escherichia coli* and *Salmonella typhimurium*: RNA secondary structure and regulation of termination. *Proc. Natl. Acad. Sci. USA,* **74**: 4365–4369.

Miozzari G.F. & C. Yanofsky. 1978. Translation of the leader region of the *Escherichia coli* tryptophan operon. *J. Bacteriol.* **133**: 1457–1466.

Morse D.E. & A.N.C. Morse. 1976. Dual-control of the tryptophan operon is mediated by both tryptophanyl-tRNA synthetase and the repressor. *J. Mol. Biol.* **103**: 209–226.

Platt T., C. Squires & C. Yanofsky. 1976. Ribosome protected regions in the leader-*trpE* sequence of *Escherichia coli* tryptophan messenger RNA. *J. Mol. Biol.* **103**: 411–420.

Rose J.K. & C. Yanofsky. 1974. Interaction of the operator of the tryptophan operon with repressor. *Proc. Natl. Acad. Sci. USA,* **71**: 3134–3138.

Squires C., F. Lee, K. Bertrand, C.L. Squires, M.J. Bronson & C. Yanofsky. 1976. Nucleotide sequence of the 5′ end of tryptophan messenger RNA of *Escherichia coli. J. Mol. Biol.* **103**: 351–381.

Squires C.L., F. Lee & C. Yanofsky. 1975. Interaction of the *trp* repressor and RNA polymerase with the *trp* operon. *J. Mol. Biol.* **92**: 93–111.

Yanofsky C. 1976. Control sites in the tryptophan operon. In *Control of Ribosome Synthesis,* Alfred Benzon Symposium XI, pp. 149–163. Academic Press, New York.

Yanofsky C. & L. Soll. 1977. Mutations affecting tRNA·trp and its charging and their effect on regulation of transcription termination at the attenuator of the tryptophan operon. *J. Mol. Biol.* **113**: 1457–1466.

Other examples of regulation

Calvo J.M. & G.R. Fink. 1971. Regulation of biosynthetic pathways in bacteria and fungi. *Annu. Rev. Biochem.* **40**: 943–968.

Cashel M. 1975. Regulation of bacterial ppGpp and ppppp. *Annu. Rev. Microbiol.* **29**: 301–318.

Ihler G. & D. Nakada. 1970. Selective binding of ribosomes to initiation sites on single stranded DNA from bacterial viruses. *Nature,* **228**: 329–242.

Kano-Sueoka T. & N. Sueoka. 1969. Leucine tRNA and cessation of *Escherichia coli* protein synthesis upon phage T2 infection. *Proc. Natl. Acad. Sci. USA,* **62**: 1229–1236.

Miller J.H. & W.S. Reznikoff. 1978. *The Operon.* Cold Spring Harbor Press, New York.

Summers W.C. 1972. Regulation of RNA metabolism of T7 and related phages. *Annu. Rev. Genet.* **6**: 191–202.

Topic 20
Regulation of Gene Expression in Eukaryotes

OUTLINE

General aspects of eukaryotic gene regulation
  potential sites for regulation of enzyme synthesis
  nonhistones and the regulation of transcription
Regulation of gene expression in lower eukaryotes
  the galactose fermentation genes of yeast
  the genes for aromatic amino acid biosynthesis in
    Neurospora
  regulation of quinic acid metabolism in *Neurospora*
Regulation of gene expression in higher eukaryotes
  regulation of enzyme synthesis by hormones
  model for steroid hormone action
Long-term genetic regulation in higher eukaryotes
  definitions of development and differentiation
  general aspects of development and differentiation.

GENERAL ASPECTS OF EUKARYOTIC GENE REGULATION

Potential sites for regulation of enzyme synthesis

We know a lot about the regulation of gene expression in prokaryotes but relatively little about it in eukaryotes. As will be discussed, the operon system of gene regulation has proved to be inapplicable to eukaryotic organisms. Eukaryotes are more complex than prokaryotes with regard to their cellular organisation, in that their cells have compartments as a result of extensive membranous structures. Of particular interest here is the nucleus which contains most of the genetic material. This compartmentation has consequences for the regulation of gene expression and the following is a list of the sites at which regulatory processes act, or might act, to control enzyme synthesis in animal cells.

1. The chromatin. Chromatin is a complex between DNA, histones and nonhistones. The organisation of these components plays an important role in determining whether or not a region of the DNA can be transcribed.

2. Heterogeneous nuclear RNA. The primary transcripts of genes coding for proteins are much longer than the mature mRNA. Most of these heterogeneous nuclear RNA (hnRNA) molecules are modified (5′ capping and 3′ polyadenylation) and processed in the nucleus to produce the 'mature' mRNA molecule that enters the cytoplasm. Control can be exerted at either the modification or processing steps and this can modulate the amount of functional mRNA produced. In fact, there is a lot of evidence indicating that most of the hnRNA that is produced is degraded, and only a relatively small amount is processed to mRNA that is found in the cytoplasm.

3. The nuclear membrane. Once the mature mRNAs have been produced, they are transported to the cytoplasm. The mechanism for this is unknown but, if a nuclear membrane property is involved, then the amount of mRNA available to the ribosomes can be altered by regulating that property.

4. The mRNA molecule. When the mRNA molecule enters the cytoplasm, it is sometimes associated with proteins and hence is unavailable for translation. The amount of mRNA available for translation or the timing of mRNA translation can therefore be controlled by altering the ratio of available to unavailable mRNA of a particular type. Further, the amount of enzyme synthesised can be affected by the lability of the mRNA itself, which is presumably a function of the nucleotide sequence of the mRNA.

5. Effector molecules. Regulation of the transcriptional and posttranscriptional events in the nucleus may be related to the types and concentrations of effector molecules (e.g. inducers, activators, repressors) that are transported into the nucleus from the cytoplasm. Controlling factors that alter the synthesis of these molecules and/or their transport into the nucleus can therefore affect the amount of enzyme ultimately synthesised.

In addition, other subtle control devices may be operating in the cell to regulate the amount of mRNA available for translation, and at this time we certainly know very little about the regulatory signals for any of this. It is clear also that translational control is operant within eukaryotic cells. For example, the translation of a mRNA molecule may be affected by whether the ribosomes are membrane-bound or not, by specific factors which inhibit or stimulate ribosomes, by the availability of amino acyl-tRNAs, or by the accessibility of initiation sequences of mRNAs to ribosomes.

Nonhistones and the regulation of transcription

There are two classes of proteins associated with the DNA in chromatin; histones and nonhistones. The control of gene transcription must ultimately reside in the nucleotide sequences of the DNA so that the appropriate effector molecules can interact and control the amount and type of RNA produced. In addition, chromosomal proteins play a role in determining whether or not a region of DNA can be transcribed. As we have discussed in an earlier Topic, the histones are arranged in a regular fashion along the DNA and thus it is unlikely that they play any specific role in the regulation of gene expression. Histones *in vivo* have been shown to be acetylated, phosphorylated or methylated, but it is not known how these modifications relate to the transcriptional activity of chromatin.

On the other hand there is a lot of evidence implicating nonhistones in the regulation of gene expression in eukaryotes. For example, nonhistone proteins have tissue specificity and DNA binding specificity; they are present in higher amounts in transcriptionally-active tissues as compared to inactive cells; they are much more diverse than histones, and certain specific classes of nonhistone proteins have, in fact, been linked with the induction of gene activity. In addition, if chromatin from transcriptionally-active and inactive tissues is dis-

sociated into DNA, histones and nonhistones, it can be determined by reconstitution experiments which component confers the capacity for transcription. These have shown that the nonhistones are indeed the components that determine whether DNA can be transcribed. Thus it is currently believed that nonhistone proteins, presumably in response to specific signals, play a central role in the basic regulation of gene transcription in eukaryotes. At this stage, though, we are still in the model-building stage for considering how nonhistones act specifically at the molecular level and it is anticipated that, as new techniques are brought to bear in this area, we will rapidly obtain a lot more information about eukaryotic gene regulation.

In the two sections that follow, we will present something of what is known about the regulation of enzyme synthesis in both lower and higher eukaryotes. The examples chosen for these sections involve short-term gene regulation, that is, changes in gene expression that result from environmental changes. Following these sections we shall briefly consider long-term regulation in higher eukaryotes; that is, how gene expression is regulated as an organism develops and differentiates.

REGULATION OF GENE EXPRESSION IN LOWER EUKARYOTES

The demonstration that gene regulation in prokaryotes often involves operons that are controlled in ways analogous to the lactose operon of *E. coli* prompted researchers to investigate whether or not operons existed in eukaryotes. Indeed much of the early model building concerning the regulation of enzyme synthesis in eukaryotes was influenced by the regulatory models of enzyme synthesis in bacteria. However, as we shall see, enzyme synthesis is regulated in different ways in eukaryotes.

Eukaryotes have many basic similarities to prokaryotes, for example the processes of DNA replication, transcription and translation are more or less the same. However, eukaryotes are vastly more complex with discrete cellular compartments (nuclei, mitochondria, chloroplasts, etc.) that determine the organisation of the processes. In this regard the lower eukaryotes, and particularly the fungi, have proved to be useful model systems for the study of gene regulation since they are typically eukaryotic in their cellular structure and genetic organisation, yet they are microorganisms that can be handled in ways very similar to bacteria. Further, these organisms are simple and live in environments that are subject to rapid changes. Like bacteria, then, lower eukaryotes must be able to adapt rapidly at the gene expression level when such changes occur.

Yeast and *Neurospora* will be discussed in this section. These organisms have a genome complexity of approximately ten times that of *E. coli*. Early studies concentrated on determining whether or not operons exist in these fungi. Since they are highly amenable to genetic and biochemical investigations, it is relatively easy to isolate mutants affecting enzyme function and also the regulation of enzyme synthesis. This of course parallels the approach of Jacob and Monod in their studies of the regulation of the *E. coli* lactose operon. In general the fungal studies showed that, contrary to the findings in prokaryotes, genes with related function are *not* closely linked but rather tend to be scattered over the chromosomes in the genome. Nonetheless, it has been shown in both lower and higher eukaryotes that there is co-ordinate synthesis of all the enzymes in a particular biochemical pathway. This presumably involves a regulatory system different from that of prokaryotes. Characteristically, then, the gene products in eukaryotes consist of monocistronic mRNAs and not polycistronic mRNAs.

In some cases, though, evidence was obtained for apparent clustering of genes in fungi and this raised the possibility of the existence of operons in eukaryotes. Three representative cases will now be discussed and as we shall see, while the early information was highly suggestive of an operon organisation, all the evidence to date runs counter to that possibility.

The galactose fermentation genes of yeast
The first three enzymes for galactose fermentation in yeast are galactokinase, α-D-galactose 1-phosphate uridyl transferase (c.f. galactosaemia defect in humans; see *Human Genetics* Topic) and uridine diphosphoglucose 4-epimerase. The genes for these enzymes, as defined by mutants, are *GAL1*, *GAL7* and *GAL10*, respectively, and these are apparently closely linked on chromosome II in the order *GAL7–GAL10–GAL1*. The three genes are coordinately induced by the addition of galactose to the medium. The information so far certainly led to optimism among workers looking for operons in eukaryotes. Regulatory mutants that affected the expression of the *GAL* genes were then studied. One class of such mutants maps at a locus distinct from the structural genes and these mutants are characterised by constitutive synthesis of the three *GAL* enzymes. These mutants are recessive to the wild-type allele. By analogy with the lactose operon of *E. coli*, the locus involved was called *i* and the mutants were designated i^-. A second class of regulatory mutants carry mutations that map to a locus, *GAL4,* that is unlinked to either the *i* locus or the *GAL* structural genes. These *gal4* mutants are pleiotropically negative in that they are uninducible by galactose. A third class of regulatory mutants maps immediately adjacent to the *GAL4* locus at the *gal81* region and these result in a constitutive production of galactose-fermenting enzymes. The *GAL81* mutants resemble o^c mutants of the lactose operon in that they behave as cis-dominants in diploids.

From the data, H. Douglas and D. Hawthorne proposed a model for the regulation of expression of the *GAL* genes and this is shown in Fig. 20.1. Here the *i* gene produces a repressor that represses the expression of the *GAL4* gene by interacting with the adjacent *gal81* region if galactose is absent. If galactose is added, the repressor is inactivated and the *GAL4* gene can then be transcribed. Since *gal4* mutants are pleiotropically negative, the *GAL4* product is presumably a positive effector molecule that is required for the expression of the three *GAL* structural genes. How and where the *GAL4* effector interacts with the *GAL* gene cluster is not known. In other words the i^-, *gal81* relationship resembles the repressor–operator relationship of bacterial operons. Whether this is formally the case is a question for debate. The existence of i^s (super-repressible) mutations

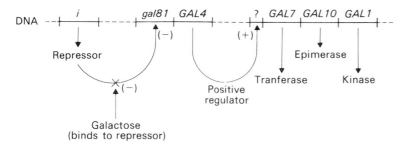

Fig. 20.1 The Douglas-Hawthorne model for the regulation of expression of the galactose fermentation genes in yeast.

certainly supports the repressor concept in the model. Still unknown, though, is the exact nature of the *i* and *GAL4* gene products at the molecular level, how the *i* product interacts with the *gal81* region, and whether a polycistronic mRNA is produced from the *GAL* structural genes. More information is needed, therefore, to determine if the *GAL* system is indeed an operon.

The genes for aromatic amino acid biosynthesis in *Neurospora*

Another potential candidate for an operon in eukaryotes were the genes for the early steps of the pathway for aromatic amino acid (phenylalanine, tyrosine and tryptophan) biosynthesis in *Neurospora crassa*. These have been studied by N. Giles, M. Case and their colleagues for many years. The enzymes involved and the genes coding for them (as defined by mutants) are shown in Fig. 20.2.

Of particular significance was the discovery that a multi-enzyme aggregate (molecular weight 230 000 daltons) contained five different enzyme activities. These five activities are coded for by the so-called *arom* gene cluster of five adjacent genes *aro2*, *aro4*, *aro5*, *aro9* and *aro1*. Genetic studies showed that mutations in a particular gene either affected the individual enzyme activity or caused the loss of two or more of the activities of the complex. These pleiotropic mutants were clearly reminiscent of nonsense mutants in operons of bacteria where polar effects result during the translation of a single polycistronic mRNA. It was suggested, therefore, that the *arom* gene cluster coded for a polycistronic mRNA, with transcription commencing at the *aro2* gene. This, then, supported the possibility that the *arom* gene cluster was an operon. However, recent information has dashed

that hope. F. Gaertner has shown that the so-called *arom* gene cluster is actually a *single* structural gene which codes for a single polypeptide of molecular weight 115 000 daltons. This dimerizes to produce an enzyme that has the five enzyme activities we have been discussing. The separate polypeptides that were found in early investigations have been shown to be artifacts of the cellular fractionation techniques where the pentafunctional polypeptide is cleaved by endogenous protease activity. Thus the *arom* system is not an operon but a fusion of five ancestral genes into one. This cluster-gene presumably has one promoter region. This is not the only example of a multi-functional polypeptide in eukaryotes — a number of other examples are known, particularly in the lower eukaryotes. The point it illustrates, of course, is that when one breaks open the cell and examines the contents, the results one obtains do not necessarily reflect the situation *in vivo*.

Regulation of quinic acid metabolism in *Neurospora*

The first three enzymes for the catabolism of quinic acid (QA), an aromatic compound, are quinic acid dehydrogenase, dehydroquinase and dehydroshikimic dehydrase (Fig. 20.3). The structural genes for these enzymes, *qa-3*, *qa-2* and *qa-4*, respectively are tightly linked. (Note that both this catabolic pathway and the aromatic amino acid biosynthetic pathway just described have a dehydroquinase involved. There is no overlap between the pathways, however, since the multi-functional polypeptide of the biosynthetic pathway serves to channel the intermediates effectively.)

The synthesis of the three genes is inducible by quinic acid. Tightly linked to the three structural

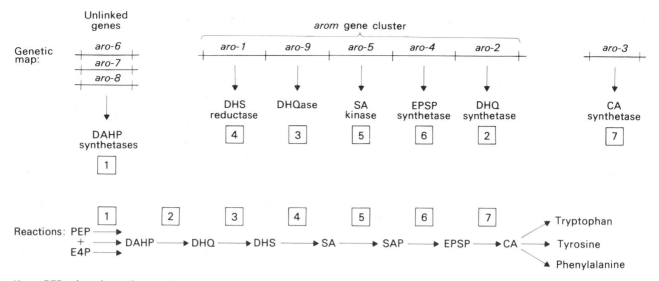

Fig. 20.2 The *arom* gene cluster of *Neurospora crassa* showing the enzymes encoded by the structural genes and the reactions they catalyse in the aromatic amino acid biosynthetic pathway (boxed numbers). (After the work of N. H. Giles and his colleagues.)

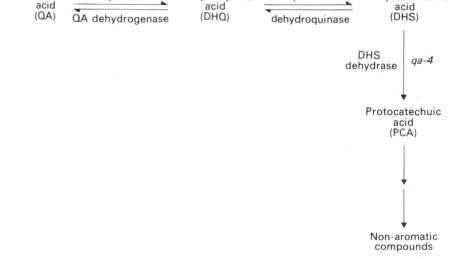

Fig. 20.3 The catabolic quinate-shikimate pathway showing the reaction steps, the enzymes involved and the structural gene loci for the enzymes in *Neurospora crassa*. The pathway is induced when aromatic compounds reach a high level in the cell and the enzymes produced break them down to non-aromatic compounds. (After the work of N. H. Giles and his colleagues.)

genes is a gene, *qa-1*, which apparently encodes a regulatory factor that acts in a positive way to activate transcription of the three structural genes of the cluster when quinic acid is present as inducer. In the absence of quinic acid, the genes are not transcribed. Since temperature-sensitive *qa-1* mutants have been found, the regulatory factor is most likely a protein. In addition there are constitutive alleles of the *qa-1* locus which are trans-dominant to the wild-type allele. These presumably are able to activate transcription of the structural genes without prior interaction with the inducer, quinic acid. Before this system can be considered an operon, though, controlling regions need to be found analogous to bacterial operators. These would be identified by mutations adjacent to the structural genes that are constitutive and cis-dominant. Also, a polycistronic mRNA for the genes would have to be defined.

In summary, then, in most instances genes with related functions in eukaryotes tend to be un-linked, though coordinately regulated. In lower eukaryotes in particular, where more extensive genetics has been possible, there are a number of examples of clustered genes, but there is no definitive example of a bacterial-like operon. At least in some instances the 'gene cluster' may actually be a single gene that codes for a multiple-functional polypeptide.

REGULATION OF GENE EXPRESSION IN HIGHER EUKARYOTES

Higher eukaryotes are characterised by the differentiation of cells into tissues, organs, etc., that have specific functions. In this they differ markedly from the comparatively undifferentiated lower eukaryotes. In the following we will concentrate on animal systems, and in particular the vertebrates.

With the greater complexity of cell specialisation in higher eukaryotes come different problems in terms of the regulation of gene expression. For example, the specialised cells in these organisms are not subjected to drastic changes in the environment as is the case for lower eukaryotes and prokaryotes. This results from the fact that animals have homeostatic mechanisms which maintain relatively constant extracellular and intracellular environments. This is mediated by the blood, which has a fairly constant composition that is maintained by a variety of mechanisms. In vertebrates, for example, this is controlled by hormones. Thus animal cells are generally not exposed to large changes in the concentration of metabolites or substrates and therefore there is less need for rapid changes in the rates of enzyme synthesis. Characteristically, then, such changes are less frequent and of less magnitude than in lower eukaryotes or bacterial cells. For example, the enzyme ornithine decarboxylase is one of the most rapidly responding enzymes and this exhibits a maximum increase of only ten- to twenty-fold in four hours when induced. Contrast this with the 1000-fold induction of β-galactosidase within minutes in *E. coli*.

Before discussing the role of hormones in the regulation of enzyme synthesis, it must be stated that there are indeed enzyme induction and repression mechanisms that operate in higher eukaryotes that are similar to those in prokaryotes. Owing to the low number of regulatory mutants available in animal cells, there are relatively few systems that have been investigated in this regard. By contrast, the actions of hormones on gene expression have been studied extensively and, in the following, some of the information that has been obtained will be considered.

Regulation of enzyme synthesis by hormones

A hormone may be defined as an effector molecule produced in low concentrations by one cell which evokes a physiological response in another cell. In vertebrates, a large number of classes of molecules have been shown to have hormonal activity, including polypeptides, amino acids, fatty acid derivatives and steroids. Some of the hormones act directly on the cell's genome whereas others act at the cell surface, thereby activating membrane-bound adenyl cyclase to produce cyclic AMP (cAMP; 3',5'-cyclic adenosine monophosphate) from ATP (Fig. 20.4). The cAMP acts as a 'second messenger' to evoke the intracellular effects observed following hormonal release.

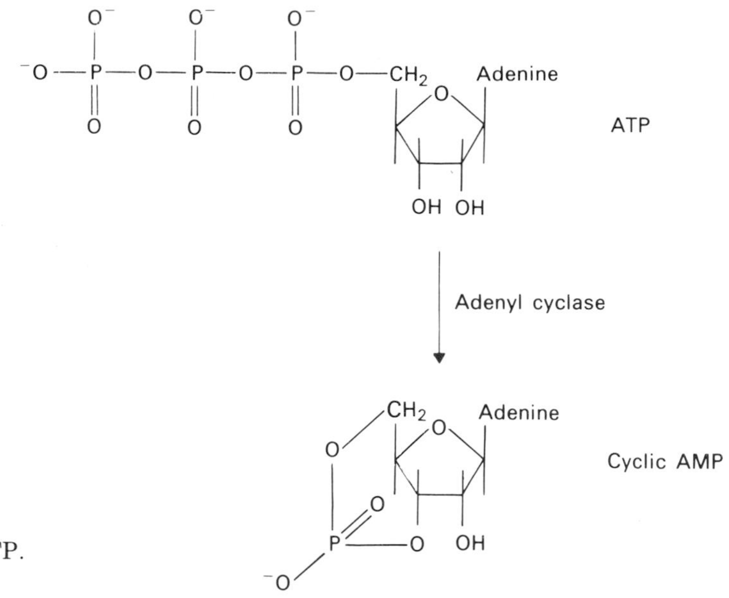

Fig. 20.4 The production of cyclic AMP from ATP. The reaction is catalysed by adenyl cyclase.

Hormones act on specific target tissues which possess receptors capable of recognising and binding that particular hormone. For most of the polypeptide hormones, the receptors are on the cell surface whereas the receptors for steroid hormones are in the cytoplasm.

Model for steroid hormone action
Steroid hormones are biosynthetically derived from sterols, which occur only in eukaryotic cells. Examples of steroid hormones are given in Fig. 20.5. These hormones apparently act at the level of genetic transcription. When a target cell interacts with the hormone, a number of events are proposed to occur. Firstly, the steroid hormone is taken up by the cell and then binds to a specific cytoplasmic receptor molecule. Then the steroid-receptor complex migrates into the nucleus, but the mechanism for this is not known. Once in the nucleus, the complex binds to acceptor sites on the chromatin thus activating the initiation of transcription at those sites. The end result is the production of specific new RNA species in the cell.

The general model just described appears to pertain to the action of a number of steroid hormones, for example, oestrogen, progesterone, aldosterone and glucocorticoids. One of the hormones of the latter class, hydrocortisone, has been studied extensively with regard to its induction of the liver enzyme, tyrosine aminotransferase (TAT). It has been shown that the receptor for this hormone is a protein, called binder II. Further it has been shown that this protein, in a complex with the hormone, can enter the nucleus where it is proposed there are 2000–10 000 acceptor sites. The extent of enzyme induction, then, appears to reflect the degree of saturation of these sites. There is good evidence from antibiotic studies that the hormone acts to stimulate transcription of mRNA molecules, although some researchers believe the hormone may exert its action at the translational level.

More generally speaking, the mechanism by which the steroid-receptor complex elicits a specific response remains to be elucidated. It seems that the receptors bind directly to exposed regions of DNA although it is not known if this is specific or non-specific. In the binding reaction, the non-histone chromatin proteins play a major role, while the histones apparently are not involved. Possibly the specificity of hormone action is keyed to the nonhistone proteins that can interact with the receptor. The interaction of steroid-receptor complex with the acceptor sites probably serves to melt

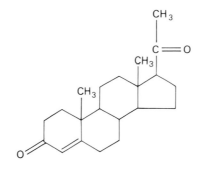

Fig. 20.5 Structures of some mammalian steroid hormones. All share a common four-ring structure. The small differences in the side groups are responsible for marked differences in the physiological effects in the animal. (a) Hydrocortisone — this is one of the glucocorticoid hormones produced by the cortex of the adrenal gland, and acts primarily to regulate carbohydrate and protein metabolism. (b) Aldosterone — this is a mineralocorticoid hormone secreted by the adrenal cortex and it acts to regulate salt and water balance. (c) Testosterone — this is made by the testes and is responsible for the production and maintenance of male sexual characteristics, and for the stimulation of sperm production. (d) Progesterone — this hormone is produced by the ovary and the placenta and, with oestrogen, is needed to prepare and maintain the uterine lining for embryo implantation and for the ensuing pregnancy.

the DNA thus exposing initiation sites for RNA polymerase to bind and commence RNA synthesis.

In summary, hormones act to integrate metabolism in higher eukaryotes. In some cases (e.g. in the liver) the coordination of metabolic activities involves the combined actions of several hormones. It is generally accepted that hormones act at the transcriptional level, although there is some controversy on this point.

LONG-TERM GENETIC REGULATION IN HIGHER EUKARYOTES

The examples of the previous section all show short-term regulation of gene expression in higher eukaryotes, that is where adjustments are made in cellular activity in response to environmental changes (e.g. hormone release). There are, however, two properties of higher eukaryotes (and some lower eukaryotes) that reflect the long-term regulation of gene expression, namely development and differentiation. These processes are really outside the area of genetics and in the areas of developmental biology and embryology, and therefore only a very gereral discussion of them will be given in this text.

Definitions of development and differentiation
Development is essentially the processes of growth and differentiation that characterise the life-cycle of an organism. Development can be observed as phenotypic changes in the organism and results from the interaction of the genome with the cytoplasm, with the internal cellular environment and with the external environment. In other words the genome carries the potential for the adult organism, but the final product results from complex gene–environment interactions.

In general, development is an irreversible or virtually irreversible process. We can consider development to involve at least three interacting processes:

1. The replication of the genetic material.
2. The growth of the organism as a result of cellular metabolic activity.
3. Cellular differentiation by which genetically-identical cells diverge in their structure and function to give rise to organised tissues, which in turn associate to form organs.

Differentiation is the formation of different types of cells and tissues from a zygote by the specific regulation of gene activity in temporal and spatial ways.

General aspects of development and differentiation

There are a number of general attributes of development and differentiation that can be related to the genetic concepts that have been presented in this text.

Nuclear DNA remains constant

Early models for gene involvement in development included one where there was a programmed loss of nuclear DNA as the organism developed, or rather that the development processes that occurred were the result of losses of particular genes in an ordered sequence. That is not true. Rather, cells of differentiated tissues contain the same genomic content of DNA as the fertilised egg (although some differentiated cells may be polyploid). One elegant experiment that showed this to be the case was performed by J. Gurdon. He transplanted a nucleus from the gut cell of a tadpole of *Xenopus laevis,* the South African Clawed Toad into an unfertilised egg of that organism from which the nucleus had been removed. The result was that the egg, once stimulated, developed into a normal adult toad. Thus the differentiated cells of the tadpole were totipotent, that is they contained all of the genetic information required for the egg to develop and differentiate into an adult organism.

The DNA is transcribed in a programmed way

All of the available evidence shows that development and differentiation involve a detailed pro-gramme of transcription of the DNA, which occurs in response to specific activator and repressor molecules. Two lines of evidence to support this will be considered here:

a. As mentioned previously, it is possible to quantify the extent to which RNA isolated from a cell hybridises to the nuclear DNA. In general, the experiments involve RNA and DNA molecules that are radioactively labelled with different isotopes. A refinement of this technique is competitive DNA : RNA hybridisation where unlabelled RNA is first allowed to hybridise with the DNA before the radioactive RNA is added. If the RNAs are from the same tissue, then the unlabelled RNA should effectively block all of the DNA sites to which the labelled RNA can bind and this would be detected as 100% competition when the radioactivity is measured. If the RNAs are from two different tissues, however, the effect on the amount of radioactive RNA that will bind to the DNA will depend on how many of the RNA species were synthesised in common by both the tissues. One can do this experiment, then, using mRNAs isolated from different tissues, e.g. lung, liver, kidney, muscle, of the same organism. Results of such an experiment show that there is limited competition between the mRNAs of the tissues in the hybridisation thus leading to the conclusion that differentiated cells reflect differences in the gene transcription activity. This correlates well with other studies showing different spectra of enzymes and relative differences in enzyme amounts and activities in different tissues of the same organism. These differences must, of course, reflect differential gene activity.

b. In certain insects such as *Drosophila,* the chromosomes of the larval salivary gland cells undergo polytenisation. That is to say, the chromosomes replicate up to a thousand times but without there being cell division. The replicated chromosomes remain together as the so-called polytene chromosomes which show characteristic banding patterns. A diagrammatic representation of a polytene chromosome is shown in Fig. 20.6. The bands are thought to represent the coding sequences of genes, while the function of the interband region is not known. The DNA is continuous along the length of

10 μ

Fig. 20.6 Diagrammatic representation of part of a polytene chromosome of the fruit fly, *Drosophila melanogaster*. The bands (solid and dotted) may be seen under the light microscope and are thought to represent the genes. The diagram actually shows one end of the 414 μm long polytenised x-chromosome. There are 1024 bands on this chromosome.

these chromosomes. In *Drosophila,* there are three larval stages each separated by a moulting event. The last larval stage is followed by pupation. This is an interesting model system, therefore, for studying gene activity (since the genes are visible, if indeed the bands are genes) during development. In fact, specific bands 'puff' in a particular pattern related to the time of larval development. The puffs are localized loosenings of the compact polytene chromosome that occur so that RNA polymerase can initiate transcription. Indeed it can be shown that RNA is being actively synthesised in the puffs and thus they are visual evidence of gene activity.

More importantly, the puffing patterns are repeatable, and, as has been stated, they are tissue and developmental stage specific.

Gene–cytoplasm interactions
The main point here is that the gene activities of a cell are affected by the cytoplasm. Thus when certain genes are turned on during differentiation, particular proteins are synthesised, some of which have a regulatory role in maintaining the differentiated state of the cell. For example, in Gurdon's transplantation experiment discussed earlier, we made the point that the nucleus carried all the genetic information necessary for the egg to develop into an adult. However, the fact that the egg cell behaved as an egg cell and not as a tadpole gut cell, is an example of how the activity of the nuclear genome is controlled by the cytoplasmic state.

In conclusion, development and differentiation involve long-term regulation of gene expression. Our discussion here has only scratched the surface of the information available for these processes. Nevertheless, we are still a long way from understanding them from a detailed molecular point-of-view, and we have a lot to learn about nuclear–cytoplasmic, cell–cell, and cell–environment interactions as they relate to developmental processes.

REFERENCES

Britten R.J. & E.H. Davidson. 1969. Gene regulation for higher cells: a theory. *Science,* **165:** 349–357.
Brown D.D. & I.B. Dawid. 1968. Specific gene amplification in oocytes. *Science,* **160:** 272–280.
Brown D.D. & I.B. Dawid. 1969. Developmental genetics. *Annu. Rev. Genet.* **3:** 127–154.
Burgoyne L., M.E. Case & N.H. Giles. 1969. Purification and properties of the aromatic (*arom*) synthetic enzyme aggregate of *Neurospora crassa. Biochim. Biophys. Acta,* **19:** 452–462.
Calvo J.M. & G.R. Fink. 1971. Regulation of biosynthetic pathways in bacteria and fungi. *Annu. Rev. Biochem.* **40:** 943–968.
Case M.E. & N.H. Giles. 1971. Partial enzyme aggregates formed by pleiotropic mutants in the *arom* gene cluster of *Neurospora crassa. Proc. Natl. Acad. Sci.*

USA, **68:** 58–62.
Case M.E. & N.H. Giles. 1975. Genetic evidence on the organisation and action of the *qa-1* gene product: a protein regulating the induction of three enzymes in quinate metabolism in *Neurospora crassa. Proc. Natl. Acad. Sci. USA,* **72:** 553–557.
Case M.E. & N.H. Giles. 1976. Gene order in the *qa* gene cluster of *Neurospora crassa. Mol. Gen. Genet.* **147:** 83–89.
Clever U. 1968. Regulation of chromosome function. *Annu. Rev. Genet.* **2:** 11–30.
Davidson E.H. 1968. *Gene Activity in Early Development.* Academic Press, New York.
Davidson E.H. & R.J. Britten. 1973. Organization, transcription and regulation in the animal genome. *Quart. Rev. Biol.* **48:** 565–613.
Davidson E.H. & R.J. Britten. 1979. Regulation of gene expression: possible roles of repetitive sequences. *Science,* **204:** 1052–1059.

Douglas H.C. & D.C. Hawthorne. 1966. Regulation of genes controlling synthesis of the galactose pathway enzymes in yeast. *Genetics*, **54**: 911–916.

Douglas H.C. & D.C. Hawthorne. 1972. Uninducible mutants in the *gal i* locus of *Saccharomyces cerevisiae*. *J. Bacteriol.* **109**: 1139–1143.

Giles N.H., M.E. Case, C.W.H. Partridge & S.I. Ahmed. 1967. A gene cluster in *Neurospora crassa* coding for an aggregate of five aromatic synthetic enzymes. *Proc. Natl. Acad. Sci. USA*, **58**: 1453–1460.

Gurdon J.B. 1968. Transplanted nuclei and cell differentiation. *Sc. Am.* **219**: 24–35.

Gurdon J.B. 1974. *The Control of Gene Expression in Animal Development*. Harvard University Press, Cambridge, Mass.

Hautala J.A., J.W. Jacobson, M.E. Case & N.H. Giles. 1975. Purification and characterization of catabolic dehydroquinase, an enzyme in the inducible quinic acid catabolic pathway of *Neurospora crassa*. *J. Biol. Chem.* **250**: 6008–6014.

Jacobson J.W., J.A. Hautala, M.E. Case & N.H. Giles. 1975. Effect of mutations in the *qa* gene cluster of *Neurospora crassa* on the enzyme catabolic dehydroquinase. *J. Bacteriol.* **124**: 491–496.

Lodish H.F. 1976. Translational control of protein synthesis. *Annu. Rev. Biochem.* **45**: 39–72.

Matsumoto K., A. Toh-E & Y. Ushima. 1978. Genetic control of galactokinase synthesis in *Saccharomyces cerevisiae*: evidence for constitutive expression of the positive regulatory gene *gal 4*. *J. Bacteriol.* **134**: 446–457.

Metzenberg R.L. 1972. Genetic regulatory systems in Neurospora. *Annu. Rev. Genet.* **6**: 111–132.

O'Malley B.W., H.C. Towle & R.J. Schwartz. 1977. Regulation of gene expression in eukaryotes. *Annu. Rev. Genet.* **11**: 239–275.

Pitot H.C. & M.B. Yatvin. 1973. Interrelationships of mammalian hormones and enzyme levels in vitro. *Physiol. Rev.* **53**: 228–325.

Revel M. & Y. Groner. 1978. Post-transcriptional and translational controls of gene expression in eukaryotes. *Annu. Rev. Biochem.* **47**: 1079–1126.

Stein G.S., T.C. Spelsberg & L.J. Kleinsmith. 1974. Nonhistone chromosomal proteins and gene regulation. *Science*, **183**: 817–824.

Tomkins G.M. & D.W. Martin. 1970. Hormones and gene expression. *Annu. Rev. Genet.* **4**: 91–106.

Walker P.R. 1977. Regulation of enzyme synthesis in animal cells. *Essays in Biochem.* **13**: 39–69.

Yamamoto K.R. & B.M. Alberts. 1976. Steroid receptors: elements for modulation of eukaryotic transcription. *Annu. Rev. Biochem.* **45**: 721–746.

Topic 21
Population Genetics

OUTLINE
Definitions
Hardy-Weinberg equilibrium
Formalisation of the Hardy-Weinberg Law
Applications of the Hardy-Weinberg Law
Factors affecting genetic equilibrium
  preferential mating
  selection against certain genotypes in reproduc-
    tion
  mutation
  migration
  genetic drift
Conclusions.

Up to this point in the text we have discussed the structure and function of genes as they have been studied in the laboratory. There, the organisms under investigation are often pure-breeding strains so that differences seen in experiments are the result of the experimental treatments rather than the result of genotypic and hence phenotypic differences. Further, in laboratory experiments matings can be carried out with organisms whose genotypes are known. Indeed, from studies of this kind has come our basic conceptual understanding of the transmission of genetic material from one generation to the next.

By contrast the world outside the laboratory is very different. For instance, natural populations of an organism do not generally mate in the ordered way 'preferred' by the geneticist. Also, in the 'wild', the relative frequencies of alleles at a locus may vary over a wide range, whereas in the laboratory these frequencies are usually 'set' at values convenient to the investigator. In this Topic we shall examine some of the basic principles of population genetics, that is the study of genes in natural (and sometimes laboratory) populations of organisms.

DEFINITIONS

Population
A Mendelian population is a group of interbreeding individuals. The largest possible population of a particular organism is the *species*. Of particular importance here is the gene exchange within the population from generation to generation. There are a number of factors that affect the genetic constitution of a population, including selection for particular alleles, mutation, migration into and out of the population of individuals, and genetic drift. These factors will be discussed later.

Allele frequencies
To analyse the genetics of populations, one must determine the frequencies of the alleles present so that changes can be detected over time and, hence, through evolution. By convention the frequency of the dominant allele of a pair is termed p and that of the recessive allele is called q. By definition $p + q = 1$.

HARDY-WEINBERG EQUILIBRIUM

Let us consider an hypothetical case of two alleles A and a in a diploid organism. In a population of 300 individuals let us suppose that there are 148 AA, 125 Aa, and 28 aa individuals. From these values we can compute the frequencies of the A and a alleles. The frequency of the A allele is $[(2 \times 148) + 125]/600 = 0.7$. (Here we are merely counting all the A alleles in the AAs, i.e. 2×148, and the A alleles in the Aa individuals. 600 is the total number of alleles in 300 diploid individuals.) Similarly, the frequency of the a allele is $[(2 \times 28) + 125]/600 = 0.3$. If we now allow (or in this ca⸮

Allele of		Frequency of pairing		Progeny		Number in population of 300
♀	♂			Genotype	Frequency	
A	A	0.7 × 0.7		AA	0.49	147
A	a	0.7 × 0.3	⎫			
a	A	0.3 × 0.7	⎬	Aa	0.42	126
a	a	0.3 × 0.3	⎭	aa	0.09	27

Frequency of *A* in progeny = 0.7

Frequency of *a* in progeny = 0.3

Fig. 21.1 Demonstration that random mating in a population will maintain genetic equilibrium if that population is in genetic equilibrium. The allele frequencies remain the same in the parental and progeny generations, as do the relative genotype frequencies.

require) this population to mate at random, so that all possible pairings occur, we can predict the distribution of the three genotypes in the next generation from the calculated allele frequencies as shown in Fig. 21.1.

As the figure shows, the relative numbers of the three genotypes closely match those of the parental generation. Note further that the frequency of the *A* allele is still 0.7 and that of the *a* allele is still 0.3. These relationships will hold at every successive generation if random mating is maintained.

But it may be argued that the two sets of figures match closely since we calculated the allele frequencies from the original set of numbers. We can prove that it was not the case by considering a second hypothetical population with 300 individuals, 190 of which are *AA*, 40 *Aa*, and 70 *aa*. In this population also the allele frequency of *A* is 0.7 and of *a* 0.3. If this population were mated at random, then, the *progeny* genotypes would have the same relative frequencies that we calculated in Fig. 21.1, that is frequencies quite different from those of the parents in this second population.

To return to the first population, we showed that the allele and genotype frequencies remained the same as a consequence of random mating. In other words the population is in genetic equilibrium. This basic tenet was described independently in

1908 by G.H. Hardy and W. Weinberg and it has become known as the Hardy-Weinberg Law or the Hardy-Weinberg equilibrium (HWE).

For a population to remain at genetic equilibrium there are a number of requirements:

1. Mating must be random through the population. This implies that all gametes are equally viable and able to participate in the fertilisation event.
2. Mutation must not occur or, if it does, the forward (*A* to *a*) and back (*a* to *A*) mutation rates must be the same.
3. The generations cannot overlap in the breeding sense.
4. There is no migration occurring to shift the relative frequency of alleles and genotypes.
5. The population is of infinite size.

All this would seem a tall order, but indeed there are human populations that have been studied that are at genetic equilibrium for particular alleles. It is worthwhile to note here, too, that if a population is not in Hardy-Weinberg equilibrium, then it only requires one generation of random mating to establish an equilibrium which successive 'rounds' of random mating will then maintain. This was demonstrated for the second hypothetical population discussed above.

FORMALISATION OF THE HARDY-WEINBERG LAW

We start out with a gene pool of a randomly mating population at Hardy-Weinberg equilibrium of p A alleles and q a alleles. By definition $p + q = 1$, as we have said. If these alleles pair at random we have the situation shown in Fig. 21.2, that is the frequencies of the three genotypes will be p^2 $AA + 2pq$ $Aa + q^2$ aa (a binomial distribution) and the allele frequencies are p $A + q$ $a = 1$.

	p A	q a
p A	p^2 AA	pq Aa
q a	pq Aa	q^2 aa

Genotype frequencies: p^2 $AA + 2pq$ $Aa + q^2$ $aa = 1$

Allele frequencies: p $A + q$ $a = 1$

Fig. 22.2 For a population at Hardy-Weinberg equilibrium with pA alleles and qa alleles, random pairing of alleles will give a genotype distribution of p^2 $AA + 2pq$ $Aa + q^2$ $aa = 1$.

We can now formally prove that if we get this distribution in one generation, then we shall get the same distribution in the next generation if random mating has occurred. This is shown in Fig. 21.3 where all possible matings are considered in terms of the frequencies in which they will occur by random mating in the populations.

APPLICATIONS OF THE HARDY-WEINBERG LAW

For populations that are in genetic equilibrium, the Hardy-Weinberg Law is very useful in making predictions, for example, of genotype frequencies (and therefore phenotype frequencies) in ensuing generations. For instance, let us consider a human population where the inability to taste PTC (see p. 182) is caused by homozygosity for a recessive allele, t. Tasters are either TT or Tt in genotype since the T allele is dominant to the t allele. In the population 70% of the individuals are tasters and 30% are non-tasters.

As a first step we must calculate the frequencies of the T and t alleles. The frequency distribution of genotypes is:

p^2 $TT + 2$ pq $Tt + q^2$ tt.

q^2 (the frequency of nontasters) $= 0.3$

$\therefore q = \sqrt{0.3}$

$= 0.55$, the frequency of allele t.

Since $p + q = 1$

$\therefore p = 0.45$, the frequency of allele T.

Once we know the allele frequencies, we can calculate the genotype frequencies:

i.e. $TT = p^2$

$= (0.45)^2$

$= 0.2$

$Tt = 2$ pq

$= 2 (0.45)(0.55)$

$= 0.5$

and $tt = 0.3$, from before.

Now we can pose a theoretical question and use the calculated allele and genotype frequencies to solve it. The question is: what will be the frequency of non-taster children from marriages of parents neither of whom are non-tasters? Random mating is assumed here. First we will solve the problem using genotype frequencies we have calculated and then we shall derive a generalised formula for a problem of this kind.

Using genotype frequencies, we must consider all possible pairings and the frequencies of their occurrences. This is shown in Fig. 21.4. Of the four types of marriages, only one, $Tt \times Tt$, can give rise to tt (non-taster) children and such children would be expected to constitute one fourth of the children of these marriages (according to basic Mendelian segregation). Thus the frequency of non-taster children among progeny of all marriages

Mating ♀ ♂	Frequency of mating		Progeny distribution AA	Aa	aa
$AA \times AA$	$p^2 \times p^2$	$= p^4$	p^4	–	–
$AA \times Aa$ $Aa \times AA$	$p^2 \times 2pq$ $2pq \times p^2$	$= 4p^3q$	$2p^3q$	$2p^3q$	–
$Aa \times Aa$	$2pq \times 2pq$	$= 4p^2q^2$	p^2q^2	$2p^2q^2$	p^2q^2
$AA \times aa$ $aa \times AA$	$p^2 \times q^2$ $q^2 \times p^2$	$= 2p^2q^2$	–	$2p^2q^2$	–
$Aa \times aa$ $aa \times Aa$	$2pq \times q^2$ $q^2 \times 2pq$	$= 4pq^3$	–	$2pq^3$	$2pq^3$
$aa \times aa$	$q^2 \times q^2$	$= q^4$	–	–	q^4

∴ Frequency of progeny is:

$$AA = p^4 + 2p^3q + p^2q^2 = p^2(p^2 + 2pq + q^2)$$

$$Aa = 2p^3q + 4p^2q^2 + 2pq^3 = 2pq(p^2 + 2pq + q^2)$$

$$aa = p^2q^2 + 2pq^3 + q^4 = q^2(p^2 + 2pq + q^2)$$

Cancelling the term in brackets, the distribution is:

$$p^2\,AA + 2pq\,Aa + q^2\,aa$$

Fig. 21.3 Algebraic demonstration that, for a population in genetic equilibrium, random mating will produce a progeny population that has the same relative frequencies of *AA*, *Aa*, and *aa* genotypes as the parental generation.

between taster parents is:

(Relative frequency of $Tt \times Tt$ marriages) × (Probability of non-taster child from $Tt \times Tt$ marriage)

$$= \frac{0.25}{0.49} \times 0.25$$

$$= 0.128$$

In other words 128 out of every 1000 children from such pairings should be non-tasters.

Now we can derive a generalised formula for problems like this. Fig. 21.5 shows the frequencies of occurrence of marriages where both parents have at least one dominant allele. From this we see that there are p^2q^2 progeny with the recessive phenotype. The frequency of recessive phenotype

Parental genotypes		Frequency of pairing
♀	♂	
TT	*TT*	0.2 × 0.2 = 0.04
TT	*Tt*	0.2 × 0.5 = 0.10
Tt	*TT*	0.5 × 0.2 = 0.10
Tt	*Tt*	0.5 × 0.5 = 0.25
	Total =	0.49

∴ Proportion of *Tt* × *Tt* pairings $= \dfrac{0.25}{0.49}$

$= 0.51$

Fig. 21.4 Calculation of the proportion of marriages in which both parents are heterozygous tasters (*Tt*) among all marriages in which both parents are tasters. The population on which the calculation is based has a distribution of 0.2 *TT*, 0.5 *Tt*, and 0.3 *tt* individuals, i.e., the frequency of the *T* allele is 0.45 and that of the *t* allele is 0.55.

progeny from such pairings, then, is:

$$\frac{p^2 q^2}{p^4 + 4p^3 q + 4p^2 q^2} = \frac{q^2}{p^2 + 4pq + 4q^2}$$

$$= \left(\frac{q}{p + 2q}\right)^2$$

Since $p + q = 1$, this becomes:

$$\left(\frac{q}{1 - q + 2q}\right)^2$$

$$= \left(\frac{q}{1 + q}\right)^2$$

To apply this to the taster/non-taster example where $q = 0.55$, this would give:

$$\left(\frac{0.55}{1.55}\right)^2 = 0.126$$

which fits well with the frequency calculated before.

Marriages		Frequency (if random)	Frequency of progeny phenotypes	
			Dominant	Recessive
AA × *AA*		$p^2 \times p^2 = p^4$	p^4	—
AA × *Aa*		$p^2 \times 2pq = 2p^3 q$	$4p^3 q$	$4p^3 q$
Aa × *AA*		$2pq \times p^2 = 2p^3 q$		
Aa × *Aa*		$2pq \times 2pq = 4p^2 q^2$	$3p^2 q^2$	$p^2 q^2$

Proportion of recessive phenotype progeny $= \dfrac{p^2 q^2}{p^4 + 4p^3 q + 3p^2 q^2}$

Fig. 21.5 Algebraic derivation of a generalised formula for calculating the frequency of recessive phenotype children among progeny of all marriages where each parent has at least one dominant allele.

In conclusion, for populations in genetic equilibrium, the Hardy-Weinberg Law is very useful in making predictions about the upcoming generations or about subsets thereof.

FACTORS AFFECTING GENETIC EQUILIBRIUM

Preferential mating

As has been mentioned repeatedly, the maintenance of Hardy-Weinberg equilibrium depends upon random mating in the population. Of course, in many populations mating is not random and this has an effect on genotype frequencies in the next generation, but it has no effect on allele frequencies. In other words the population will no longer be in genetic equilibrium if there is preferential mating.

We can illustrate this by using the taster/nontaster example from before where 2/10 of the population were *TT*, 5/10 were *Tt* and 3/10 were *tt*. Now let us apply the constraint that an individual of a particular genotype can only marry another

of the same genotype; *TT* with *TT*, *Tt* with *Tt*, and *tt* with *tt*. This is called *inbreeding*. In the first case only *TT* progeny will result and in the last case only *tt* progeny will be produced. However, from the middle marriage 1/4 of the progeny will be *TT*, 1/2 will be *Tt*, and 1/4 will be *tt*. With this in mind, Fig. 21.6 shows the calculation of genotype frequencies for the progeny of these preferential matings. As can be seen, the frequency of *T* remains 0.45 and that of *t* remains 0.55. However, the genotype frequencies have changed since the *Tt* × *Tt* pairing result in a dispersion of alleles throughout the three possible genotypes with the notable consequence that the frequency of the heterozygote, *Tt*, has been reduced by one half from 1/2 to 1/4. Concomitantly the frequencies of the two homozygotes have increased. This will occur at each successive generation as long as the same preferential mating constraints are in effect. Inbreeding, then, serves to result in homozygosity at all loci. When applied to laboratory organisms such as mice or rats, inbreeding produces purebreeding strains that can be used in studies where genotype constancy is important, e.g. in tests of potential carcinogens on mammals. The danger of

	Genotype frequencies:	*TT*	*Tt*	*tt*
		$^2/_{10}$	$^5/_{10}$	$^3/_{10}$
∴	Allele frequencies:	*T* = 0.45		*t* = 0.55

Progeny frequencies if only preferential mating (*TT* × *TT* ; *Tt* × *Tt*, and *tt* × *tt*) occurs:

TT	*Tt*	*tt*
$^2/_{10}$ from *TT* × *TT*	(½ × $^5/_{10}$) from *Tt* × *Tt*	$^3/_{10}$ from *tt* × *tt*
(¼ × $^5/_{10}$) from *Tt* × *Tt*		(¼ × $^5/_{10}$) from *Tt* × *Tt*
Total: 13/40	10/40	17/40

and allele frequencies are still *T* = 0.45 and *t* = 0.55

Fig. 21.6 An example of how preferential mating (inbreeding) alters the distribution of the three genotypes in one generation. The starting population here is a taster/nontaster population with 0.2 *TT*, 0.5 *Tt*, and 0.3 *tt*. In this case only individuals of like genotype mate with the consequence that the frequency of *Tt* individuals in the next generation is halved. The relative frequencies of *T* and *t* alleles is not changed by such a preferential mating regime.

this is that many loci will become homozygous for recessive deleterious genes, and this is a principal reason why cousin marriages are illegal in many countries as this is a form of inbreeding.

Selection against certain genotypes in reproduction

One of the major assumptions in the Hardy-Weinberg Law is that all individuals in the population are equally able to reproduce, with no preference of one gamete over another, and no differential mortality of the conceived offspring. Now, in natural populations, the recessive alleles have often arisen by mutations and in many instances result in the production of a defective gene product. Depending on the genes involved, this could indeed have effects on the overall ability of homozygous recessive individuals to have children. For many human traits, for example, the homozygotes for deleterious recessive alleles may be so severely affected as to die before reproductive age, and obviously this means that not all genotypes can contribute to reproduction. Similar arguments can be made for dominant mutations which may have lethal effects when homozygous. And, in addition, there are dominant and recessive mutations that have a range of effects on the ability of particular individuals carrying them to reproduce. This brings us to the concept of *fitness* which is the relative reproductive success, or the relative numbers of offspring produced by given genotypes. In the extreme, if a genotype has a fitness value of zero, then that genotype is not contributing genes to the gene pool of the next generation. We shall now consider an hypothetical example of this kind.

Suppose we have a population with genotype frequencies 0.25 AA, 0.5 Aa, 0.25 aa, where A is dominant to a. Phenotypically then, 3/4 of the population is A and 1/4 is a. Now, $q^2 = 0.25$ and thus $q = 0.5$. As $p + q = 1$, then $p = 0.5$. If aa individuals do not survive to reproductive age, or are sterile, then the effective breeding population (EBP) is composed of only AA and Aa individuals. In the EBP, then, the relative frequency of AAs is $\frac{0.50}{0.75} = 0.67$. These individuals then mate randomly and thus we can calculate the frequency

of matings and the distribution of progeny as shown in Fig. 21.7.

As can be seen, not only have the genotype frequencies shifted markedly in just one generation but the allele frequencies have changed; that of A from 0.5 to 0.665, and that for a from 0.5 to 0.335.

One might want to argue from these data that a way to remove a deleterious recessive allele from a population is to prevent the homozygous recessive individuals (if they survive to reproductive age) from participating in the procreation process. Unfortunately this does not work as with continued selection against aas, selection becomes less effective as the frequency of aas decreases (Table 21.1). In fact, even at low aa frequencies there are many a alleles to be found in heterozygous Aa individuals. Therefore the eradication of a 'deleterious' recessive allele from a population depends upon identifying the heterozygotes and restricting their contributions to the future gene pool. In human populations, this is where pedigree analysis and genetic counselling come into play. Even so, the mutation process will continue to pump deleterious alleles into the population, albeit at a low frequency.

On the other hand, deleterious dominant mutations are not 'hidden' by the heterozygotes condition as are the deleterious recessive alleles. Therefore, selection is very effective in removing deleterious dominant mutations from a population.

Mutation

Mutation is, of course, the source of all variation in a population, and it will disrupt the genetic equilibrium of a population. The rate of spontaneous mutations at a locus may only be 10^{-6} or less, but even so mutations are essential if evolution (that is adaptation to, for example, new environments) is to take place. If mutation generates an allele that is deleterious to the organism in the particular environment in which it lives, then there will be selection against the allele. In general, a balance will occur between the rate of occurrence of the mutation and the rate of loss of the mutation from the population by selection. Thus the mutation can then be distributed throughout the population by the processes of mating and recombination.

Original population: 0.25 *AA* + 0.5 *Aa* + 0.25 *aa*

∴ Allele frequencies: *p* = 0.5 , *q* = 0.5

In this population the *aa* individuals do not reproduce.

The effective breeding population, then, consists of

$\frac{0.25}{0.75}$ = 0.33 *AA* individuals and $\frac{0.50}{0.75}$ = 0.67 *Aa* individuals

These pair at random:

		Progeny frequencies		
Matings	Frequency	*AA*	*Aa*	*aa*
AA × *AA*	0.33 × 0.33 = 0.109	0.109	–	–
2 (*AA* × *Aa*)	2 (0.33 × 0.67) = 0.442	0.221	0.221	–
Aa × *Aa*	0.67 × 0.67 = 0.449	0.112	0.225	0.112
	Total = 1.000	0.442	0.446	0.112

Original population: 0.25 *AA* + 0.5 *Aa* + 0.25 *aa*

Progeny population: 0.442 *AA* + 0.446 *Aa* + 0.112 *aa*

Progeny allele frequencies: *A* = 0.665

a = 0.335

Fig. 21.7 Algebraic demonstration of the consequences of selection against recessive phenotype individuals in the population. Here we start with an hypothetical population of 0.25 *AA*, 0.50 *Aa* and 0.25 *aa*, where *aa* individuals are not part of the breeding population. Random mating of *AA* and *Aa* individuals result in a progeny population where both the allele and genotype frequencies have changed from those of the parental populations.

It may be the case that the mutation has a more favourable effect in a particular genotype and/or a particular environment and this will enable the allele to be 'fixed' in the population. This is the foundation of evolutionary change.

Let us consider mutation in a more formal sense. There are two kinds of mutation that we need to deal with; *forward mutation,* the change from the (now) wild-type form of the gene to the mutated form (e.g. *A* to *a*, for the purposes of discussion), and *back mutation,* the change from the mutated to the wild-type form (*a* to *A*). For any gene locus the rate of forward mutation is likely to be different from the rate of back mutation, and usually the former is higher than the latter. These relationships can be represented diagrammatically:

$$A \underset{v}{\overset{u}{\rightleftarrows}} a$$

where the mutation rate for *A* to *a* is *u* and that for *a* to *A* is *v*.

Now, if in one generation the frequency of *A* is *p* and that of *a* is *q*, then for the next generation the proportion of *A* alleles that mutate to *a* alleles is *pu* and the proportion of *a* alleles that mutate to *A* alleles is *qv*, since then there will be no change in

Table 21.1 Effects of constant selection against *aa* individuals on the frequency of *aa*'s in the population.* Six different 'starting' frequencies of *aa* individuals are shown.

Generation	Frequency of *aa* individuals					
0	0.990	0.750	0.500	0.250	0.100	0.010
1	0.249	0.215	0.172	0.112	0.058	0.008
2	0.112	0.100	0.086	0.062	0.038	0.007
3	0.062	0.058	0.051	0.040	0.026	0.006
4	0.040	0.038	0.034	0.028	0.019	0.005
5	0.028	0.026	0.024	0.020	0.015	0.004

* This is calculated for each generation as follows:

Effective breeding population $= p^2\ AA + 2pq\ Aa$.
Matings that give *aa* progeny are $Aa \times Aa$. In the breeding population the frequency of *Aa*'s is $\dfrac{2pq}{p^2 + 2pq}$.

$\therefore$ The frequency of $Aa \times Aa$ pairings $= \left(\dfrac{2pq}{p^2 + 2pq}\right)^2$

Dividing through by p we have $\left(\dfrac{2q}{p + 2q}\right)^2$

Substituting $1 - q$ for p, this gives $\left(\dfrac{2q}{1 - q + 2q}\right)^2$

$$= \left(\dfrac{2q}{1 + q}\right)^2$$

$$= \dfrac{4q^2}{(1 + q)^2}$$

From $Aa \times Aa$, 1/4 of the progeny will be *aa*.

$\therefore$ Frequency of *aa* here $= 1/4 \times \dfrac{4q^2}{(1 + q)^2}$

$$= \dfrac{q^2}{(1 + q)^2}$$

$$= \left(\dfrac{q}{1 + q}\right)^2$$

allele frequencies in succeeding generations. That is:

$$pu = qv$$
$$\text{so,} \quad (1 - q)u = qv$$
$$\text{i.e.} \quad u - uq = qv$$
$$\therefore \quad u = qv + uq$$
$$\text{then,} \quad u = q(u + v)$$

And thus at mutational equilibrium, the new frequency of allele a, will be:

$$q = \dfrac{u}{u + v}$$

This effect of mutation on the distribution of

genotypes in a population is called *mutation pressure*.

Let us consider an example where the forward mutation rate is four times that of the back mutation rate, i.e. $u = 4v$.

In this case the frequency of the a allele, q, at equilibrium under such mutation pressures will be:

$$q = \frac{u}{u + v} \text{ (from above)}$$

$$\therefore q = \frac{4v}{5v} = \frac{4}{5}$$

In other words an equilibrium will be reached at which no change in allelic frequencies will occur (even though mutation is occurring), when the frequency of the a allele is 0.8 and that of the A allele is 0.2. In natural populations, as we have said, it is usually the case that the forward mutation rate is greater than the back mutation rate (i.e. $u > v$) and thus mutation pressure serves to introduce mutant alleles into the population.

Before leaving this subject it is worth noting that mutations can be advantageous, disadvantageous or neutral to the organism at a given time and in a given environment. In the first two cases, the mutations are the source of variation in the population, and selection acts on them; but selection does not act on neutral mutations. Charles Darwin proposed the theory that evolution occurs through natural selection, that is, the present-day diversity of life has evolved from common ancestors rather than having a divine origin. Evolution by selection on advantageous and disadvantageous mutations is therefore usually called Darwinian evolution. It has been proposed by others that evolution might also occur by the accumulation of neutral mutations, that is without selection. This form of evolution is called non-Darwinian evolution.

Migration

Another assumption of the Hardy-Weinberg Law is that the population is a closed one in that there is no loss of individuals from it or additions of new individuals from the outside. In natural populations (except those that are geographically isolated), migration of individuals commonly occurs.

From all that we have said concerning genetic equilibrium it should be obvious that the introduction of new alleles into the gene pool by individuals entering the population and interbreeding with it will shift the equilibrium. This process is presumably essential for evolution.

Genetic drift

The theoretical population we considered to be in Hardy-Weinberg equilibrium at the outset of our discussions was of infinite size. Even though natural populations are smaller than that, many of them are large enough for random mating to occur in effect and genetic equilibrium is maintained through the generations. On the other hand, if the sample of alleles contributing to the zygotes for the next generation is not representative of the overall allelic composition of the population's gene pool, then deviations from genetic equilibrium can occur. This will be observed either as chance variations in allele frequencies in the population or possibly as chance fixation of an allele (i.e. p or $q = 1$) in the population. This phenomenon is called genetic drift and is commonly observed when a small population is involved. Therefore, genetic drift is very likely to occur when a population is drastically reduced in size for some reason (the bottleneck effect) or if a small subset of a large population breaks away and colonises a new area (the founder effect). An example of the latter is seen in the Dunker religious group that lives in Pennsylvania, USA. Over 250 years ago, 28 Dunkers migrated to the USA from Germany and formed a community that has remained more-or-less self-contained ever since. Studies have shown that the Dunkers have frequencies of blood group alleles that are quite different from those found in Germany and in the USA, whereas the general populations of these two countries have quite similar blood group allele frequencies. This, then, is genetic drift occurring by the founder effect.

CONCLUSIONS

We have seen in this Topic a little of how genes in populations are distributed and the effects of various factors, such as mutation, selection and migration, on the gene pool of a population. All of the genes of an organism, in association with the environment, are responsible for the phenotype of that organism. Therefore, as the environment changes, different combinations of genes may be more favoured and changes in gene frequencies in the population will result over many generations by the 'forces' described. This is the process of evolution, at least in simple terms. However, evolution is clearly an extremely complex process in which numerous factors are intertwined, and a lot remains to be learned about it. In this Topic we have presented a simplified view of the genetics of populations and it is hoped that the reader will extrapolate the basic concepts that have been discussed to natural populations and to the evolutionary process.

REFERENCES

Bodmer W.F. & L.L. Cavalli-Sforza. 1976. *Genetics, Evolution, and Man.* W.H. Freeman, San Francisco.

Crow J.F. & M. Kimura. 1970. *An Introduction to Population Genetics Theory.* Harper and Row, New York.

Darwin C. 1859. *The Origin of Species.* John Murray, London.

Dobzhansky T. 1947. Adaptive changes induced by natural selection in wild populations of *Drosophila. Evolution,* 1: 1–16.

Dobzhansky T. 1955. A review of some fundamental concepts and problems of population genetics. *Cold Spring Harbor Symp. Quant. Biol.* 20: 1–15.

Falconer D.S. 1960. *Introduction to Quantitative Genetics.* Oliver and Boyd, Edinburgh.

Fisher R.A. 1930. *The Genetic Theory of Natural Selection.* Clarendon Press, Oxford.

Harland S.C. 1936. The genetic conception of the species. *Biol. Rev.* 11: 83–112.

Kettlewell H.B.D. 1961. The phenomenon of industrial melanism in Lepidoptera. *Annu. Rev. Entomol.* 6: 245–262.

Lewontin R.C. 1974. *The Genetic Basis of Evolutionary Change.* Columbia University Press, New York.

Li C.C. 1955. The stability of an equilibrium and the average fitness of a population. *Am. Naturalist,* 89: 281–295.

Mather K. 1953. The genetical structure of populations. *Symp. Soc. Exp. Biol.* 7: 66–95.

Mayr E. 1963. *Animal Species and Evolution.* Harvard University Press, Cambridge, Mass.

Merrell D.J. 1953. Selective mating as a cause of gene frequency changes in laboratory populations of *Drosophila melanogaster. Evolution,* 7: 287–298.

Ohta T. 1974. Mutational pressure as the main cause of molecular evolution and polymorphism. *Nature,* 252: 351–354.

Ohta T. & M. Kimura. 1971. Functional organization of genetic material as a product of molecular evolution. *Nature,* 233: 118–119.

Powell J.R. & R.C. Richmond. 1974. Founder effects and linkage disequilibrium in experimental populations. *Proc. Natl. Acad. Sci. USA,* 71: 1663–1665.

Simpson G.G. 1953. *The Major Features of Evolution.* Columbia University Press, New York.

Wallace L.B. 1968. *Topics in Population Genetics.* Norton, New York.

Wright S. 1951. The genetic structure of populations. *Ann. Eugen.* 15: 323–354.

Index

acridines 51–3
acriflavin 202
adenine 2, 18
alkaptonuria 204–5
allele 47
allele frequency 243–53
allolactose 213
amino acid 18, 82–3
amino acyl synthetase 69
2-aminopurine (2-AP) 50
anaphase 42, 45–6
aneuploid 185–7, 190–1
antibiotic resistance 63, 135–7, 196
antibiotic selection 59
Antirrhinum 146
arabinose operon 223–5
  cyclic AMP and expression 224
  model for regulation 223–5
aromatic amino acid genes,
  Neurospora 235–6
Aspergillus 166–7, 174–9
attenuator 226–8
autosome 12, 151
auxotrophy 59

Bacillus subtilis 62, 125, 129–30
back mutation 250–2
bacterial chromosomes 10–1, 13
bacteriophage 5–10, 12, 99, 106–16,
  124–9, 132–3, 140, 230
  genetics 106–16
  lambda (λ) 10, 12, 126–9, 133
  lysogeny 124
  P1 125
  P22 124
  φ × 174 10, 114–6, 140
  prophage 124
  replication 6–7
  SP10 125
  structure 6, 9
  T phages (*see* T2, T3, T4, T5, T7)
  temperate 124
  virulent 124
bacterium 32, 118–29, 132, 213–30
  conjugation 118–24
  genetics 118
  major control of transcription and
    translation 228–30
  operon 213–28
  plasmid 132
  regulation of gene expression 213–30

bacterium (*Cont.*)
  relaxed mutant 229
  stringent control 228–9
  transduction 124–9
  transformation 129, 32
  translational control 229–30
Barr body 189–92
base 1
β-galactosidase (*see* lactose operon)
biochemical genetics 204–12
biochemical pathway 208, 210–11
broad bean 33
5-bromouracil (5-BU) 49–50

catabolite repression 220–3
  catabolite gene activator protein
    220–2
  cyclic AMP and 220–2
cell cycle 31–37
  G1 phase 31–2
  G2 phase 36–7
  mitosis 36
  mutants 37
  RNA synthesis in G1 32
  S phase 32–6
central dogma 61
centromere 11, 39
chicken 146
chloramphenicol 196
chloroplast 11, 195
  DNA 11, 195
chromatids, sister 39, 43
chromatin 13, 15, 36, 232–3
chromosome 8, 12, 31, 36, 151–3,
  185–193, 240–1
  aberrancies 185–93
  aneuploidy 185–6
  autosome 12, 151
  bacteriophage 8
  condensed 31, 36
  deficiency 186–8
  deletion 189
  duplication 186–8
  euploidy 185–6
  extended 31
  heterochromatin 189
  inversion 186–8
  Lyon hypothesis 191–3
  non-disjunction 187, 190
  polytene 240–1
  prokaryotes v. eukaryotes 8

chromosome (*Cont.*)
  rearrangement 189
  sex 151–3
  translocation 186
circular permutation 8–10
cis-trans test 211
cistron 114, 211–2
cloning DNA 132
codon 66, 82
coincidence 162–3
colinearity 211–2
complementary base-pairing 61
complementary DNA (cDNA) 139
complementation group 208
complementation test 113
conditional mutations 59, 107
conjugal plasmid 119
conjugation 118–24
  DNA transfer 121
  Hfr bacterial strains 120
  interrupted mating 123
  mapping 123
co-transduction 125–6
coupling 151
crossing-over 45, 150, 153–6, 162–3,
  167–79
  chiasmata and 45
  coincidence 162–3
  interference 162–3
cyclic AMP 220–1, 224, 237–8
cycloheximide 36, 196
cytochromes 196
cytosine 2, 19

Darwinian evolution 252
deletion 111, 189
deletion mapping 113
deoxyribonuclease 6, 66
deoxyribonucleic acid (*see* DNA)
deoxyribose 2, 19
development 239–41
  gene–cytoplasm interactions 241
  transcription and 240
differentiation 237, 239–41
  gene–cytoplasm interactions 241
  transcription and 240
Diplococcus pneumoniae 129
DNA 1–4, 10–11, 16, 23–6, 52–6, 132–41,
  195
  base ratio 2
  chloroplasts 11, 195
  complementary base-pairing 3–4

DNA *(Cont.)*
  complementary DNA (cDNA) 139
  composition 2
  denaturation 10
  density gradient analysis 25
  double helix 2, 4
  mitochondria 1, 195
  nearest neighbour analysis 23–5
  polarity 2, 4
  polymerase (*see* DNA polymerase)
  reassociation 16
  recombinant DNA 132–41
  repair of mutational damage 52–6
  repetitive sequences 16
  satellite 16
  unique sequences 16
  X-ray diffraction 3
DNA polymerase 14, 23, 28–9, 33–4,
    56, 139
  eukaryotes 33–4
  I 28–9, 56, 139
  II & III 28–9
  mitochondria 33
DNA replication 14, 20–9, 31–6, 122
  bidirectionality 34
  conservative model 26
  discontinuous synthesis model 26–8,
    32
  E. coli 34
  eukaryotes 32–6
  eukaryotes – histones 36
  eukaryotes – nucleosomes 36
  eukaryotes – replication units 34
  in cell cycle 32
  in vitro 20–5
  in vivo 25–9
  prokaryotes 31, 34
  protein synthesis involvement 36
  regulation 20
  RNA primer 28–9, 33
  RNA synthesis requirement 36
  rolling-circle model 122
  semi-conservative in eukaryotes 33
  semi-conservative model 26
DNA–RNA hybridisation 139, 240
dominance 47, 144, 146
  incomplete 146
  molecular model 146
Down's syndrome 187–9
Drosophila 57, 76, 151, 174, 205–7, 240
  gene control of eye pigments 205–7
  polytene chromosomes 240–1
Dunkers – founder effect and 252

elongation factor 91, 95
endonuclease 56, 79
endoplasmic reticulum 40, 96
enzyme 232–9
  aggregates 235
  control of synthesis 232–3

enzyme *(Cont.)*
  hormonal regulation of synthesis
    237–9
episome (*see* plasmid)
Escherichia coli 8, 56, 58–9, 118–30,
    132, 136, 213–30
  arabinose operon 223–5
  conjugation 118
  cultures 118–24
  DNA repair 56
  Hfr strains 216–7
  lactose (*lac*) operon 213–23
  mapping methods 118–24
  partial diploid 216–7
  recombination genes 56
  transduction 125–9
  transformation 129–30
  translational control 229–30
  tryptophan operon 225–8
euchromatin 16
eukaryote 8, 11–12, 40, 232–41
  cell structure 40
  cellular compartmentation 232
  differentiation 239–41
  gene regulation 232–41
    in eukaryotes 233–9
    homeostasis 237
    hormones 237–9
    long-term genetic regulation
      239–41
    sites for gene regulation 232–3
    steroid hormones 238–9
eukaryotic chromosomes 11–12
euploid 185–6
evolution 252
excision repair 54–6
exonuclease 9, 28, 56, 79, 94
extrachromosomal genetics 195–202
  characteristics 196–8
  maternal inheritance 196
  plasmid 195
  traits 198–202

F factor 119
F plasmid 135, 216–7
F' plasmid 216–7
filtration enrichment 59
forward mutation 250–2
founder effect 252
frameshift mutation 100
fungus 164–79, 199–202
  extrachromosomal traits 199–202
  life cycles 164–7
  meiotic genetic analysis 167–74
  mitotic genetic analysis 174–9
  respiratory deficiencies 199–202
  tetrad analysis 167–74
    tetratype ascus 170–4

G1, G2 phase (*see* cell cycle)

galactosaemia 185
galactose genes, yeast 234–5
gene 111, 113, 150–63, 167–74, 186–8,
    204–41, 243–53
  biochemical pathway and 208, 210–1
  cistron 211
  colinearity 211–2
  complementation group 208
  complementation test 113
  control of metabolism 204–5
  coordinate expression in operons 215
  deficiency 186–8
  duplication 186–8
  frequency 243–53
  function 204–12
  hormonal regulation of expression
    237–9
  linkage 150–63
  mapping methods, diploids 156–62
  mapping methods, haploids 167–74
  mapping by tetrad analysis 167–74
  operon 213–28
  regulation of expression 213–41
  regulation of expression, eukaryotes
    232–41
  regulatory 215–6
  regulatory mutants 218–9
  repressor 214–5
  steroid hormones and gene expression
    238–9
  structural 215–6, 227–8
  unit of function 113
  unit of mutation 111
  unit of recombination 111
genetic code 66, 82, 85, 87, 93–95,
    99–104, 195, 223
  codon 66
  degenerate 103
  evidence for codon assignments 101
  initiation codon 87, 94, 195
  mutations and 104
  start and stop codons 103
  table 102
  termination codons 93, 223
  triplet 99
  tRNA binding technique 103
  universal 103
  'wobble' 103
genetic drift 252
genetic engineering 132
genetic equilibrium 243–52
  applications 245–8
  factors affecting 249–52
  genetic drift and 252
  Hardy–Weinberg Law 243–8
  migration and 252
  mutation and 24, 249–524
  mutation pressure and 252
  preferential mating and 248–9
  selection and 249

genetic fine structure 108–12
genetic material 1–7, 8ff
genotype 47
Golgi apparatus 40
guanine 2, 18, 229
    ppGpp and pppGpp 229

haemophilia 184–5
Hardy–Weinberg equilibrium (see genetic equilibrium)
HeLa 76–8
heterochromatin 16, 189
heterogeneous nuclear RNA (see messenger RNA)
heterozygous 47
Hfr bacteria 120, 216–7
histone 13–6, 232–3
    genes 16
    modification and gene expression 233
    nucleosome involvement 15–6
homothallism 166–7
homozygous 47
hormone 237–9
    model for steroid action 238–239
    regulation of gene expression 237
host-controlled modification and restriction 132–3
human 181–93, 204–5 (see also eponymous syndromes)
    alkaptonuria 204–5
    aneuploidy 185–7, 190–1
    chromosomal aberrancies 185–93
    galactosaemia 185
    gene control of metabolism 204–5
    genetic counselling 193
    genetic diseases 184–93
    haemophilia 184–5
    inversion 186–8
    Lyon hypothesis 191–3
    pedigree analysis 181–5
    superfemale 191
    traits 184–5
    translocation 186
    translocation and Down's syndrome 189, 191–2
    Y-linked trait 184
Huntington's chorea 185
hydroxylamine 51–2

incomplete dominance 146
induced mutation 47–52
inducer 232
initiation codon 87, 94, 96, 103
initiator tRNA 86–7, 195
inosine 103
interference 162–3
interphase 45
    between meiotic divisions 45
interrupted mating 123

intervening sequences 140
inversion 186–8
iojap 198–9

karyotype 11–2, 14
kinase 19
Klinefelter's syndrome 191

lactose operon 213–23
    catabolite repression 220–2
    constitutive mutants 217–20
    control site sequences 222–3
    cyclic AMP and expression 220–2
    inducer 213
    initiation of translation 221–3
    model for control 215–6
    mutations 216–20
    operator 215–8
    polar effect of nonsense mutation 216–7
    positive control 220–2
    promoter 215–6, 219–20
    repressor 214–5, 218–9
lambda (λ) 10, 12, 126–9, 133
    chromosome 10, 12
    sticky ends 10, 12
    transduction with 126–9
ligase (see polynucleotide ligase)
linkage 150–63
Lyon hypothesis 191–3
lysogeny 124
lysosome 40

maize 198–9
maternal effect 196–8
maternal inheritance 196
meiosis 43–6, 144, 187, 190
    non-disjunction 187, 190
    tetrad stage 46
Mendel 142–9
    First Law 143–5
    life history 142
    Second Law 146–8
Meselson & Stahl experiment 25–6
messenger RNA (mRNA) 65–9, 94, 139, 215–6, 226, 232
    end modifications 67–8, 94
    eukaryotic 66
    gene regulation and 232
    heterogeneous nuclear RNA 69, 139, 232
    non-coding sequences 68
    polycistronic mRNA 215–6, 226
    pre-mRNA 68
    processing 68–9
    prokaryotic 66
    stability 66, 94
metaphase 42, 45
missense mutation 104

mitochondria 11, 40, 195–6
    cytochromes 196
    DNA 11, 195
    initiator tRNA 195
    protein synthesis 195–6
    ribosomes 195–6
mitosis 31, 36–7, 39–43
    mammalian cells 37
    nuclear membrane 37
    protein synthesis 37
    RNA synthesis 36
mitotic genetics 174–9
    Aspergillus 174–9
    crossing-over 174–9
    Drosophila 174
    gene mapping 177–9
    haploidisation 176–7
mouse 32–3
mutagen 47–53
    2-aminopurine (2-AP) 50
    5-bromouracil (5-BU) 49–50
    acridines 51–3
    hydroxylamine 51–2
    nitrous acid 50–1
    ultraviolet (UV) light 48
    X-ray 48
mutagenesis 47
mutation 47–8, 52–60, 100, 104, 111, 205, 208–11, 217–20, 228–9, 244, 249–52
    auxotroph 228–9
    base-pair transition 47–8
    base-pair transversion 47
    biochemical mutants 56, 59, 208–11
    cold-sensitive 59
    deletion 52–3, 104, 111
    genetic equilibrium and 24, 249–52
    heat-sensitive 59
    in operator 217–8
    in promoter 219–20
    in regulatory genes 218–9
    induced 47–52
    insertion 52–3, 104
    polar effect of nonsense mutation 216–7
    repair 52–6
    selection procedures 55–60
    suppressor 100
    temperature-sensitive 59
    tritium suicide 59–60
    visible 56–7, 205
mutation pressure 252

nearest neighbour analysis (see DNA)
Neurospora 39, 57–60, 78, 165–74, 199–200, 208–11, 234–7
    biochemical mutants 208–11
    genetic analysis 167–74
    heat-sensitive mutants 59–60
    life cycle 165–6

Neurospora (Cont.)
  mating types 165
  poky mutant 199–200
  quinic acid gene regulation 235–7
  tritium suicide 59–60
nitrous acid 50–1
non-disjunction 187, 190
nonhistones 13, 232–3
nonparental ditype ascus 170–4
nonsense mutation 104, 216–7
  polar effect 216–7
nuclear membrane 8
nucleic acid (*see also* RNA and DNA)
  1–5
  UV absorbance 5
nucleoid 11
nucleolar organiser 76
nucleolus 39–42
nucleoside 18
nucleosomes 15, 36
nucleotide 1, 18
nucleus 8, 40

one-step growth 106
operon 213–28
  arabinose catabolism 223–5
  attenuator regulation 226–8
  catabolite repression 220–2
  coordinate gene expression 215–6
  cyclic AMP and expression 224
  glucose sensitivity 220–3
  inducible 213, 223
  lactose catabolism 213–23
  model for function 215–6
  operator 215–6
  promoter 215–6, 219–20
  regulatory mutations 218–9
  repressible 225–8
  tryptophan biosynthesis 225–8
ordered tetrads 167–70
overlapping genes 116

parental ditype ascus 170–4
pedigree analysis 181–5
peptide bond 82
peptidyl transferase 91
petite strains 200–2
phage (*see* bacteriophage)
phenotype 47
ϕ × 174 10, 114–5, 140
  chromosome 10
  genetic organisation 115
  mapping using restriction
    endonucleases 140
  overlapping genes 116
photoreactivation 54
photosynthesis 195
Physarum polycephalum 32
Pisum sativum 143

plasmid 119, 132, 135–6, 216–7
  F 119, 135, 216–7
  F′ 216–7
Pneumococcus 5, 57
poky 199–200
polycistronic mRNA 215–6, 226
polynucleotide ligase 56
polynucleotide phosphorylase 101
polypeptide (*see also* protein) 85
polysomes (polyribosomes) 93–4
polytene chromosomes 240–1
population genetics 243–53
  founder effect 252
  genetic drift 252
  Hardy–Weinberg equilibrium 243–4
positive control 220–2
Principle of Independent Assortment
  146–8, 150
Principle of Segregation 143–5
prokaryote 8, 213–30
  gene regulation 213–30
  translational control 229–30
promoter 63, 215–6, 219–20, 225–8
  mutations in 219–20
  recognition by RNA polymerase 63
prophase 41–5
protease 235
protein 82–6, 96
  direction of synthesis 85–6
  peptide bond 82
  secretion mechanism 96
  signal hypothesis 96
  structure 84
protein synthesis (*see* translation)
prototrophy 59
pseudo-wild 99
purine 1–2, 18–20
  biosynthesis 18, 20
puromycin 36
pyrimidine 1–2, 18–21, 54
  biosynthesis 19–21
  dimers 54

recessiveness 47, 144
recombinant DNA 132, 135–40, 195
  applications 140
  cloning 136–40
  cloning vehicle 132, 135–6
  plasmid 132, 135–6
recombination 45, 107–10, 150–7,
  162–3, 167–79
  chiasmata and 45
  coincidence 162–3
  frequency 150–1
  in phage T4 107–10
  interference 162–3
  meiotic 45
  mitotic 174–9
  physical exchange and 155–7

regulation 213–41
  activated tRNA and 227–8
  arabinose operon 223–5
  aromatic amino acid genes,
    Neurospora 235–6
  attenuation in operons 226–8
  bacterial 213–30
  cyclic AMP and 237–8
  development 239–41
  differentiation 239–41
  effector molecules 232–3
  eukaryotes 232–41
  galactose catabolism in yeast 234–5
  gene–cytoplasm interactions 241
  hormonal control of gene expression
    237–9
  in eukaryotes 233–9
  lactose operon 213–20
  long-term 239–41
  model for lactose operon 215–6
  quinic acid genes in *Neurospora*
    235–7
  role of steroid hormones 238–9
  transcription 228–9
  translational control 229–30
  tryptophan operon 227–8
regulatory mutants 218–9
relaxed mutant (*see* stringent control)
repair 52–6
  enzymes involved 56
  excision repair 54–6
  genes involved 56
  photolyase 54–5
  photoreactivation 54
replica plating 59
replication (*see* DNA replication)
repressor 218–9, 225–6, 232
  regulation of operons 225–6
  mutation effects 218–9
repulsion 151
restriction endonuclease 132–4
reverse transcriptase (*see* RNA-
    dependent DNA polymerase)
reversion 49
rho factor 65
ribonuclease 71, 74
  processing role 74
ribonucleic acid (*see* RNA)
ribose 2, 18–9
ribosomal RNA (rRNA) 16, 66, 72–9,
  195
  genes 16, 74–6, 195
  methylation 75–7
  mitochondria 195
  pre-rRNA 74–6
  processing 74–6, 79
  prokaryotic v. eukaryotic 66, 72
ribosome 40, 66, 72–9, 87, 93, 195–6,
  215–6, 221, 223
  70S 72–4

ribosome (Cont.)
  80S 72
  binding site in lactose operon 221–3
  mitochondria 195–6
  nucleolar organiser and 76
  polysome 93
  protein synthesis role 87
  proteins 72–3, 76–7, 195–6
  regulation of synthesis 79
  RNA content 66
  rRNAs 72
  self-assembly 74
  subunits 72
  synthesis 74–9
rifampicin 63
RNA (see messenger RNA, ribosomal
    RNA, transfer RNA)
  composition 2
RNA polymerase 14, 28–9, 61–6, 215,
    225
  binding to promoter 63
  conformational changes 64
  core enzyme 63
  E. coli 63
  eukaryotes 65–6
  inhibition 65
  prokaryotes 63–5
  sigma factor 63–4
RNA primer (see DNA replication)
RNA processing (see messenger RNA,
    transfer RNA, ribosomal RNA)
RNA synthesis 19–20, 32, 61–79
  chromatin structure and 66
  in cell cycle 32
  inhibitors 63
  one strand copied 62
  regulation 20
  rho factor 65
  termination 65
RNA-dependent DNA polymerase 139

S phase (see cell cycle)
Salmonella typhimurium 124
satellite DNA 16
Schizosaccharomyces pombe 32–3
sea urchin 32
selection 249
sex chromosome 11, 151–3, 189–93
  aberrancies 189–93
  Lyon hypothesis 191–3
sex linkage 151–3
sigma factor 63–4
signal hypothesis 96
spontaneous mutation 47–8
stringent control 228–9
  ppGpp, pppGpp and 229
  relaxed mutant 229
  stringent factor 229
superfemale 191
suppressor mutation 100

T phages 5, 8–10, 99, 106–13, 132, 230
  chromosomes 8–10
  DNA replication 8
  genetic material 8
T2 phage 8–9, 132, 230
  life cycle 9
  translational control 230
T3 phage 8
T4 phage 5, 8–9, 99, 106–13, 230
  complementation test 113
  conditional mutations 107
  fine structure analysis 108–12
  life cycle 106–7
  mapping using deletions 112
  recombination 107–10
  rII mutant strains 99, 108, 113
  structure 5, 9
  translational control 230
  visible mutations 107
T5 phage 8
T7 phage 8
telophase 42, 45–6
temperature-sensitive mutants 59–60,
    71
terminal redundancy 8–10
termination codon 93, 103, 215, 223
termination factors 93–5
testcross 145, 148, 150, 159–62
tetrad analysis 164–74
  gene mapping 167–74
  gene–centromere distance 167–70
  ordered tetrad 165, 167–70
tetratype ascus 170–4
thymine 2, 19
  dimers 54
tobacco mosaic virus – chromosome 10
transcription (see RNA synthesis)
  in development & differentiation 240
  promoter 215–6
  regulation 228–9
  regulation by nonhistones 233
transduction 118–29
transfer RNA (tRNA) 16, 66, 69–72, 86,
    94, 195, 221, 225
  activated 225
  amino acyl synthetase 69
  amino acylated 85
  codons and 69
  genes 16, 71
  initiator 86, 94, 195, 221
  pre-tRNA 71
  processing 71
  role in translation 89
  structure 70
transformation 5–6, 118–24, 129, 132,
    136–7
transition mutation 47–8
translation 36, 61, 82–97, 195, 221–3,
    229–30
  amino acyl synthetases 85

translation (Cont.)
  cellular compartmentation and 96
  codon 84
  elongation 89–93, 95
  elongation factor 91, 95
  endoplasmic reticulum and 96
  eukaryotes 94–7
  genetic code and 84
  inhibitors 36
  initiation 86–9, 94–5
  initiation in lactose operon 221–3
  initiator tRNA 86, 94, 195
  mRNA end modifications and 94
  peptide bond formation 91–2
  peptidyl transferase 91–2
  polysomes 93
  prokaryotes 86
  regulation 229–30
  relation to transcription 93
  ribosome binding site 87–8
  signal hypothesis 96
  termination 93–6
  translocation 91
translocation 186, 189, 191–2
transversion mutation 47
trisomy 187
tryptophan operon 225–8
  activated tRNA and regulation
    227–8
  attenuator site 226–8
  promoters 225–8
  regulation 227–8
  tryptophan as co-repressor 225
Turner's syndrome 190–1

ultraviolet light 48, 54
  pyrimidine dimer induction 54
uracil 2, 19

virus 8
visible mutations 107

wobble 103

X chromosome 11, 189–93
  aberrancies 189–93
  Lyon hypothesis 191–3
Xenopus laevis 32, 76–8
X-ray 48

Y chromosome 11
yeast 32, 39, 57–9, 76–8, 164, 167,
    200–2, 234–5
  galactose fermentation genes 234–5
  genetic analysis 167
  petite mutant strains 200–2